AF801515

Springer Theses

Recognizing Outstanding Ph.D. Research

Aims and Scope

The series "Springer Theses" brings together a selection of the very best Ph.D. theses from around the world and across the physical sciences. Nominated and endorsed by two recognized specialists, each published volume has been selected for its scientific excellence and the high impact of its contents for the pertinent field of research. For greater accessibility to non-specialists, the published versions include an extended introduction, as well as a foreword by the student's supervisor explaining the special relevance of the work for the field. As a whole, the series will provide a valuable resource both for newcomers to the research fields described, and for other scientists seeking detailed background information on special questions. Finally, it provides an accredited documentation of the valuable contributions made by today's younger generation of scientists.

Theses are accepted into the series by invited nomination only and must fulfill all of the following criteria

- They must be written in good English.
- The topic should fall within the confines of Chemistry, Physics, Earth Sciences, Engineering and related interdisciplinary fields such as Materials, Nanoscience, Chemical Engineering, Complex Systems and Biophysics.
- The work reported in the thesis must represent a significant scientific advance.
- If the thesis includes previously published material, permission to reproduce this must be gained from the respective copyright holder.
- They must have been examined and passed during the 12 months prior to nomination.
- Each thesis should include a foreword by the supervisor outlining the significance of its content.
- The theses should have a clearly defined structure including an introduction accessible to scientists not expert in that particular field.

More information about this series at http://www.springer.com/series/8790

John Callum Alexander

Surface Modifications and Growth of Titanium Dioxide for Photo-Electrochemical Water Splitting

Doctoral Thesis accepted by
Imperial College London, UK

Springer

Author
Dr. John Callum Alexander
Imperial College London
London
UK

Supervisors
Prof. Neil McN Alford
Department of Materials
Imperial College London
London
UK

Prof. Geoff H. Kelsall
Department of Chemical Engineering
Imperial College London
London
UK

ISSN 2190-5053 ISSN 2190-5061 (electronic)
Springer Theses
ISBN 978-3-319-34227-6 ISBN 978-3-319-34229-0 (eBook)
DOI 10.1007/978-3-319-34229-0

Library of Congress Control Number: 2016939052

© Springer International Publishing Switzerland 2016
This work is subject to copyright. All rights are reserved by the Publisher, whether the whole or part of the material is concerned, specifically the rights of translation, reprinting, reuse of illustrations, recitation, broadcasting, reproduction on microfilms or in any other physical way, and transmission or information storage and retrieval, electronic adaptation, computer software, or by similar or dissimilar methodology now known or hereafter developed.
The use of general descriptive names, registered names, trademarks, service marks, etc. in this publication does not imply, even in the absence of a specific statement, that such names are exempt from the relevant protective laws and regulations and therefore free for general use.
The publisher, the authors and the editors are safe to assume that the advice and information in this book are believed to be true and accurate at the date of publication. Neither the publisher nor the authors or the editors give a warranty, express or implied, with respect to the material contained herein or for any errors or omissions that may have been made.

Printed on acid-free paper

This Springer imprint is published by Springer Nature
The registered company is Springer International Publishing AG Switzerland

Supervisor's Foreword

Titanium dioxide is an extraordinarily versatile material. We know it best as a fine submicron powder used in paint pigment or in sunscreens, but its electronic, dielectric, and electrochemical properties are such that TiO_2 continues to attract huge research interest. Its ease in changing valency means it can be a very good insulator, but it can easily be reduced to make it an oxide capable of conducting electricity. As an insulator, TiO_2 and compounds containing TiO_2 can be formed into microwave dielectric resonators for mobile communications, but the need for alternative energies and fuels has brought a resurgence of interest in this material.

Over the past decade, there has been sustained interest in TiO_2 and its diverse applications. Of particular note has been the concerted effort to understand the photo-electrochemical properties of TiO_2 for applications such as water splitting, a major focus of the thesis that follows.

Recently, the use of TiO_2 as a substrate for metal nanoparticles and plasmonic applications has received serious attention. A range of apparently plasmon-driven phenomena have been observed and studied, such as water splitting and other redox reactions, for which theoretical models have not yet been demonstrated.

Other applications of TiO_2 include as follows: as a transparent conducting oxide (TCO) for electronics applications and as the active material in memristors, for which understanding of the rich defect chemistry of TiO_2 is crucial. Understanding the electronic structure of TiO_2 has therefore been the focus of a significant theoretical effort by computational chemists using techniques such as density functional theory. Experimental confirmation of modelling has been limited by the availability of high-quality synthetic surfaces with controlled defects and impurities.

The work reported in this thesis was carried out as part of the Engineering and Physical Sciences Research Council Programme Grant in 'Nanostructured Functional Materials for Energy Efficient Refrigeration, Energy Harvesting and Production of Hydrogen from Water', which brought materials scientists together with (computational) chemists and physicists to understand the links between the structure and the functional performance of materials and devices, including TiO_2.

The thesis provides, in a single source, a wide-ranging yet extremely thorough introduction to the growth, structure, defect chemistry, and functional applications of TiO_2, providing a comprehensive reference for researchers looking to work with this material.

In particular, the theory and governing equations of the titanium dioxide–water interface have been covered in greater-than-usual depth, providing a consistent frame of reference for those analysing the structure of the interface.

The chapters on thin-film growth provide a guide for the controlled growth of oriented epitaxial and polycrystalline thin films, for use in surface science studies supporting theoretical work.

Furthermore, the critical discussion of the possible plasmonic water splitting mechanisms and the impact of the defect chemistry on the electronic and photo-electrochemical performance of TiO_2 provides researchers with numerous avenues for the further investigation. The value of this thesis derives from the joint application of materials and electrochemical analysis, considering the functional properties of titanium dioxide in the light of its structural properties and defect mechanisms, and providing a different perspective on this well-known and studied material.

London, UK
April 2016

Prof. Neil McN Alford

Abstract

This study investigates photo-anodes based on titanium dioxide (TiO_2) that can be used to produce hydrogen by the photo-electrochemical decomposition of water.

TiO_2 is a wide bandgap semiconductor that absorbs only the UV region of the solar spectrum. Sensitization of TiO_2 to visible light by the addition of gold nanoparticles (AuNPs) was studied. AuNPs sustain localized surface plasmon resonance (LSPR) that results in the absorption of light at the resonant energy. The evidence for water splitting by Au-TiO_2 systems is discussed critically. Fabrication of arrays of AuNPs was done by annealing sputtered gold thin films, micellar nanolithography, and nanosphere lithography. The optical characteristics and photo-electrochemical 'water splitting' performance of AuNP-coated rutile (110) electrodes were determined. Nb-doped crystals coated in AuNPs of ca. 20 nm exhibited a small photocurrent that was not present with the bare rutile electrode. Reduced undoped rutile (110) with AuNPs did not exhibit the 'plasmonic photocurrent'. Some Nb-doped electrodes did not exhibit an effect. Batches of Nb-doped and reduced rutile were examined using voltammetry and impedance spectroscopy, and it was found that the 'inactive' Nb-doped TiO_2 was partially reduced.

Thin films of TiO_2 were fabricated by pulsed laser deposition (PLD) onto amorphous and single crystal substrates. The effect of growth conditions on the phase and orientation of the film was studied, and procedures to grow anatase films oriented with (100), (001), and (101) were developed. The temperature and heating regime of TiO_2 films fused silica affected the orientation of film growing. Nb doping of the films also affected the temperature of the anatase–rutile phase transition and the orientation of the films, acting to stabilize anatase at higher temperatures. Surprisingly, highly doped films were found to be non-conductive. The importance of the oxygen partial pressure in producing conductive films for use as electrodes is discussed.

Acknowledgements

There are numerous people who I wish to acknowledge for their assistance during this Ph.D.

Professor Neil McN Alford for providing me with the opportunity to return to academia from industry to undertake research in a completely new subject area.

Professor Geoff H. Kelsall for his patience with my impromptu visits and his encyclopaedic knowledge of electrochemistry.

Dr. Anna-Karin Axelsson for daily input and direction on both plasmonic water splitting and thin-film deposition.

Dr. Steve Dennison, Dr. Anna Hankin, and Dr. Chin Kin Ong for teaching me the basics of all the electrochemical measurement techniques used in this thesis and for discussions on photo-electrochemistry.

Dr. Derfogail Delcassian for micellar lithography, Dr. Jing Pang for nanosphere lithography, and Dr. Kai Wang for reduction of TiO_2 crystals.

Manisha Chhikara for atomic force microscopy of TiO_2 thin films.

I acknowledge the Engineering and Physical Sciences Research Council (EPSRC) for funding on Programme Grant EP/G060940/1 'Nanostructured Functional Materials for Energy Efficient Refrigeration, Energy Harvesting and Production of Hydrogen from Water'.

Declaration

I hereby declare that this thesis is my own work. To the best of my knowledge, the arguments made and results contained within are my own and have not been published by another person. The intellectual content of this thesis is my own, with the exception that I acknowledge that I have accepted suggestions and modifications to the content and format of the text and figures of this thesis.

This thesis does not contain material that has been previously accepted for the award of any qualification at Imperial College London or any other educational institution.

Where previously published results and arguments have been included, I have referenced the work and identified the authors.

I have acknowledged any experimental work by others who have contributed materially to the results presented in this thesis.

The copyright of this thesis rests with the author and is made available under a Creative Commons Attribution Non-Commercial No Derivatives License. Researchers are free to copy, distribute, or transmit the thesis on the condition that they attribute it, that they do not use it for commercial purposes, and that they do not alter, transform, or build upon it. For any reuse or redistribution, researchers must make clear to others the license terms of this work.

John Callum Alexander

Contents

List of Figures

List of Tables

Chapter 1
Introduction

1.1 Scope of Research

This thesis describes investigations into the use of titanium dioxide (TiO_2) as a photo-anode for photo-electrochemical water splitting, otherwise termed the production of solar hydrogen.

The thesis reports on the results of four strands of work related to the surface modification and growth of TiO_2 electrodes. The work can be grouped as studies on rutile (110) single crystals, and the growth of TiO_2 thin films by pulsed laser deposition (PLD) as follows:

- The surface modification of rutile (110) single crystals by the addition of gold nanoparticles (AuNPs).
- Studies on Nb-doped and reduced rutile (110) single crystals.
- Growth of oriented TiO_2 thin films, for use as electrodes, on amorphous substrates like fused silica, including the effect of Nb doping.
- Growth of oriented TiO_2 thin films on single crystal substrates.

1.2 Structure of Thesis

This chapter summarizes the scope of research, structure and contents of the thesis, and resulting publications and conference attendance.

In Chap. 2—Literature Review, the global energy context for solar hydrogen is outlined. The basic principles of photo-electrochemical water splitting cells and material are introduced, including the use of titanium dioxide as a water splitting material. The fundamental problem with TiO_2 as a solar hydrogen material i.e. its wide band gap, is discussed, and the idea of sensitizing TiO_2 to visible light by using AuNPs is introduced. Studies using AuNP-TiO_2 electrodes for water splitting

© Springer International Publishing Switzerland 2016

J.C. Alexander, *Surface Modifications and Growth of Titanium Dioxide for Photo-Electrochemical Water Splitting*, Springer Theses,
DOI 10.1007/978-3-319-34229-0_1

and other photo-electrochemical reactions are reviewed and the various mechanisms of 'plasmonic water splitting' proposed in the literature are discussed critically. This is followed by a short review of TiO_2 thin films fabricated by pulsed laser deposition (PLD), which was used to identify starting conditions for the experimental work. Finally, a detailed review of the defect chemistry of TiO_2 with Nb is presented. The review was carried out to try to understand the differences between doped and reduced TiO_2, and the lack of conductivity in doped thin films.

In Chap. 3—Principles of Photo-Electrochemical Cells, the theory of semiconductor-liquid interfaces is described in greater detail than Chap. 2, and the relevant equations are presented. The structure of the space charge layer in the semiconductor is derived as a function of potential. Characteristic parameters such as the space charge width constant and the capacitance are defined, and key terms like the flat band potential are defined. A simple model for the photocurrent is also presented. The theory and models outlined are used as background to the electrochemical analysis methods described in Chap. 4. The information is also to provide the necessary background to the results and discussion of the voltammetry and impedance spectroscopy presented Chaps. 5 and 6.

In Chap. 4—Experimental Methods, the range of fabrication and characterization methods are detailed. Metal nanoparticle array fabrication techniques are introduced; annealing sputtered gold thin films; micellar nanolithography; and nano-sphere lithography. The thin film deposition techniques of sputtering for metal films and pulsed laser deposition are introduced. Materials characterization techniques are discussed including; X-ray Diffraction (XRD), Raman spectroscopy, UV-Visible (UV-Vis) spectroscopy, and atomic force microscopy (AFM). Finally, electrochemical methods including voltammetry, chronoamperometry and impedance spectroscopy are introduced.

In Chap. 5—Results: Plasmonic Photocurrent with AuNP-TiO_2, the results of experiments on the fabrication of AuNP arrays on single crystals of rutile (110) and their performance as water splitting photo-anodes are presented. This includes optical and electrochemical measurements on several types of AuNP array on both Nb-doped and reduced rutile.

In Chap. 6—Results: Electrochemistry of TiO_2—Rutile (110), the differences between two batches of Nb-doped rutile and reduced rutile are investigated in greater detail using impedance spectroscopy and voltammetry. This chapter provides deeper understanding of the behaviour of the TiO_2 crystals used to form the AuNP covered electrodes discussed in Chap. 5, and studies the qualitative differences between those doped with Nb and those reduced.

In Chap. 7—Results: TiO_2 thin films grown on Fused Silica, the effects of deposition conditions on the phase and orientation are presented. The effect of heating during deposition versus deposition at room temperature with post-annealing is also presented. Finally, the effect of Nb doping on both the phase and orientation of the film is investigated. The lack of conductivity of the doped films is also discussed, in light of the review of the defect chemistry of Appendix A.

In Chap. 8—Results: TiO_2 thin films grown on single crystals, the effect of both deposition conditions and substrate lattice on phase and orientation are studied. In addition, the epitaxial relationships between the TiO_2 film and the single crystal substrate are determined and discussed.

1.3 Publications and Conferences

Publications produced and conference presentations given during the course of this thesis are:

- A. Hankin, J.C. Alexander, G.H. Kelsall, Constraints to the flat band potential of hematite photo-electrodes, *Phys. Chem. Chem. Phys.*, 2014,16, 16176–16186 [1].
- J.C. Alexander, A.-K. Axelsson, N. Alford, Anatase TiO_2 thin films for photo-electrochemical energy and solid-state electronics applications: the effect of substrate on film orientation, at EcoMatech 2013, Bled, Slovenia, September 2013 (oral presentation).
- J.C. Alexander, A.-K. Axelsson, G.H. Kelsall, N. Alford, Gold Nanoparticles on TiO_2: mechanism of plasmonically-enhanced water splitting, Materials Research Society Spring 2013, San Francisco, USA, April 2013 (oral presentation).

Additional publications based on the results of each of Chaps. 5–8 are in preparation, but will not be submitted before this thesis. Working titles are:

- Plasmonic water splitting with AuNPs on Nb-doped rutile (110).
- The effect of reduction of TiO_2 on the semiconductor-liquid interface and photo-electrochemical performance of photo-anodes.
- Control of phase and orientation of TiO_2 films on fused silica by thermal processes and doping.
- Deposition of (100) and (001) oriented anatase films on single crystal substrates.

Reference

1. Hankin, A., Alexander, J.C., Kelsall, G.H.: Constraints to the flat band potential of hematite photo-electrodes. Phys. Chem. Chem. Phys. **16**, 16176 (2014)

1.2 Structure of Thesis

In Chap. [illegible] Results, [illegible] deposition [illegible] and [illegible] [illegible] simple [illegible] and discuss[illegible]

[illegible]

Chapter 2
Literature Review

This chapter discusses the use of titanium dioxide as the active material in photo-electrochemical cells for the electrolysis of water, with the aim of producing a source of renewable hydrogen. Recent reports of sensitizing TiO_2 to visible light using localized surface plasmon resonance (LSPR) on metal nanoparticles are discussed.

The choice of titanium dioxide as a material for water splitting was due to its stability as an n-type semiconductor, the extensive literature on its photo-electrochemical behaviour and, in the wider project at Imperial College London, the development of a parallel stream of computational descriptions of the surfaces of rutile and anatase (EPSRC Programme Grant on Nanostructured Functional Materials for Energy Efficient Refrigeration, Energy Harvesting and Production of Hydrogen from Water).

The global context driving interest in photo-electrochemical hydrogen production by water splitting is introduced in Sect. 2.1, followed by a basic overview of photo-electrochemical cells (PECs), with a more detailed treatment in the following chapter Principles of Photo-Electrochemical Cells. The main requirements for a water splitting material are introduced in Sect. 2.3, and the use of metal oxides is introduced in Sect. 2.4.

In Sect. 2.5, the concept of 'plasmonic water splitting' is introduced, followed by a brief outline of plasmonic resonance on metal nanoparticles in Sect. 2.6. Fabrication methods for metal nanoparticles are discussed in Chap. 4, Experimental Methods. The bulk of the literature review covers the use of gold nanoparticles (AuNPs) to "enhance" the performance of TiO_2 (both rutile and anatase). Various experimental reports are discussed in Sect. 2.7 followed by a critical discussion of the proposed mechanisms of "plasmonic enhancement" in Sects. 2.8–2.10. The published data reviewed and discussed in this chapter led to the experimental work reported in Chaps. 5 and 6 of this thesis. The purpose of this work was to detect the reported effect, and to try to determine and understand the mechanisms involved.

In Sect. 2.11, the literature review also includes a brief summary of reports on the growth of oriented and epitaxial thin films of anatase and rutile for a range of

© Springer International Publishing Switzerland 2016

J.C. Alexander, *Surface Modifications and Growth of Titanium Dioxide for Photo-Electrochemical Water Splitting*, Springer Theses,
DOI 10.1007/978-3-319-34229-0_2

purposes. These were used as a starting point for determining the growth conditions for rutile and anatase thin films, as reported in Chaps. 7 and 8 of this thesis, aiming to provide high quality surfaces to support theoretical models based on hybrid density functional theory (h-DFT), undertaken in the same Programme Grant.

Finally, Sect. 2.12 presents a review of the theoretical defect chemistry of TiO_2, aiming to understand why heavily doped anatase and rutile films grown during this project did not show the levels of conductivity reported by other authors. The interpretation of this theory in the context of the doped films is reported in Chap. 7, TiO_2 films on fused silica.

In Chap. 4, Experimental Methods, a range of experiments are outlined in response to this review of the literature.

2.1 Global Energy Drivers for Solar Water Splitting

During the 21st Century, the world population is expected to increase to a peak of 9 billion, while demand for energy will also increase as people aspire to a Western standard of living [1, 2]. Against a background of increased energy demand, reducing carbon dioxide emissions to prevent climate change is a significant global challenge [3]. For example, the UK has committed to 80 % reduction compared with 1990 levels by 2050 [4]. Low-carbon and renewable energy sources, such as wind and solar power, will be crucial in meeting this target. One of the major drawbacks of renewable energy sources is intermittency: solar technologies such as photovoltaic (PV) cells are affected by weather and the fact that the sun only shines on half of the surface of the Earth at any one time. As such, without energy storage technology, power output is unlikely to match demand [5, 6].

Conversion of electrical energy into a 'solar fuel' is one possible approach to using energy storage to mitigate this intermittency. One of the main candidates for a solar fuel is hydrogen (H_2), which can be produced by the electrolysis of water. Hydrogen is a versatile fuel—it can be used in both fuel cells and combustion engines—with the only combustion products being water vapour and heat [7]. As such, it also presents an alternative to hydrocarbons for transport, provided hydrogen storage can be made more efficient and economic. The wide-scale adoption of H_2 as a fuel for transport and energy storage is known as the Hydrogen Economy [8]. Currently, most H_2 is produced by steam reforming of natural gas, which produces CO_2 and largely negates the benefits on offer. Renewable and carbon-free H_2 production is possible by using PV generated electricity to power a separate electrolyser (alkaline or PEM).

An alternative approach is to simultaneously harvest sunlight and produce hydrogen in a photo-electrochemical cell (PEC) employing materials such as TiO_2 and Fe_2O_3 as photo-anodes. The potential benefit is decreased material and equipment cost, as the area of solar collector also acts as the electrode (although this has yet to be demonstrated in practice). As H_2 would compete with battery technology for electricity storage, another potential market for H_2 is as a chemical feedstock for ammonia production and/or to help reduce CO_2 to a fuel in an Artificial Leaf.[1]

At present, there are no commercialized PEC systems for photo-electrolysis of water (water-splitting), although devices with up to 16 % solar-to-hydrogen (STH) efficiency have been produced by employing a $GaInP_2$/GaAs multi-junction cell [9]. However, devices using these rare III–V semiconducting materials are prohibitively expensive and unstable in aqueous environments, so are unsuitable for large-scale commercialization. Although Si is also unstable in oxidizing environments, even in water, a promising Si-based device has been reported [10] with efficient Co catalysts, forming a stable photo-anode for water splitting. The fundamental advantage of Si is that, with a band gap of ca. 1.1 eV, it is a good absorber of sunlight, in principle enabling high efficiencies. The challenges are in catalysing the reactions, and material stability.

However, devices employing cheap, earth-abundant materials that are stable in water, such as metal oxides, suffer from low efficiencies, due to large band gaps, conduction band edge energies that will not reduce H^+ to H_2, and poor electron-hole (e^-h^+) lifetimes [11]. The focus of the work reported in this thesis was to investigate routes to improving the understanding and performance of metal oxides, especially titanium dioxide, for water splitting applications. In the following discussion, brief reference will also be made to hematite (Fe_2O_3), which was also studied during the research period.

2.2 Photo-Electrochemical Cells (PECs)

The basic principles of PECs are discussed, followed by consideration of the materials TiO_2 and Fe_2O_3 for successful water splitting. The explanations that follow are based on photo-anodes, at which the oxidation of water occurs by the oxygen evolution reaction (OER). A photo-anode is formed by an n-type semiconductor, such as TiO_2 or Fe_2O_3 in contact with a redox couple in electrolyte. Further detail on the structure of the semiconductor-liquid interface and the terminology used may be found in the Chap. 3 of this thesis, Principles of Photo-electrochemical Cells.

2.2.1 Overview

The components of a basic n-type photo-electrochemical cell (PEC) are illustrated in Fig. 2.1. A PEC for water splitting consists of a semiconductor photo-anode, an aqueous electrolyte, a counter electrode, and an ohmic contact that transports

[1]See http://www3.imperial.ac.uk/energyfutureslab/research/grandchallenges/artificialleaf.

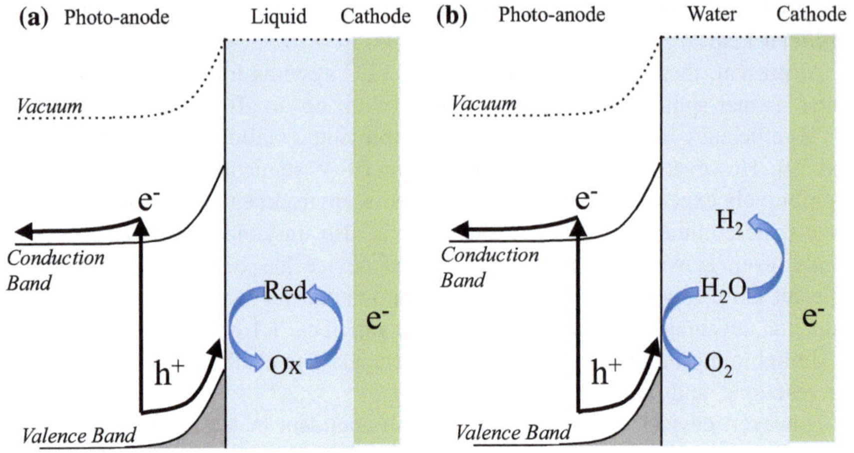

Fig. 2.1 Basic schematic diagram of (**a**) a regenerative PEC and (**b**) a hydrogen producing PEC. Image based on publication [12]

electrons from the semiconductor to the counter electrode (or cathode), where they reduce protons/water to hydrogen. In the case of a photo-anode, the electron and hole are generated by the absorption of a photon. In Fig. 2.1a, a reductant (Red) is oxidized at the photoanode of a regenerative PEC by a hole (h^+) and becomes an oxidant. The oxidant (Ox) diffuses to the counter electrode (cathode or photo-cathode) and is reduced by an electron (e^-) to its original state i.e. the process is regenerative. In the case of a water splitting cell (Fig. 2.1b), the oxidation of water/OH^- results in the generation of oxygen, which leaves the PEC. Likewise, the reduction of water/H^+ results in the generation of hydrogen, which leaves the PEC. The reaction is not regenerative; water is consumed in the production of hydrogen.

Practically, the O_2 and H_2 must be generated and separated in the cell to avoid the spontaneous reverse reaction. In a PEC with a photoactive anode (as in the case of TiO_2 and Fe_2O_3), the H^+ ions (protons) produced by OER reaction must be transported away from the photo-anode by a conductive electrolyte to the cathode for reduction. A typical means of ensuring separation is the use of a proton exchange membrane (PEM) which allows protons to pass, but prevents electrons and OH^- ions doing so. Alternatively, if the photocathode was the driving force, then the OH^- ions must be transported to an anode for oxidation [13, 14]. In some water splitting 'Z-schemes', both a photo-anode and a photo-cathode are employed together to boost the fraction of light absorbed [13, 15] and to provide sufficient electrical potential to drive water splitting.

Materials produced and investigated in this project were for use as photo-anodes in a PEC. This means that the practicalities of PEC operation need to be borne in mind when proposing new material systems. These considerations include: efficiency targets for a real-world technology; possible efficiency limits in proposed systems; and electron and hole potentials relative to the potentials of redox couples. Other issues such as charge separation, lifetime and mobility also need to be considered.

2.3 Requirements for Economical PEC Water Splitting

To assess the suitability of materials and systems for economically feasible water splitting, it is necessary to be efficient, cheap and robust, which may be summarised in scientific terms as: generate sufficient photo-potential to split water; generate sufficient photocurrent to efficiently produce hydrogen; be stable, i.e. not be decomposed by electrons, holes, hydroxide ions or protons; and do so cheaply [15]. The first two requirements are discussed below in more detail in relation to TiO_2 and Fe_2O_3.

2.3.1 Solar to Hydrogen Efficiency

The generally accepted solar-to-hydrogen (STH) efficiency target is ca. 10 %. The goal is to compete with the closest technological competitor: photovoltaic cells coupled to electrolysers, as highlighted e.g. by a US Department of Energy report [11, 15]. The ca. 10 % efficiency goal is based on multiplying the efficiencies of a typical PV cell and a typical electrolyser. For electrolysis, 1.6 V is a typical practical potential difference, and 1.23 V is the reversible potential difference, giving an (voltage) efficiency of 77 % (1.23 V/1.6 V). A typical commercial PV module efficiency is on the order of 15 % (electrical power output to solar power input). A photo-electrochemical device would need to meet this efficiency at the same or lower cost to be competitive with PV/electrolysis as a reference technology. The voltage of 1.23 V is the reversible potential for water splitting. Practical devices require higher voltages, as discussed further below.

As an example, a STH of 10 % efficiency for a typical solar flux, P_{solar}, of 1000 W m^{-2} corresponds to a photocurrent of 8.13 mA cm^{-2} or 81.3 A m^{-2}. This is based on the ratio of Gibbs free energy of the hydrogen ($\Delta G^{\circ} = -237.178$ kJ mol^{-1}) generated at a rate R_{H_2} to the incoming solar irradiance [11, 16]. It also assumes zero external bias applied to the cell. The STH efficiency is given by Eq. (2.1).

$$STH\ \% = \Delta G^{\circ} R_{H_2} / P_{solar} Area_{cell} \tag{2.1}$$

Given that the production of each Mole of H_2 is directly related to the current, this is equivalent to:

$$STH\ \% = 1.23 J_{photo} \tag{2.2}$$

where J_{photo} is the photocurrent density. As each H_2 molecule requires two electrons (see hydrogen evolution reaction, Chap. 3 Eq. 3.2), the molar production rate is $R_{H_2} = j_{photo}/2F$, where F is the Faraday constant (F = 96,485 C mol^{-1}), the charge on N_{Av} electrons [14, 16]. At 10 % STH this corresponds to quantities of about 150 μmol h^{-1} cm^{-2} of hydrogen and 75 μmol h^{-1} cm^{-2} of oxygen, as the ratio of production rates should be 2:1 according to the overall water splitting reaction.

$$H_2O \xrightarrow{hv} H_2 + \frac{1}{2}O_2 \tag{2.3}$$

A photo-anode needs to approach these efficiency values (or have the potential to) to be considered viable. However, few studies report their results in such a manner that the STH efficiency can be established, due to uncertainties of sample size, illumination intensity, and applied external bias.

2.3.1.1 Efficiency Limits

The main efficiency losses at various stages of solar water splitting, from initial photon absorption to the production of hydrogen, can be categorized as follows [14, 17]:

- Absorption of photons—how much of the solar spectrum that can be absorbed is limited by a material's band gap;
- Thermalisation of electrons as the energy of photons in excess of the band gap energy is lost by interaction with the lattice;
- Recombination of charge carriers in the bulk, at surfaces, and traps;
- Overpotential of electrons/holes—excess potential not utilised, but dissipated as heat;
- Resistive losses—due to resistivity of materials, interfaces, and electrolyte.

Osterloh et al. [17] noted that a single photocatalytic material device has a theoretical limit of less than 10 %—principally due to issues of bandgap, overpotential and the Shockley-Queisser limit, which is based on radiative, recombinative, and spectral losses. The overpotential required for a significant H_2 production rate is suggested to be about 0.6 V (0.3 V each for reduction and oxidation), giving 1.83 V as a minimum practical band gap and practical potential difference [9, 17]. A tandem device is more flexible as it adds the potential of two cells in series, but again the limit is of the order of 15 % from the need to absorb twice as many photons per electron. Multi-junction devices have reached efficiencies approaching 16 % through the use of multiple layers with graded band gaps [9, 18]. In both tandem and multi-junction devices, the top layer material has the highest bandgap, with each subsequent layer or cell having a smaller bandgap. For example, a top-cell made of TiO_2 could absorb UV light, and a bottom-cell made of Fe_2O_3 could absorb visible light. In multi-junction cells, the efficiency increases with the number of junctions.

2.4 Materials: Metal Oxides for Water Splitting

The materials TiO_2 and α-Fe_2O_3 (for comparison) are now discussed in the context of the main material requirements for efficient water splitting as previously outlined. General material properties such as physical and electronic structure, or synthesis methods are not discussed here. These can be found in a range of reports [18, 19]. The basic requirements for water splitting can then be borne in mind when reviewing the effect of "plasmonic" metal nanoparticles in later sections.

2.4.1 *Bandgap and the Solar Spectrum*

The fundamental property of a semiconductor for solar cell applications is its bandgap. The generally accepted values of the indirect band gaps for rutile and anatase TiO_2 are 3.0 and 3.2 eV respectively [19, 20]. In terms of the wavelength of a photon, this corresponds to 413 and 388 nm. By comparison, the band gap of α-Fe_2O_3 is in the region of 2.0–2.2 eV (620–564 nm) depending on material preparation [12, 20]. This immediately gives an upper limit to the efficiency of a solar cell. Figure 2.2 shows the standard Air Mass 1.5 solar spectrum used by the solar cell industry for terrestrial systems and shows how the 1000 W m^{-2} irradiance is distributed over the range of photon wavelengths.

A modified version of the spectrum in Fig. 2.2 shows the proportion of total energy in the solar spectra from photons of energy equal to, or greater than the bandgap of a material. The range of band gaps for TiO_2 and Fe_2O_3 are shown.

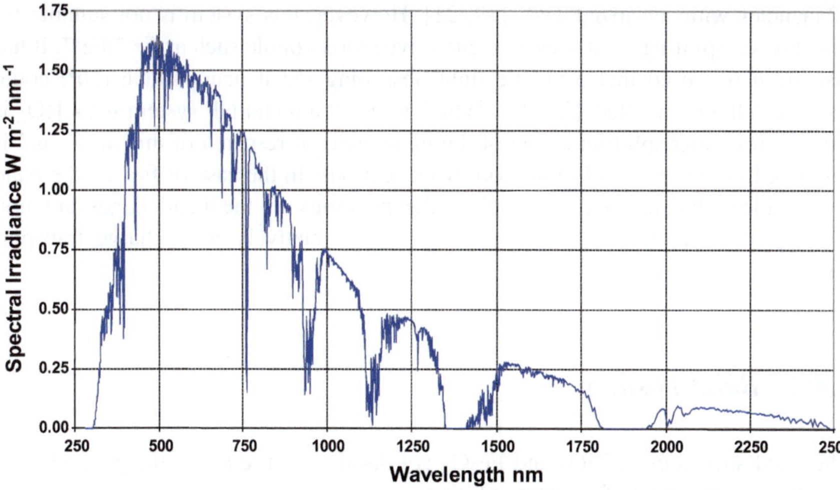

Fig. 2.2 Standard solar reference spectra—ASTM G173-03 (2012) used for solar cell design

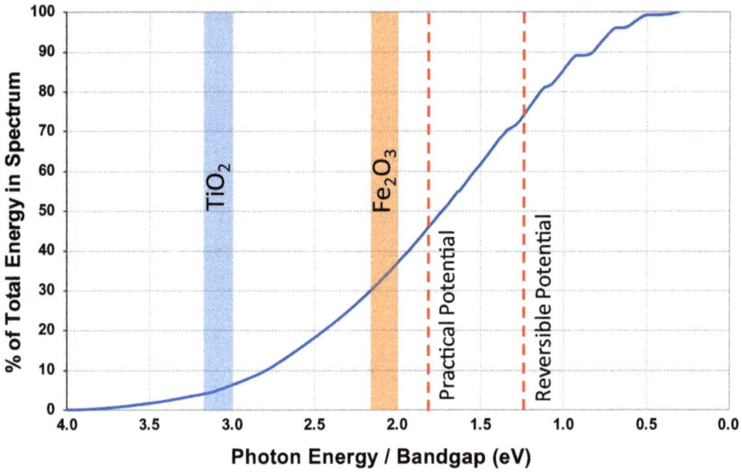

Fig. 2.3 Modified reference spectra showing cumulative solar energy with decreasing bandgap. The band gaps of TiO_2 and Fe_2O_3 are shown along with the reversible potential for water splitting and a practical potential

Based on Fig. 2.3 the maximum percentage of absorbable solar energy is between 4–6 % for TiO_2, and 30–40 % for α-Fe_2O_3. Taking thermalisation, recombination and resistive losses into account, the actual efficiency will be much lower. For TiO_2, it is clear that 10 % STH efficiency is unachievable unless it is used alongside another material or sensitized to photons into the visible range (400–700 nm), for example with a light absorbing dye as in a dye-sensitized solar cell (DSSC), in which electrons are injected into the TiO_2 by the dye. The efficiency of a TiO_2 based DSSC is above 10 % indicating the potential for moderately high efficiencies with sensitized TiO_2 [12, 21]. However, this system is not suitable for direct water splitting, as it uses a regenerative redox couple such as $Fe^{3+}|Fe^{2+}$. It has also been proposed that a visible light absorbing metal nanoparticle (NP) could reproduce the same effect [21, 22]. Whether or not a metal NP system with TiO_2 is suitable for water splitting is one of the main areas of research of this thesis and is discussed in greater detail in the following sections. In the case of Fe_2O_3, the band gap is ideal, but as discussed below the positions of the band edges are not. Additionally, Fe_2O_3 has been reported to have relatively poor charge transport characteristics [22, 23].

2.4.2 *Band Position*

The band structures of TiO_2 and Fe_2O_3 are discussed in terms of the general band structure requirements for water splitting.

Xu and Schoonen [23, 24] compiled a comprehensive list of the band positions of oxides based on experimental flat band data and calculation. Both TiO_2 and Fe_2O_3 have valence band edges significantly positive of the oxidation potential of water, so they should both sustain the oxygen evolution reaction (OER) under illumination. The conduction band of TiO_2 is just positive of the H^+/H_2 redox potential, suggesting that reduction of H^+/water and therefore unbiased water splitting is not possible. The position of the TiO_2 conduction band is sometimes reported as being slightly negative [24, 25]. This may be due to differences between anatase and rutile and other factors such as particle size and the surface of the material studied.

Fe_2O_3 also has a conduction band that is too positive to reduce H^+/water, so while it can absorb a greater proportion of the solar spectrum, it requires a significant bias in a hydrogen-producing PEC. The need to input energy decreases the achievable STH efficiency. Cesar et al. [25, 26] used two series-connected TiO_2 based DSSCs in a tandem cell configuration with a Fe_2O_3 anode to achieve water splitting estimated at 2.1 % STH efficiency from purely solar sources. This is a relatively high efficiency for a metal oxide based PEC. However, using three cells together in a system is complex and unlikely to be economic.

It is noted that another metal oxide, $SrTiO_3$, with a bandgap of 3.2 eV (the same as anatase TiO_2) is the only oxide to have demonstrated full water splitting at zero bias [23, 26]. This can be attributed to the position of the conduction band being sufficiently negative of the reduction potential of water. As with TiO_2, the large bandgap severely limits efficiency. However, it could offer a more promising starting material than TiO_2, as it removes the need for bias. Nevertheless, the focus of this thesis is on aspects of TiO_2 related to water splitting.

In Fig. 2.4, the calculated conduction and valence band edges of a range of metal oxides are given, relative to the reduction and oxidation energies of water at pH 0. The data for TiO_2 and Fe_2O_3 are highlighted.

2.4.3 *TiO_2 and Fe_2O_3 for Water Splitting*

The first use of TiO_2 for photo-electrochemical water splitting is generally credited to Fujishima and Honda, who used a (001) rutile single crystal [27, 28]. Earlier work by Boddy, occasionally referenced, was an electrochemical investigation of TiO_2 only [11, 28, 29].

There have since been numerous reports of water splitting by TiO_2 studying effects of doping, particle size, synthesis and more. All have in common an extremely low STH efficiency, usually $\ll 1$ %. Many comprehensive reviews are available [11, 29, 30]. While the internal quantum efficiency of TiO_2 photo-anodes may be improved, the STH conversion efficiency is fundamentally limited by the large bandgap to well below an economical level. These studies in just TiO_2 are not

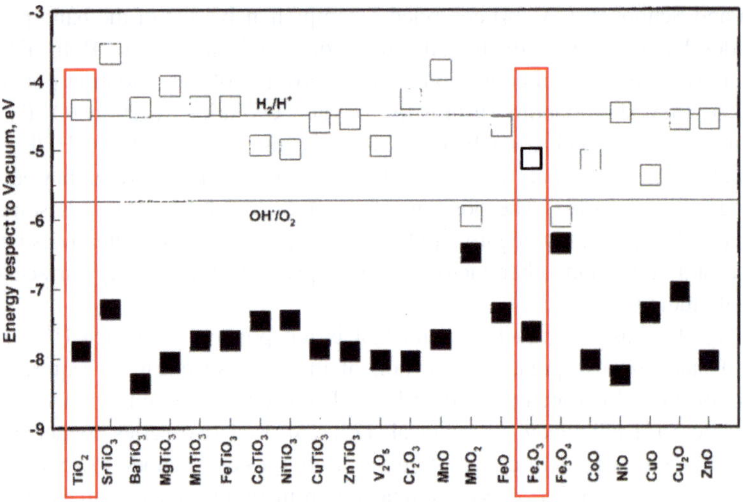

Fig. 2.4 Calculated energy positions (versus vacuum) of the conduction and valence band edges at pH 0 for selected metal oxide semiconductors. Taken from Xu and Schoonen [23, 27]. Reprinted with permission from Mineralogical Society of America

elaborated on here. The focus will be on systems functionalized with metal nanoparticles as a far less studied topic.

Neither TiO_2 nor Fe_2O_3 alone are suitable for achieving efficient water splitting; both require some bias or need to be used together in a tandem cell. However, they are worth studying as they are already fairly well documented and can be used as starting materials for modelling, to understand the various aspects of interesting surface phenomena, such as plasmonically-enhanced water splitting.

2.4.4 Other Water Splitting Materials

While TiO_2 is the focus of this thesis, there are numerous other possible materials for water splitting, such as $SrTiO_3$ as mentioned in Sect. 2.4.2. Tungsten trioxide (WO_3), which is stable in acidic solutions, is another well-known metal oxide and has even been used with Fe_2O_3 to improve charge separation by formation of a heterojunction [30, 31].

Oxynitrides are also being investigated as a potential materials group, due to the less positive position of their valence band. The nitrogen *2p* orbitals are less negative (versus SHE) than the O orbitals, decreasing the band gap of the oxynitride compared to the metal-oxide. Essentially, this decreases the bandgap, allowing a greater proportion of the solar spectrum to be absorbed, which seems to be a promising avenue of investigation, although efficiencies remain low [22, 31].

2.4.5 *Improving Solar to Hydrogen Efficiency*

One approach to achieve increased efficiency of water splitting, while still using the cheap and stable material TiO_2, might be sensitization to visible light: increasing how much of the energy in the spectrum is absorbed. As mentioned previously, dyes may achieve this, but would need to be developed to be stable in aqueous solution and resistant to oxidation. Another alternative, explored in this thesis is the use of surface plasmonic resonance on metal (gold) nanoparticles to absorb visible light. The general principal behind light absorption by metal NPs is outlined briefly below, followed by a review of key publications on TiO_2|Au systems used for solar energy, water splitting, and other redox reactions. The Fe_2O_3/Au system has also been investigated [12, 22], but so far there is little in the literature on this combination, particularly for water splitting.

2.5 Plasmonic Water Splitting

The general idea of plasmonic photocurrent is that the addition of metal nanoparticles that support localized surface plasmon resonance (LSPR) can lead to the absorption of parts of the solar spectrum that the substrate material does not absorb. This is analogous to the dye-sensitization of TiO_2 in DSSCs [12, 32, 33]. Using a noble metal particle such as gold as the sensitizer may result in a more stable system in the oxidizing environment required for oxygen evolution in water splitting, as it is harder to oxidize gold compared to an organic dye. The intention in loading TiO_2 with gold nanoparticles is to extend the photocurrent generation of a TiO_2 based photo-electrochemical cell (PEC) from the UV into the visible region of the solar spectrum. The basic principle is illustrated in Fig. 2.5. In more detail, the concept of plasmonic enhancement of TiO_2 is as follows.

Titanium dioxide absorbs only UV light ($\lambda < 400$ nm), which makes up only 5 % of the energy in the solar spectrum. The majority lies in the visible and infrared part of the spectrum ($\lambda > 400$ nm). A gold nanoparticle demonstrating localized surface plasmon resonance will interact strongly with visible light, scattering and

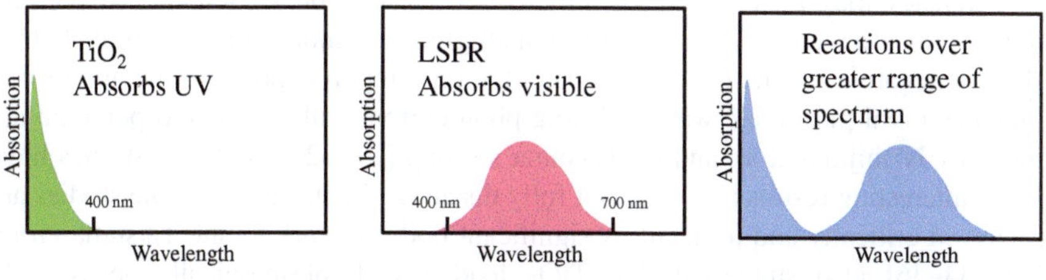

Fig. 2.5 Principle of using LSPR on a metal nanoparticle to extend the absorption of TiO_2 into the visible spectrum

absorbing around the resonant energy of the plasmon. The target is that the AuNP–TiO_2 system would therefore result in absorption of light from a larger portion of the spectrum. This is shown schematically in Fig. 2.5. Furthermore, it is hoped that the AuNP and TiO_2 will couple favourably and produce a photocurrent, enhanced relative to the TiO_2 alone, that will drive chemical reactions such as the oxidation of water.

In fact, enhanced photocurrents with visible light have been observed experimentally. Although, in most cases, the photocurrents generated at visible wavelengths of light are negligible. Any photocurrent generated at visible wavelengths corresponding to the LSPR, that cannot be attributed to the substrate material (TiO_2) is referred to in this thesis as 'plasmonic photocurrent'. The term 'plasmonic enhancement' is often used in the literature. But this phrase is suggestive of the enhancement of a pre-existing photocurrent by the action of the plasmon. The phenomena of interest here is the generation of a photocurrent that is not possible for TiO_2 alone, due to the energy of photons being insufficient to generate electron-hole pairs, i.e. sub-band gap photons. Hence, the use of 'plasmonic photocurrent' in this thesis to distinguish the effects from each other.

This section examines the literature of 'plasmonic enhancement' and 'plasmonic photocurrent' of titanium dioxide by metal nanoparticles for photo-electrochemical water splitting. The focus is on studies with gold nanoparticles rather than silver as the plasmon resonance of silver is in the UV region of the spectrum, while the resonance of gold is in the visible. As the aim is to sensitize the TiO_2 to visible light, silver is not considered suitable. Furthermore, silver will rapidly oxidize in aqueous solution. The review covers existing studies of metal nanoparticles on TiO_2 and includes a critical discussion of the possible physical mechanism that could lead to an enhancement of performance. While the review is focussed primarily on water splitting for hydrogen production, other photo-electrochemical and photocatalytic applications are also discussed, as this is a relatively new area. Selected studies based on silver rather than gold nanoparticles are also discussed when relevant. The review concludes with suggestions for a course of investigation, which formed the results reported in this thesis.

Increasing the efficiency of metal NP and TiO_2 systems in practice will be dependent on being able to optimize the processes at work. To optimize a process requires a precise understanding of the mechanism at work, which is yet to be determined unambiguously. It is possible that once identified, the theoretical limit of efficiency may prove to be too low to indicate economic viability. At present, most studies have shown either no or negligible plasmonic photocurrent. Others show a negative impact overall. To date, the only reports known to the author that indicate a real plasmonic water splitting photocurrent with real-world potential are those by Nishijima et al. and by Tatsuma's group [21, 32–35]. These studies have some interesting results that were not fully discussed by the authors. The studies are reviewed critically and in detail. A significant body of work by the Tatsuma group [21, 34–36] also suggested that TiO_2 loaded with plasmonically active gold nanoparticles can give good visible light performance in a PEC.

One of the main objectives of this thesis was to gain an understanding of the mechanism of plasmonic water splitting and to use this to fabricate photo-anodes that could be used to test and optimize the process. Hence, the critical review of the mechanisms presented in the literature.

The various proposed mechanisms discussed in detail during the course of this chapter are:

- 'Hot' electron injection or direct electron transfer (DET)
- Plasmonic Resonant Energy Transfer (PRET);
- Local heating
- Various effects of the 'near-field' around the nanoparticle (of which PRET could be considered one)

The following sections introduce plasmonic effects on nanoparticles, studies on water splitting and other photo-electrochemical systems involving plasmonic enhanced TiO_2, and a critical discussion of the mechanisms that may be operative.

2.6 Plasmonic Nanoparticle Arrays

Illuminated metal nanoparticles (NP) may exhibit Localized Surface Plasmon Resonance (LSPR): oscillations of the electron plasma about the ion cores that affect the absorption and scattering of light. The electric field around the nanoparticles is also concentrated due to the small size of the nanoparticles: the 'lightning-rod' effect. The electric field local to the nanoparticle is often referred to as the 'near-field' [36]. The effect of the electric field of incident light on the electrons in the nanoparticle is shown schematically in Fig. 2.6. The action of the electric field (denoted by E) causes the electrons at the surface of the metal nanoparticle to oscillate relative to the positive ions of the metal atoms. The electrons oscillate collectively at the frequency of the incident photon, resulting in

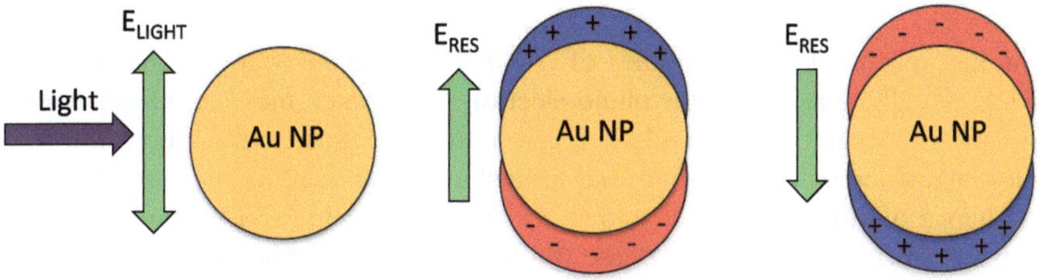

Fig. 2.6 Oscillation of electrons in a metal nanoparticle due to oscillating electric field of incident light, E_{LIGHT} results in electric field of localized surface plasmon resonance E_{RES} [36, 38]

positive and negative poles on the surface of the nanoparticle. For a spherical nanoparticle, a dipole is formed with an electric field E_{RES}. For other geometries (rectangles, triangles), multiple poles are possible [37].

For nanoparticles deposited on a substrate, there can also be a plasmonic effect at the interface called a Surface Plasmon Polariton (SPP). This is a wave that travels along the interface, and that can be used as a waveguide. The effect of the near-field of both LSPs and SPPs on the substrate is unknown. It may alter the absorption or band structure of the supporting material, it may influence charge separation or transfer, or have little effect.

Systems incorporating 'plasmonic' metal nanoparticles and semiconductors have been investigated as a means to improve the efficiency of both photovoltaic (PV) and photo-electrochemical cells. While the mechanisms behind the enhancement of PV, such as increased path length due to scattering [38, 39] and increased coupling of light into the substrate [39, 40] are reasonably well understood, the effects in photo-electrochemical and photocatalytic systems including TiO_2 are not.

The detailed theory of plasmonics is available from many sources [36, 40–42], but is not presented here. The focus of this review is not on how a plasmon is excited, or modelling the absorption, but on the effect in photo-electrochemical systems.

Studies have shown that the number of resonances, resonant wavelengths, and intensity of plasmonic resonance is highly dependent on the size and shape of the nanoparticle, as well as the refractive index of the environment [32, 36, 41, 42]. It is reasonable to suppose that metal nanoparticle arrays of gold or other metals may be designed to absorb light optimally at any wavelength or range of wavelengths desired, such as the solar spectrum. These arrays can be fabricated on suitable TiO_2 substrates. What is unknown is how to design the Au-TiO_2 system as a whole, to optimize the generation of photocurrent under visible light.

2.7 TiO_2-AuNP Systems for Water Splitting

The key results of several studies of AuNP enhanced TiO_2 are presented below. While not all are related to the photo-electrolysis of water, they are informative for this thesis. Nishijima's study [32, 33] is introduced first, as it is both the most relevant and the most interesting, and formed a starting point for the work described in Chap. 5 of this thesis. Other studies are introduced as necessary to aid discussion of issues such as the effect of phase (rutile or anatase), doping, nanoparticle array architecture, and applications. Finally, the proposed mechanisms are discussed in the light of these studies and other 'plasmonic' photo-electrochemical studies involving different oxides and applications.

2.7.1 *Nishijima et al. [32, 33]*

Nishijima et al. [32, 33, 42] investigated photo-anodes comprised of arrays of Au nano-rods (AuNRs) and nano-discs (AuNDs) on niobium doped (0.05 % by weight) rutile single crystals with a (110) orientation. The study reports incident current to photon efficiencies (IPCE) of 5–10 % at the peaks, and internal quantum efficiencies (IQE) of 10–15 %. To our knowledge, these figures are orders of magnitude higher than any others reported for the Au-TiO_2 system. The authors suggest that injection of hot electrons from the AuNR/AuND into the conduction band of the TiO_2 is responsible for the photocurrent.

The nanostructures were deposited by electron-beam lithography, which allows precise control over size and shape, but is limited to coverage of small areas: 1 mm^2 in this instance. The AuNRs were 110×240 nm^2 and AuNDs were 125×125 nm^2 as shown in Fig. 2.7. Both sets of structures were 40 nm thick with a 2 nm Ti layer between the gold and TiO_2, which is a typical approach used to improve the adhesion of gold to a substrate in physical vapour deposition. The more recent of the two studies by the same authors uses AuNRs of similar dimensions. For the AuNRs, the UV-Vis-NIR extinction spectra (Fig. 2.8a, black line) demonstrated two peaks, at 650 nm and 1000 nm, which are attributed to LSPR on the short and long axes of the rods. Light polarized transversely or longitudinally with respect to the AuNRs showed only one of the peaks (Fig. 2.8a, red and blue lines). It is noted that this measurement was of extinction, which included absorption and scattering/reflection. However, the symmetric AuNDs displayed

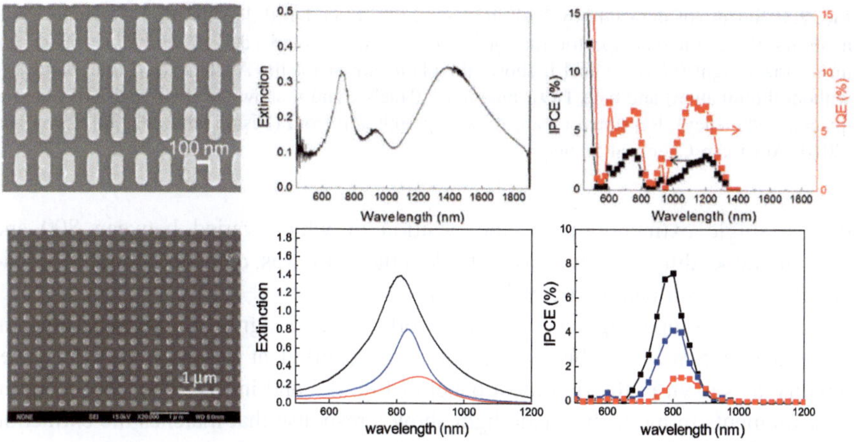

Fig. 2.7 Results of the studies by Nishijima et al. [32, 42]. The *top row* shows the characteristics of nano-rods, while the *bottom row* shows nano-disks. The *left-hand panels* are SEM images, the *central panel* is the optical extinction, and the *right-hand panel* is the incident-photon-to-current-efficiency (IPCE). The *curves* in the *bottom row* correspond to varying separation of the nano-disks. Reprinted (adapted) with permission from Nishijima et al. [32]. Copyright (2010), American Chemical Society

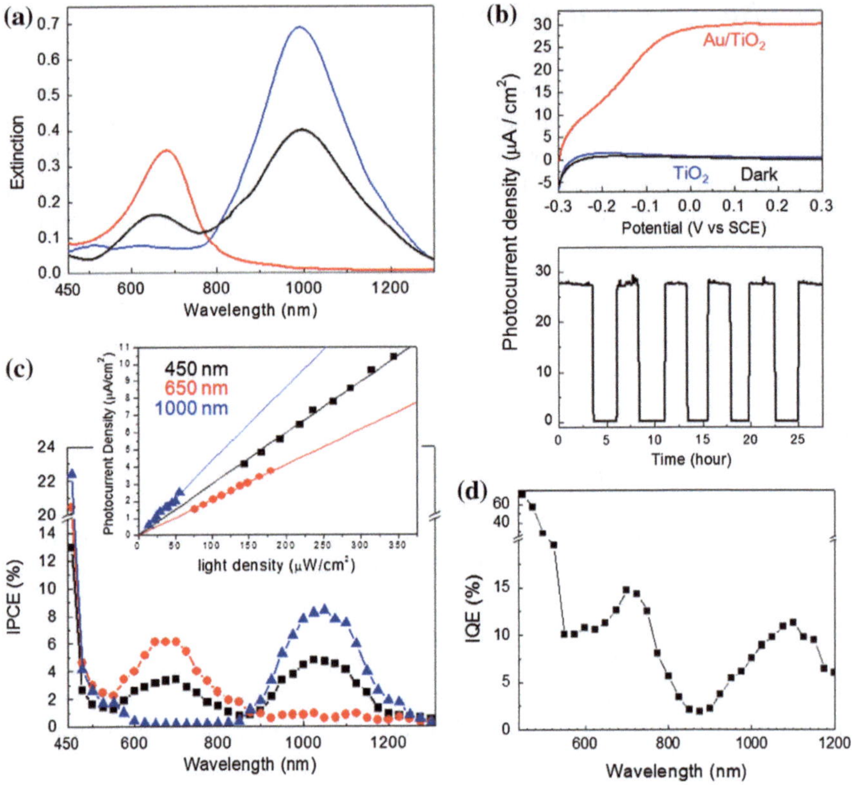

Fig. 2.8 Results of the study by Nishijima et al. [33] reproduced directly from the paper. Panel **a** shows the extinction spectra for the nano-rods with polarized (*red and blue lines*) and un-polarized light (*black*). Panel **b** shows the photocurrent for the Au/TiO_2 (*red line*, *black line* without illumination) and bare TiO_2 (*blue line*). Panels **c** and **d** show the IPCE and IQE (internal quantum efficiency). Reprinted (adapted) with permission from (Nishijima et al. [32]. Copyright (2010) American Chemical Society

only a single extinction peak, the position of which varied between 800 and 850 nm, depending on how densely packed the array was, demonstrating plasmonic coupling between nanoparticles [32, 42].

The most compelling aspect of this work is that a significant photocurrent (an IPCE above 5 % at visible wavelengths) is reported at visible and NIR wavelengths, as shown in the rightmost panel of Fig. 2.7 and in Fig. 2.8c. Photocurrent measurements in monochromatic light show a response that matches the extinction spectra closely. There is a slight shift in the position of both the visible and the near-infra-red (NIR) peak. This could be attributed to the fact that the resonance for absorption and scattering in LSPR on a metal nanoparticle may occur at slightly different wavelengths [33, 42]. Monochromatic light was used with a cut-off filter to remove any residual UV to ensure that all current measured was attributable to visible light. This was identified as a potential issue during the work reported in this thesis; a monochromator can produce higher order harmonics. For example, light monochromated to 700 nm can have a 350 nm component, which will directly excite electron-hole pairs in TiO_2 leading to spurious results. The type of cut-off filter used in the reviewed publication was not identified.

Under illumination by 500–1300 nm light (of unspecified intensity), a photocurrent of 30 $\mu A\ cm^{-2}$ was achieved. On illumination by monochromatic light at 450, 650, and 1000 nm wavelengths, there appeared to be a linear increase of photocurrent with increasing intensity (inset in Fig. 2.8c). Illumination intensity varied from 10–350 $\mu W\ cm^{-2}$, resulting in photocurrents of up to 10 $\mu A\ cm^{-2}$. As presented, the measured photocurrent seemed to increase with photon energy; the intensity of the shorter wavelength light was higher.

In the insert of Fig. 2.8c, extrapolated linear fits of the photocurrent for 450, 600, and 1000 nm light indicates that a higher photocurrent could be produced by the lowest energy photons. If this is a process with an energy threshold, then it is reasonable, as the 1000 nm incident beam would contain more photons per Watt per unit area. However, the 650 nm illumination produced a lower current than the 450 nm, which is not congruous with a simple picture of an energy threshold. The authors of the study did not make these observations.

Furthermore, it is noted in this thesis that a photocurrent is generated at photon energies below the reversible energy for water splitting of 1.23 eV, which is equivalent to a wavelength of 1008 nm. This might indicate that the energy of multiple photons is required to produce the observed photocurrent.

The experiment was carried out in water at pH 7 with a 100 mM $KClO_4$, with a platinum wire counter electrode and a saturated calomel reference electrode (SCE). The photocurrent onset seemed to start at −0.3 V and incident photon charge carrier efficiency (IPCE) measurements were made at +0.3 V (SCE), which corresponded respectively to −0.056 and 0.544 V versus SHE (based on SCE being +0.245 V vs. SHE). At pH 7, the equilibrium potentials for hydrogen and oxygen are at −0.413 V (SHE) and +0.817 V, respectively. Measurements were also made in KCl with 1×10^{-3} mol dm^{-3} of 1,2-benzenediol as an electron donor, resulting in photocurrents greater by a factor of 1.4. The equilibrium potential of 1,2-benzenediol (presumably oxidized to CO_2) was given as +0.32 V (SHE), which is much less positive than that for oxidation of water. Such a small increase is somewhat surprising given the usual exponential effect of increasing overpotential, unless adsorbed reactants and products inhibited reaction rates.

It was also noted by Nishijima et al. that the photovoltage of the AuNR patterned sample increased compared to that of the bare(un-patterned) TiO_2, which could indicate more free electrons in the conduction band, due to increased absorption of light. The authors attribute the detected current to the oxidation of water, suggesting the possibility of utilizing visible light in a water splitting system based on TiO_2. This was confirmed in their later study by detection of generated hydrogen and oxygen by gas chromatography—mass spectroscopy (GS–MS) [33, 43].

One detail that is not mentioned in the paper is the optical extinction spectra of the bare TiO_2 substrate. While the doping level of the TiO_2 substrate with Nb was only 0.05 wt%, this would have resulted in significant absorption in the visible and

IR spectrum compared to undoped single crystal TiO_2, as found in attempts to reproduce the result of this paper and discussed in Chap. 5.

Based on this result, it seems to be possible to oxidize water efficiently with visible light using gold nanoparticles on TiO_2, specifically if using 0.05 wt% Nb-doped rutile (110). Whether the effect can occur with other crystallographic orientations, dopants and doping levels, or on other phases like anatase is undetermined. The following reviews of papers try to address these questions. It is noted that the authors of the study did not discuss whether or not the doping of the TiO_2 substrate was an important feature.

2.7.2 *Liu et al. (2011)*

Liu et al. [43, 44] also studied plasmonic enhancement of water splitting using gold nanoparticles on TiO_2. The study was based on anatase TiO_2 nanotubes formed by anodizing Ti foil in a NH4F electrolyte (Fig. 2.9). Then 5 nm of Au was deposited using electron beam evaporation, which formed densely packed islands rather than a continuous film. The authors highlighted a decrease in photocurrent under UV light, due to the dense gold film preventing light from reaching the TiO_2.

Under illumination by monochromated light, there is a twofold enhancement of photocurrent at 532 nm and a peak enhancement of sevenfold near 650 nm (Fig. 2.9f). However, these enhancements are of a small photocurrent density of the order of hundreds of nano-amperes (not given as a current density, Fig. 2.9e. In this

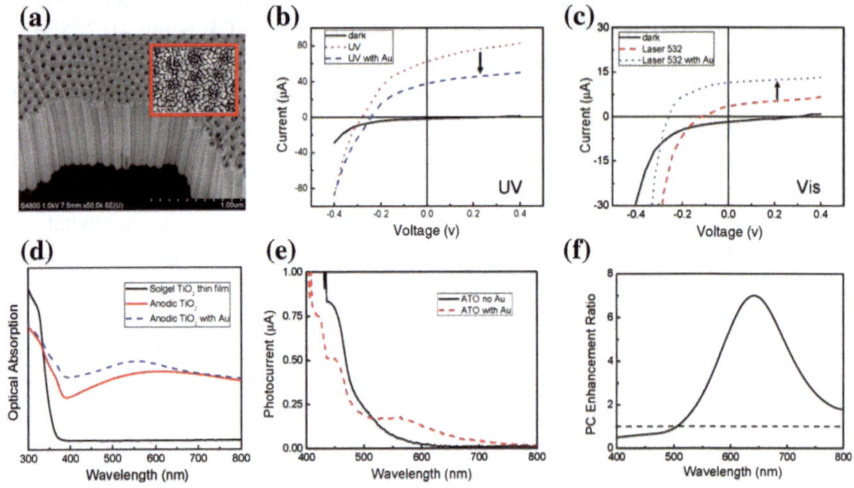

Fig. 2.9 Effect of gold nanoparticle film (*inset*) on anatase TiO_2 nanotubes under UV light (**b**) and visible illumination (**c**). Absorption, photocurrent over the visible part of the spectrum and the enhancement ratio are shown in (**d**), (**e**), and (**f**). Reprinted with permission from Liu et al. [43]. Copyright 2010 American Chemical Society

case, the plasmonic effect may have been an enhancement of a pre-existing photocurrent or the dark current. What is clear is that the TiO_2 nano-tube film had significant absorption of visible light compared to un-doped TiO_2 synthesized by a sol-gel process, which was explained by accidental doping by N and F present in the electrolyte during the anodizing process (Fig. 2.9d).

Repeating the photocurrent experiment with a sol-gel TiO_2 film (presumably also anatase) rather than anodized Ti gave no photocurrent under visible light, either with or without gold particles. The authors suggested that enhancement of absorption by nanotube films due to their doping was responsible, rather than injection of electrons, which they argued should have occurred with both the sol-gel film and the nanotube film. This conclusion suggested that doping could also have been responsible for the photocurrent reported in Nishijima's paper, based on Nb-doped TiO_2 having significant absorption in visible light (see Chap. 5 of this thesis).

The two studies introduced so far showed that it may be possible to use TiO_2 with gold nanoparticles to split water using visible light. However, it is equally possible to fabricate materials that decrease the overall efficiency of the system.

2.7.3 *Naseri et al. (2010)*

Naseri et al. [44, 45] studied anatase TiO_2 films loaded with AuNPs that exhibited visible light absorption due to SPR, resulting in doubling of photocurrents with Au present, but of magnitudes less than 0.4 A m^{-2}, which corresponded to a STH efficiency of <0.5 % under 0.1 W cm^{-2} light or 4 A $W^{-1.}$ This was attributed to (assumed rather than measured) water splitting in the visible part of the spectrum. It was not clear whether the source spectra included UV light, so it is hard to confirm whether photocurrents were due to visible light absorption or UV.

2.7.4 *Chandrasekharan and Kamat (2000)*

Chandrasekharan and Kamat [45] authored the earliest paper reporting attempts to exploit visible light absorption of AuNPs to enhance the photo-electrochemical performance of TiO_2. They did not discuss water splitting, nor assigned reactions, nor identified a redox couple explicitly, but as they used 0.05 M NaOH electrolyte, it is assumed that no other redox species were added. The TiO_2 was produced by a sol-gel process, followed by annealing at 400 °C, which should have produced an anatase film, although this was not characterised. The gold nanoparticles were formed by immersing the electrode in a toluene solution containing 5 nm gold colloids that adsorbed onto the TiO_2 film surface and aggregating as 20–70 nm clusters (Fig. 2.10). The Au/TiO_2 film exhibited a SPR at 546 nm (red-shifted from Mie Theory), but the spectral response of photocurrents was shown not to be

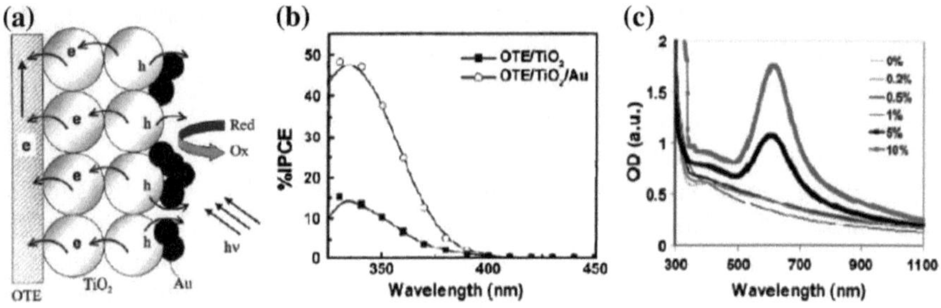

Fig. 2.10 Effect of gold nanoparticles on photocurrent of TiO_2 PEC. Schematic diagram of Au nanoparticles on TiO_2 film (**a**), measured photocurrent (**b**), and film absorbance (**c**) [43, 45]. Reprinted with permission from Chandrasekharan et al. [45]. Copyright 2000 American Chemical Society

influenced by the addition of the Au particles, i.e. there was no photocurrent response to visible light. However, photocurrents were enhanced threefold and the photovoltage also increased. It was suggested that the AuNPs increased charge transfer rates between the TiO_2 and electrolyte by decreasing overpotentials. While the catalytic process was not identified, an example might be $Au_2O_3/Au(OH)_3$, formed by hole-driven oxidation following photon absorption by TiO_2, acting as a catalyst for water/OH^- oxidation:

$$Au_2O_3 \rightleftarrows 2Au + 6OH^- + 6h^+_{VB,TiO_2} \tag{2.4}$$

where,

$$E_{Au_2O_3/Au}(SHE)/V = 1.363 - 0.0592\ \text{pH} \tag{2.5}$$

Another possibility is that the nanoparticles scattered the incident UV light, increasing the optical path length of the material, so increasing rates of electron-hole pair generation.

Based on the small photocurrent densities reported by Liu et al. [21, 43] and the results in Chap. 5 of this thesis, it is possible that there was a small visible light photocurrent present in the study by Chandrasekharan and Kamat, but that it was not detected. As the following studies by the Tatsuma group showed, it is possible to have a measurable and significant plasmonic enhancement of photo-electrochemical reaction rates on anatase TiO_2.

2.7.5 *Tatsuma Group*

Tatsuma's group at the University of Tokyo, Japan, have reported several studies of the photo-electrochemistry of Au-TiO_2 systems. Though not necessarily all for aqueous solutions, they are critically reviewed for this thesis when relevant information was identified.

Tian and Tatsuma [21] investigated nano-porous anatase TiO_2 films loaded with Au or Ag nanoparticles in a photo-electrochemical cell with aqueous 0.05 M NaOH solution and also with an $I^-|I_3^-$ couple for measurement short-circuit currents. The $I^-|I_3^-$ couple was likely used due to being more thermodynamically and kinetically facile than oxidizing hydroxide ions to oxygen. The redox potential also sits between the conduction and valence bands, permitting the reaction to occur without additional bias. The application of bias would render the short circuit current meaningless. An anatase precursor sol was spin-coated onto ITO and annealed at 500 °C. The resulting anatase film was then immersed in an ethanol solution containing Au colloids of 5–20 nm diameter to coat the surface with Au particles. The Au coated TiO_2 exhibited LSPR at 542 nm, which was red-shifted from the 530 nm of the colloidal solution, due to the higher refractive index of the TiO_2 substrate. The response of the system to monochromatic light from 400 to 700 nm followed the absorption/extinction spectra closely (Fig. 2.11). There was a

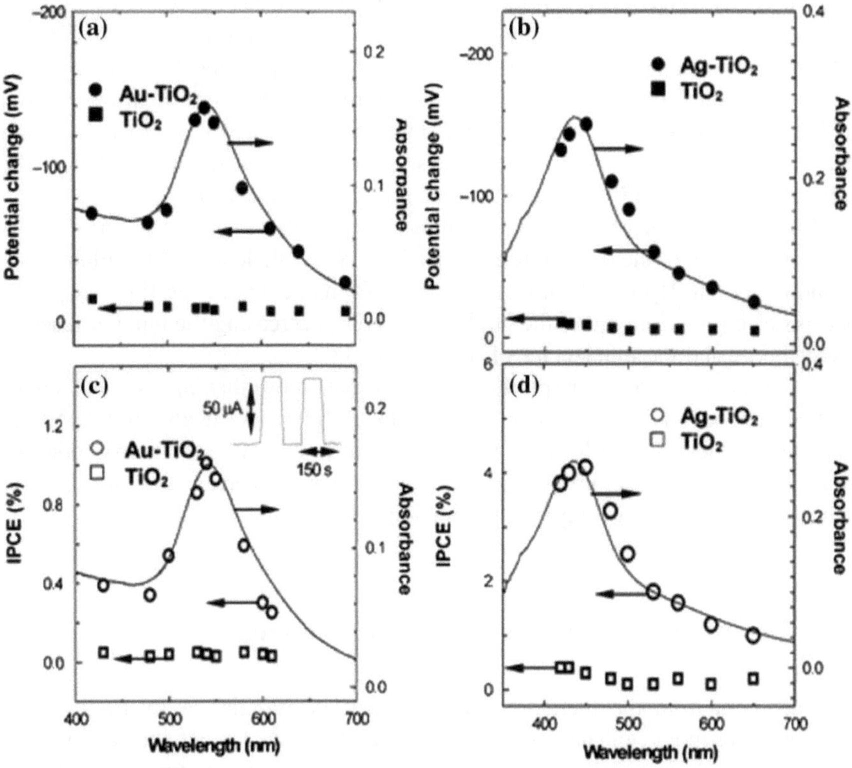

Fig. 2.11 Results reported by Tian and Tatsuma [21, 34]. The open circuit potential and IPCE are shown against the absorbance for both gold- (*left-hand images*) and silver- (*right-hand images*) NP coated anatase TiO_2. Reprinted (adapted) with permission. Copyright © 2004, Royal Society of Chemistry

measurable short-circuit current under illumination that peaked at 1 % IPCE at 540 nm. Likewise, the open circuit potential became more negative in the presence of the gold, peaking at −140 mV relative to that of bare TiO_2. A similar behaviour was seen with silver nanoparticles in place of gold.

The authors described this behaviour as 'remarkable' and attributed the effect to the plasmon absorption of the gold nanoparticles. They proposed that the electric field in the TiO_2 space charge layer influenced the oscillating electrons in the metal nanoparticles, drawing the electrons into the conduction band of the TiO_2 substrate, and leaving positively charged metal nanoparticles that oxidized hydroxide ions to oxygen.

Tian and Tatsuma [34, 45] continued the work by investigating the effect of changing the redox couples, noting that the system with I^- as an electron donor was not entirely stable. TiO_2 films of 4 μm thickness were obtained by spin coated sol-gel. In contrast to the previous work, the gold nanoparticles were formed by photocatalytic reduction. Films were immersed in 5×10^{-3} mol dm^{-3} of $HAuCl_4$ for 30 min and then irradiated with UV light to form <50 nm diameter nanoparticles. The most promising redox couple tested was $Fe^{2+}|Fe^{3+}$, with the optimum ratio of $FeCl_2$–$FeCl_3$ identified to be 0.1 and 0.05 M, respectively, in a solution of N_2-saturated acetonitrile and ethylene glycol in a 60/40 volume ratio. An IPCE of 12 % at a wavelength of 560 nm was reported, corresponding to an energy conversion efficiency of 1.27 % with illumination of 10 mW cm^{-2}. With the addition of 0.2 M 4-nitrobenzoic acid, the IPCE increased up to 26 %. Methanol, ethanol and other hydrocarbons were also investigated to establish the feasibility of a 'photovoltaic fuel cell', but the IPCE was found to be much lower, perhaps due to gold being a poor alcohol oxidation catalyst. It was also seen that the presence of O_2, a by-product of the oxidation, decreased the IPCE efficiencies. The solution was found to contain H_2O_2, which is a product of O_2 reduction, suggesting that the O_2 accepts electrons from either the gold and/or TiO_2, decreasing the number reaching the cathode, and the efficiencies of the circuit.

While this is not a water splitting application, it shows that an anatase TiO_2-Au system can be driven by visible light to do useful work. It also highlights the potential problem of simultaneous O_2 reduction at an O_2 evolving photo-anode. In a water splitting cell, this unwanted side-reaction would decrease the electron concentration available for hydrogen production.

2.8 Mechanisms of Plasmonic Enhancement

It is apparent from the literature that there are uncertainties as to the processes leading to plasmonic enhancement of photo-electrochemical reactions or generation of plasmonically-induced visible light photocurrent. Several common explanations are presented below and are discussed critically. Reports on plasmonic enhancement of both TiO_2 and other metal oxides and semiconductors are considered. The phenomenon of interest is how the Au-TiO_2 system can produce photocurrent under visible light. Increased photocurrent at UV wavelengths, at which TiO_2 is already photoactive, as seen by Chandrasekharan and Kamat [12, 45], is not considered in detail.

2.8.1 *Direct Electron Transfer (DET)*

The most common mechanism proposed in the literature is that of electron injection also referred to as direct electron transfer (DET), which is proposed to act in a similar manner as dye-sensitized solar cells [12, 34]. The originators of this mechanism appear to be Tian and Tatsuma [32, 34]. They propose that:

1. A photon is absorbed by a surface plasmon on a metal nanoparticle;
2. Electrons in the metal nanoparticle oscillate close to the interface with the TiO_2;
3. The electric field in the space charge layer of the TiO_2 causes electrons in the metal nanoparticle to be injected into the TiO_2 conduction band;
4. The positively charged metal nanoparticle oxidizes the electron donor in solution.

The suggested steps in the process are illustrated schematically in Fig. 2.12. The effect of the plasmon is to give an electron sufficient energy to overcome the Schottky barrier at the interface between Au and TiO_2.

Nishijima et al. [32, 33] also proposed the injection of an electron as the principal mechanism in a water splitting process, although they do note that there may also be some non-linear effects due to the electric field enhancement operative. They suggested that an electron is promoted from the filled d-band of the AuNP into the conduction band of TiO_2, followed by electron donation from H_2O to the AuNP to compensate the electron loss. In more recent papers [33, 46], they expanded on this concept. Later, the authors considered the requirement for a hole

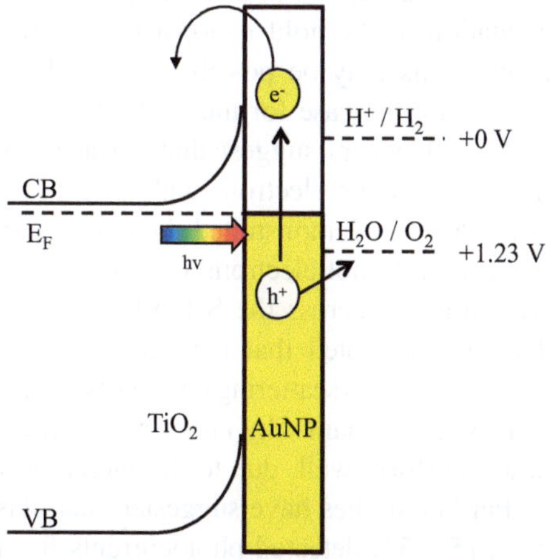

Fig. 2.12 Schematic of electron injection from a metal nanoparticle into the conduction band of TiO_2

to drive the oxidation. It was noted that for water oxidation to occur, four holes of sufficient potential are required for each molecule of oxygen. They suggested that the injection of electrons from the gold would leave a net positive charge on the metal nanoparticle that oxidizes H_2O or OH^- and that the positive charge could be considered as a de-localized hole at the surface.

The suggestion of a positively charged hole in a metal could be considered controversial. The electronic structure of metals is conventionally described in terms of the free electron gas [46–48]. To suggest that there is a long-lived hole with a defined energy and sufficient potential to accept an electron is a novel and unproven concept. However, Petek et al. [47–49] have measured the generation of electron-hole pairs in metals such as gold and copper at ultrafast timescales. While an extremely short-lived hole may be possible, it is not certain whether it is possible for these holes to have a sufficient lifetime to accept an electron from a solution species. Typically, the hole is filled by a cascade of electrons from higher to lower energies within the nanoparticle, for example by Auger recombination. This is also unlikely given the typical femtosecond timescale (10^{15} Hz) of the plasmon oscillation. Furthermore, the potential of any hole is undetermined, but would have to be positive of the water oxidation potential.

On the physical scale of energy, and at pH 0, water reduction occurs at about −4.5 eV (versus vacuum), and oxidation at −5.73 eV. The work function of gold is around 5.4 eV [49, 50]. Converting to the absolute potential scale, this suggests that the Fermi level of gold at +0.9 V (SHE) is not positive enough to oxidize water, which is at +1.23 V (SHE). If the work function of gold is unaffected by pH (and adsorbed ions), then the potentials of water oxidation and the Fermi level of gold become equal at pH 5.6. At pH 5.6 and above, it should be possible for water to be oxidized by a hole at the potential of the Fermi level. However, if the gold is in nanoparticle form, it might be possible that the potential of the TiO_2 substrate determines the potential of the Fermi level of gold.

Details of how the injection of the electron into the TiO_2 might proceed are also uncertain. Nishijima et al. proposed the electron is excited from the d-band to the s, p band in gold, and is then swept into the TiO_2. Again, this suggests a long-lived excitation in the gold nanoparticle, as if it were a molecule with discrete energy levels. This may be possible with gold nanoparticles below 10 nm [51]. This is clearly not the case for the Nishijima study.

Recent papers suggest that surface plasmons can indeed decay into ‘hot electrons’. These are electrons with an excess of energy above the Fermi level. Giugni et al. [50, 52] demonstrated the conversion of SPPs on the tip of a gold plasmonic waveguide to hot electrons with an apparent efficiently of 30 %. The hot electrons were injected across the Schottky barrier formed between the gold tip and GaAs. The authors noted that the decay of a plasmon into a hot electron, rather than radiative decay (scattering), might be enhanced by a rough Schottky barrier contact. This might explain why larger nanoparticles (such as those fabricated by Nishijima et al.) perform well, due to the increased likelihood of defects in the contact area.

Further studies have suggested that plasmons decay into ‘hot electrons’. Knight et al. [52, 53] detected photocurrents from arrays of AuNPs deposited on Si, with photocurrents corresponding to the absorption of photons with energies smaller than the band gap of Si. This was attributed to hot electron transfer into the semiconductor. Mubeen et al. [53, 54] coated AuNRs with TiO_2, Pt for water reduction and a Co oxidation catalyst. They observed both hydrogen and oxygen evolution under visible light, and attributed this to hot electron transfer. The results of experiments by Mukherjee et al. suggested that these hot electrons can also transfer into a Fesbach resonance of hydrogen molecules on gold nanoparticles, causing the dissociation of H_2 into atomic hydrogen [54, 55].

How the hot electrons are generated by the decay of the surface plasmon is not clear at this stage, and is only speculated. It could be hypothesized that an electron oscillating around the ionic core due to an applied electric field has increased potential and kinetic energies (forced-damped harmonic oscillation), which could raise the potential of some electrons above the Fermi level, to an energy at which they can be injected into the conduction band by the electrostatic field, or by tunnelling. This picture of a distribution of electron energies, above the usual Fermi-level, was used by Cushing et al. [55, 56] in their investigation of the photocurrent enhancement mechanism of Cu_2O films loaded with gold nanoparticles. The plasmonic photocurrent was attributed to a plasmonic resonant energy transfer (PRET, discussed in Sect. 2.8.2) mechanism between the Au and Cu_2O. However, follow up papers on Ag–Cu_2O core-shell structures by the same authors [52, 53, 56], reported a mixture of behaviour that was attributed to both PRET and DET, depending on the wavelength of the incident light.

In a talk at the Materials Research Society Spring 2013 symposium, Peter Nordlander[2] suggested that, rather than decaying to a distribution of energies, the plasmon may decay into a single hot electron with energy above the Fermi level of the metal equal to the energy of the exciting photon (no discussion was given regarding the generation of a hole). Given that a surface plasmon is often visualized as the collective oscillation of all free electrons at the surface of a nanoparticle, it is surprising that, rather than a distribution of energies, that the energy should be transferred to a single electron.

Recently, Sundararaman et al. [57] predicted the decay of plasmons and generation of both hot electrons and hot holes in metals by means of a quantized plasmon model and electronic states calculated with density functional theory (DFT). The authors predict a distribution of hot electrons above the Fermi level and a distribution of holes below the Fermi level. This appears to be an important step in beginning to explain how a plasmon can decay and result in the reduction or oxidation of a species in solution. Manjavacas et al. [58] have also estimated hot carrier generation using Fermi's Golden Rule. Govorov et al. [59] have investigated, theoretically, the injection of hot electrons from metal nanoparticles into semiconductor substrates by quantum mechanical methods. Overall, it appears that direct electron injection is beginning to develop a theoretical basis to support the observations of experimental work.

[2]To the knowledge of the author of this thesis, there has not been a related publication.

A number of studies [52, 53] suggest that the emission of hot electrons follows the Fowler formula (a form of which is given in Eq. (2.6)). The proposal is that the rate of photoelectrons with sufficient energy to overcome the Schottky barrier is given by a relation of the form:

$$g = C_f \frac{(h\nu - \phi_{SB})^n}{h\nu} \tag{2.6}$$

where g is a generation rate of electrons, C_f is the Fowler emission coefficient, n = 2 for most metals, and ϕ_{SB}, the Schottky barrier height. The study by Mubeen et al. [53] claimed that this equation fits data well. When the current increased at wavelengths corresponding to plasmonic resonance, the divergence from the above fit was attributed to an enhancement due to the plasmonic resonance [52, 60].

At energies above the plasmonic resonance, the photocurrent should be ascribed to the usual generation of electron-hole pairs, not hot electrons. Over the region of the plasmonic resonance, the fit of Eq. (2.6) to a typical current-potential curve for plasmonic enhancement is poor. At lower energies, the fit clearly diverges from the data, so while the fit follows the general trend, it is suggested here that the fit does not capture the effects of the plasmon resonance.

To accurately fit the plasmonic photocurrent data reported in this thesis, the Fowler Eq. (2.6) as presented above would need to be modified to account for the plasmonic enhancement, perhaps in the scaling constant C_f. It could be hypothesized that, based on equations for thermionic emission as presented by Murphy and Good [61] and Fowler and Nordheim and others [62, 63], the effect of the plasmon resonance is similar to either the temperature term of the electric field term. Greater photon energy is analogous to higher temperature. However, as the photocurrent has a maximum at a particular energy, rather than simply increasing with photon energy, it may be better to formulate this in terms of the electric (near) field, which is maximized at the plasmon resonance and may be high at the surface of a nanoparticle.

2.8.2 Plasmonic Resonant Energy Transfer (PRET)

Resonant Energy Transfer (RET) involves coupling between dipoles of two materials such as a nanoparticle and a molecule or nanowire of different materials by Coulomb forces. This can allow efficient energy transfer from one material to another, say between nanoparticles [64, 65] or from a plasmon to a fluorescent molecule. However, it is noted that the materials must have similar response to excitation, suggesting that an excitation produced by visible-light in one material cannot couple to a material with a significantly larger energy requirement for excitation. For example it is noted that ZnO and TiO_2 can couple to enhance fluorescence [64, 66] and also CdTe with plasmons in Au [66, 67]. Cascading from high to low energy is also seen from ZnO to an organic molecule via FRET;

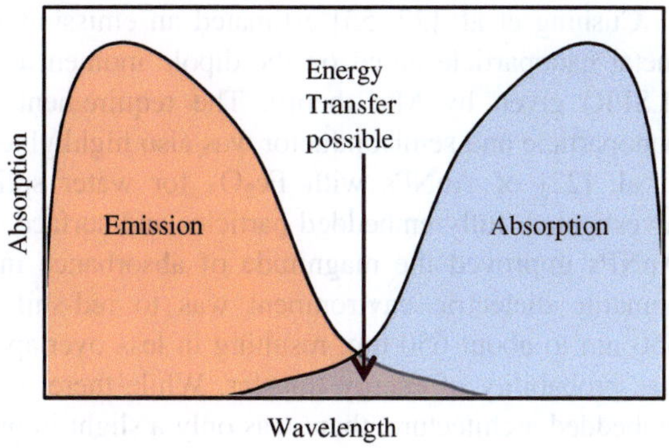

Fig. 2.13 Schematic of energy transfer between two materials, in which transfer is possible only if the absorption spectrum (*blue*) of the receiving material overlaps the emission spectrum (*red*) of the material emitting the photon

however, the reverse is not seen [67–70]. Whether this dipole oscillation energy transfer is possible for a plasmonic TiO_2 system requires further investigation, as a standard TiO_2 excitation is far from that of an Au nanoparticle.

The theoretical foundations of resonant energy transfer has been described by Andrews [55, 68–70], who showed that both Resonant Energy Transfer (RET) and inverse-square radiation were the near-field and far-field asymptotic solutions of energy exchange by emission of photons. In both cases, energy transfer is possible only if there is an overlap of the emission and absorption spectra of the emitter and absorber, as shown in Fig. 2.13. In the case of near-field energy transfer, often referred to as Förster Resonant Energy Transfer (FRET), the energy transfer is by exchange of virtual photons, while in the far field the exchange occurs through a real photon.

Energy transfer from the decay of a surface plasmon to the excitation of an electron-hole pair in a semiconductor could be viewed as a form of resonant energy transfer, and has been referred to as Plasmonic Resonant Energy Transfer (PRET). The study by Cushing et al. [55] used the effect of an insulating SiO_2 spacer layer between the Ag and Cu_2O to demonstrate that energy transfer occurs even when the gold nanoparticle and Cu_2O are separated. It is noted here that the band gap of Cu_2O (ca. 2.2 eV) coincides with the resonant wavelength of AuNPs.

A standard formulation of resonant energy transfer [55] is given without proof to demonstrate the fundamental requirement for overlapping emission and absorption spectra:

$$w_{FRET} = \frac{9\kappa^2 c^2}{8\pi\tau_{A*} n^4 R^6} \int \frac{F_A(\omega)\sigma_B(\omega)}{\omega^4} d\omega \tag{2.7}$$

where, the rate of transfer w_{FRET} depends on the product of the emission $F_A(\omega)$ and the absorption $\sigma_B(\omega)$ at each frequency ω. The transition rate decreases as R^{-6}, where R is the distance between particles. If there is no overlap, the integral is zero, and the rate of energy transfer is zero. The remaining parameters are related to the material.

Cushing et al. [22, 55] estimated an emission/absorption rate for a plasmonic metal nanoparticle based on the dipole moment of the localized surface plasmon (LSPR) given by Mie theory. The requirement for overlapping absorption of nanoparticle and semiconductor was also highlighted in a recent study by Thimsen et al. [22] of AuNPs with Fe_2O_3 for water splitting. Two architectures were investigated: fully embedded particles and surface deposition. It was clear that the AuNPs improved the magnitude of absorbance in both cases. The effect of the hematite dielectric environment was to red-shift the plasmon resonance from 550 nm to about 650 nm, resulting in less overlap with the Fe_2O_3 absorption and less probability of energy transfer. While there was increased absorbance for the embedded architecture, there was only a slight improvement in the photocurrent of the embedded system at high bias. This suggested that even with an emission and absorption overlap, it is not a simple matter that the addition of metal nanoparticles will improve performance. For the surface configuration, the overall effect under AM1.5 light was negative. The authors attributed this to the formation of Schottky barriers between Au and Fe_2O_3, resulting in lower open circuit potentials than the standard Fe_2O_3-H_2O junction. They speculated that increasing the Schottky barrier height might help overcome this drawback, although do not elaborate on the reasoning.

Given the large band gap of both rutile (3.05 eV) and anatase (3.2 eV) TiO_2, it seems that resonant energy transfer from a visible light absorbing surface plasmon to TiO_2 would occur with an extremely low probability. The only way this would be likely is if multiple plasmons could decay simultaneously and pool their energy into exciting an electron hole pair in TiO_2. The concept of energy pooling has been touched on by Andrews [70] as being a real process, but there has been no proposal that this may be in effect for the Au-TiO_2 system.

For energy pooling to occur, it seems to the author of this thesis that the surface plasmons of adjacent nanoparticles should be coupled. This is more likely to occur if the nanoparticles are of uniform size, shape, and separation and are separated by a distance of less than the wavelength of the plasmon. This may explain why the nanoparticle array used by Nishijima et al. [32] performed so well. The paper also demonstrated (in the supplementary information) that the efficiency of the process was dependent on the inter-particle spacing.

2.9 Other Plasmonic Effects

2.9.1 Effect of Enhanced Near-Field on Absorption

In addition to DET and RET, another suggestion for the mechanism of plasmonic enhancement is the effect of the enhanced electric field at the surface of a metal nanoparticle supporting a surface plasmon. Liu et al. [43] note that doping materials can reduce the bandgap and allow absorption of visible light. They suggested that the electric near-field formed at the surface of the nanoparticle creates an enhancement of the electric field at the interface of the AuNP and TiO2 surface that influences charge separation and suppresses recombination, resulting in higher photocurrent. Nishijima et al. [32] also indicated that they believed there was a non-linear effect due to the enhanced electric field, in addition to electron injection, but do not expand.

It is not unreasonable to suggest that a larger electric field could enhance photocurrent generation. From Fermi's Golden rule [71], the transition rate is dependent on the square of the Hamiltonian operating on the initial and final states. Cushing et al. [55] derived a form for the transition rate in a semiconductor (2.8) that demonstrates that the rate of photon absorption, or transition rate w_{SEMI}, is proportional to the intensity of the electric field $|E_0|^2$ (the square of the field strength E_0) and as such an enhanced field should result in more electron-hole pairs, and subsequently higher photocurrent densities. The other terms are not defined here as the equation is only presented to highlight the $|E_0|^2$ dependence.

$$w_{SEMI} = \frac{e^2|E_0|^2}{4\pi^2 m^2 \hbar\omega^2}|p_0|^2\left(\frac{2m_{reduced}}{\hbar^2}\right)^{3/2}(\hbar\omega - E_G)^{1/2} \tag{2.8}$$

This argument may be true at a particular point local to a nanoparticle, but over a large area, the incoming photon flux is fixed, so the total absorption should remain constant. The effect of the enhanced electric field should be to increase the absorption coefficient, resulting in a shorter absorption depth for a given material. Exactly how the enhanced field might decrease recombination rates, as suggested by Liu et al. [43], is not clear.

Furthermore, the effect of enhancing the absorption coefficient of a material relies on that material absorbing at the plasmon resonance. For a material like TiO_2, if the photons are of below band gap energy, then enhancing the electric field cannot cause the absorption of a photon. This is shown by the term $(\hbar\omega - E_G)^{1/2}$, which only evaluates for $\hbar\omega > E_G$. However, for above band gap energy photons, it would be possible to enhance the absorption coefficient.

For these reasons, it is suggested here that enhanced absorption due to the intensification of the electric field at the surface of a metal nanoparticle cannot be the driving mechanism for the plasmonic photocurrent in Au-TiO_2 photo-anodes, as there is no overlap of the absorption spectrum of TiO_2 with the emission of gold nanoparticles in the visible spectrum.

2.10 Evidence for Direct Electron Transfer

Several sets of data are now examined that support direct electron transfer (DET) as the mechanism of plasmonic photocurrent for the Au-TiO_2 system.

2.10.1 Electrons in the Conduction Band of TiO_2

Furube et al. [72] have demonstrated that DET might take place using femtosecond transient absorption spectroscopy (TAS). They compared TiO_2 films with embedded AuNPs to films with a light absorbing ruthenium based "N-3" dye, as used in dye-sensitized solar cells (DSSCs), in which electron injection is known to occur from the dye to the TiO_2. The absorbance of a 3500 nm IR probe by electrons in the TiO_2 conduction band was measured. The system was excited by a 550 nm laser i.e. by photons with insufficient energy to directly excite electron-hole pairs in the TiO_2. Figure 2.14 shows the transient absorbance of 3500 nm light by conduction band electrons following the 550 nm excitation. The absorbance of the Au/TiO_2 system under pulsed excitation was similar to the N3 dye/TiO_2 system, although the lifetime of the excitation was shorter lived, decaying after about 1 ns. The authors inferred that this implies that electrons were injected into the TiO_2 from the AuNP, as in the dye system.

It is suggested here that the TAS measurements are not conclusive evidence of DET. While they do demonstrate the presence of electrons in the conduction band of TiO_2 following illumination with visible light, they do not give any indication of the origin of the electrons. It is inferred by the authors of the study that because the dye injects electrons, that the gold nanoparticle must also do so. It is possible that the electrons are excited directly from the valence band of the TiO_2 somehow. A possible experiment to verify this would be a TAS measurement probing for holes rather than electrons. If a population of holes are measured in the valence band of TiO_2, then it might indicate that the electrons originate within the TiO_2, while the absence of the hole signal would support DET.

The fact that the effect is not seen with ZrO_2 is attributed by the authors to the more negative potential of the conduction band of ZrO_2 compared to TiO_2. This additional potential barrier might be sufficient to prevent the transfer of electrons. Likewise, if the conduction band electrons were generated in the TiO_2 itself, the larger band gap of ca. 5 eV could also prevent this from occurring in ZrO_2.

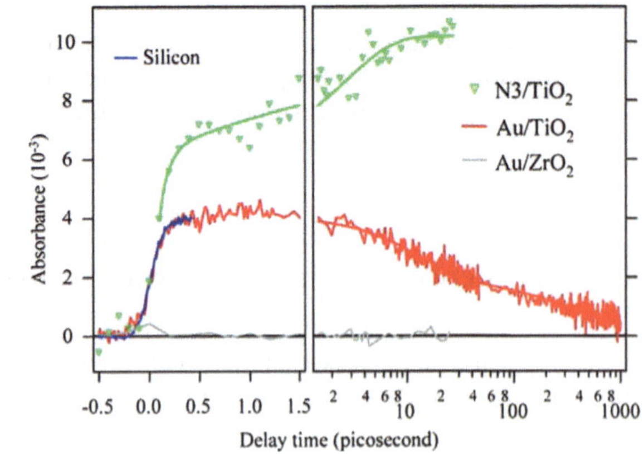

Fig. 2.14 Transient absorbance of TiO_2 films loaded with dye and AuNPs, and Au-loaded ZrO_2. Figure taken from publication [72]. Reprinted with permission from Furube et al. [72]. Copyright 2007 American Chemical Society

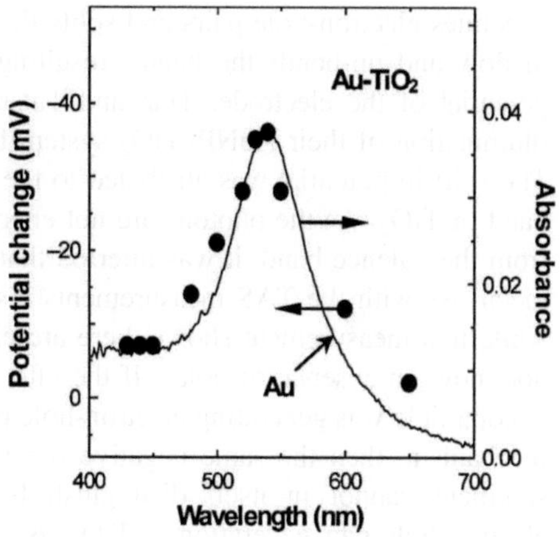

Fig. 2.15 Potential change of illuminated AuNP–TiO_2 electrode under monochromatic illumination. Figure taken from publication [21]. Reprinted (adapted) with permission. Copyright © 2004, Royal Society of Chemistry

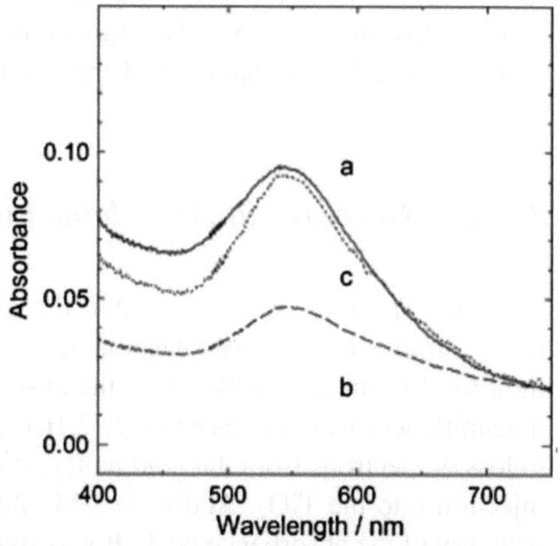

Fig. 2.16 Effect of extended period of illumination on the absorbance of Au-TiO_2 at open circuit (**a**), in the absence of an electron donor (**b**) and after the introduction of ethanol (**c**). Figure taken from publication [21]. Reprinted with permission from Tian et al. [21]. Copyright 2005 American Chemical Society

Two other sets of data used to support the DET mechanism were identified by Tian and Tatsuma [21, 34]. The first was the negative shift in open circuit potential (seemingly versus Fe^{3+}/Fe^{2+}) under visible light illumination (Fig. 2.15). The second was the decreased absorbance of the gold nanoparticles following extended illumination in white light, which was reversed by addition of an electron donor, such as ethanol (Fig. 2.16).

2.10.2 Shifting Open Circuit Potential Under Illumination

A negative shift in the open circuit potential under monochromatic illumination was detected in the anatase AuNP–TiO_2 systems studied by Tian and Tatsuma [21] (Fig. 2.15) and supports DET. In a conventional PEC, illuminating the junction generates electron-hole pairs and splits the quasi-Fermi levels. This allows current to flow and un-bends the bands, resulting in a negative shift in the open circuit potential of the electrode. Tian and Tatsuma saw a similar shift of potential on illumination of their AuNP–TiO_2 system by visible light in multiple experiments. The shift in potential was attributed to the presence of electrons in the conduction band of TiO_2. As the photons are not energetic enough to promote these electrons from the valence band, it was inferred that the electrons must have come from the metal. As with the TAS measurements discussed previously, it could be said that while this measurement shows there are electrons in the conduction band, it does not show an absence of holes. If the effect of the plasmon resonance on the gold nanoparticle was generating electron-hole pairs in the TiO_2 by some other unknown mechanism, then the same negative potential shift would be expected; this measurement cannot in itself distinguish between electron transfer from Au and electron-hole pair generation in TiO_2. As shown in Fig. 2.15 the potential change follows the absorbance of the plasmon, with the largest shift at the plasmon resonance. This suggests that the number of electrons in the conduction band was directly related to the strength of the excitation of the nanoparticles.

2.10.3 Decrease in Absorbance of Plasmon

Tian and Tatsuma also investigated the effect of an electron donor on the absorbance around the surface plasmon resonance. Following irradiation with visible light for 30 min at 10 mW cm^{-2}, the absorbance (presumably measured in terms of transmittance) was decreased (Fig. 2.16). The decreased absorbance was attributed to loss of electrons from the gold nanoparticle, due to the excitation of electrons and injection into the TiO_2. Adding 0.5 M ethanol as an electron donor resulted in the recovery of the absorbance peak. It was suggested that this regenerated the electrons in the gold nanoparticles. It was not made clear by the authors whether the Au-TiO_2 sample was immersed in water only with no control of the electrode potential.

On this basis, it might be imagined that without the addition of an electron donor, the gold nanoparticle would become depleted of electrons to the point that the absorption of light and photocurrent decreases and stops. This in turn would suggest that a system for producing hydrogen by water splitting would not be practical, as it would require a sacrificial electron donor.

The standard equilibrium potential of CO_2/ethanol is −0.346 V (SHE) at pH 7, compared to +0.817 V (SHE) for water oxidation [73]. The Fermi level of gold should be at a potential of +0.6 to +0.9 V (SHE) based on the vacuum level being at −4.5 V on the SHE scale and the work function of gold being between 5.1 and 5.4 eV. This suggests that electrons could be donated by ethanol easily, but not by water at this pH. For water to act as an electron donor to gold, the oxidation potential of water should be less positive than the potential of the gold. This will depend on both the pH and the work function of the gold nanoparticle, indicating that an investigation of the dependence of photocurrent on pH might be prudent for Au-TiO_2 systems.

The preceding paragraphs discussed the evidence for and against DET the main candidate for the plasmonic photocurrent mechanism. Generally, the evidence for Au-TiO_2 systems allows for the possibility of DET. The TAS and negative potential shift measurements are considered compatible with DET, but do not rule out any other possibilities.

2.10.4 Opposition to DET Mechanism

Liu et al. [43] refute the DET explanation, stressing that electron excitation in metals and dyes are not comparable, because in a dye molecule, there is a long-lived excitation and electron-hole pair generation, which has not been observed in gold nanoparticles. Control studies using sol gel films and ITO as the substrate for the gold nano-island film did not exhibit any visible light photocurrent. The authors reasoned that if DET was the mechanism, then the effect should work with both the sol-gel film and the ITO, but as no effect was observed, there must be some other mechanism at work. The point regarding ITO may be discounted as ITO is essentially metallic by design, so would have had no extended depletion region and electric field to effect charge separation. Any electron transferred to the ITO may easily recombine with the AuNP. The lack of photocurrent from AuNPs on anatase films prepared by a sol-gel process followed by thermal treatment is in-line with the work of Chandrasekharan and Kamat [45], and Naseri [44]. However, the studies by Tian and Tatsuma [21, 34] showed that plasmonic photocurrent with anatase and AuNPs was indeed possible (but with redox couples other than water). The lack of enhancement could perhaps have been due to the nature of the gold nano-islands, which did not demonstrate a sharp peak of plasmon resonance in absorbance, and were in addition deposited only on the top of the film. By comparison, Tatsuma's films were nano-porous and loaded with gold nanoparticles exhibiting a well-defined resonance in the absorption spectrum. This suggests that both the interfacial area and/or the strength of the resonance and the associated electric field are important factors in producing charge transfer from the AuNP to the semiconductor.

Another possibility is that in the study by Liu et al. the depletion region of un-doped anatase was significantly wider than that of the anatase nano-tubular film, due to the N and F impurities in the nanotube film decreasing the strength of the field and efficiency of charge separation. However, this would probably also apply to the films of Tian and Tatsuma.

Furthermore, there is a feature of Nishijima's study [32] that also casts doubt on the DET theory. The general picture presented is that of electron transport by the action of the electric field at a Schottky barrier between gold and TiO_2. However, in their case, there was a 2 nm thick titanium layer between the gold and TiO_2, so that the TiO_2 should have formed an ohmic contact with the Ti. While there is only

2 nm of Ti, there should be sufficient charge carrier density in both the Au and Ti to compensate for the differences in Fermi levels (work functions) between the metals and the oxide. The differences in Fermi levels at the Ti–Au interface should not present a barrier to charge movement from Au to Ti or from Ti to TiO_2. The Ti should have been screened the Au from the TiO_2, unless the Au was determining the equilibrium potential. It is noted that in the study by Knight et al. [52] that there was also a Ti layer between the AuNRs and Si substrate. This too should have prevented the formation of a Schottky barrier between the Au and Si.

Of course, the systems studied by the Tatsuma group and other groups did not include a Ti interlayer, so they would form a Schottky barrier between the anatase TiO_2 and Au. This might indicate that either a different mechanism was at work in each and every case, or that a mechanism that is not dependent on the presence of a Schottky barrier is acting.

Where the TiO_2 is directly in contact with the electrolyte, however, there should be a semiconductor-liquid junction of the Schottky barrier type, in addition to a Schottky barrier or ohmic contact at the semiconductor-metal junction. This complicates the analysis and understanding of the system significantly and is beyond the scope of this review. Further information can be found elsewhere, such as in the chapter by Allongue, in Modern Aspects of Electrochemistry [74], which discusses the effect of metal-semiconductor junctions in electrochemistry.

This is not to say that DET could not be the active mechanism. But this critical review serves to highlight the fact that it has not yet been proven conclusively to be operative. A theory of plasmonic photocurrent that can be tested rigorously is presently lacking in the literature.

2.11 Oriented Thin Films of TiO_2

Surface modification of TiO_2 by gold nanoparticles has been reviewed previously. Another aspect of this thesis is to investigate the surface of TiO_2 itself.

As described in results Chaps. 7 and 8, part of the project required the growth of epitaxial thin films of TiO_2 with different orientations, principally rutile (110), anatase (101), (100), and (001). The films were to be used to study the dependence of water adsorption and charge transfer on specific features of these surfaces, and to help to validate the theoretical modelling part of the wider research programme. In particular, a parallel project under the same Programme Grant investigated the adsorption modes of water on the rutile (110) surface using density functional theory computations [75]. For anatase TiO_2, water is also known to adsorb both molecularly and dissociatively, e.g. at oxygen bridging vacancies, depending on the polymorph and the surface exposed, in addition to factors such as temperature [76–78].

The focus of this section of the literature review is to identify a range of growth conditions to begin investigating how best to form the oriented thin films. Further publications are discussed in Chaps. 7 and 8 of this thesis reporting the growth of TiO_2 thin films by pulsed laser deposition (PLD).

The background to the need for oriented TiO_2 thin films is given in brief. It is possible to purchase high quality single crystal rutile substrates with a range of orientations including (100), (001), (110), and (011). These single crystals can also be doped with a range of other elements and can be grown using high temperature processes such as the Verneuil process. As such, it is relatively straightforward to obtain high quality 'ideal' surfaces to study and compare their properties and behaviour with computational predictions. In the case of anatase, it is not possible to grow high quality single crystals with control of the purity, as an anatase precursor would be required. To date only the synthesis of micron sized single crystals of anatase has been achieved [79]. It is likely that the high temperatures required to fuse, say an anatase powder, would result in a phase change from anatase to rutile, preventing the growth of millimetre scale single crystals. At present, most surface science studies of anatase single crystals have been carried out on naturally occurring samples of uncontrolled purity. An alternative approach is to grow epitaxial thin films of anatase on other single crystal substrates [80]. An advantage of growing anatase films is also the flexibility of doping the films during the growth process.

It is possible to deposit films of either the rutile or anatase morphology of TiO_2, and with different crystallographic orientations, by changing either the substrate [81] and/or varying the deposition temperature and pressure [82]. The initial literature search concentrated on identifying a range of suitable substrates, which act as the growth template, for films with a range of low index orientations e.g. (100), (001), (101).

Silva et al. [81] investigated the growth of TiO_2 films on a range of cubic single crystal substrates, including MgO, $LaAlO_3$, and $SrTiO_3$, on which the films were grown by pulsed laser deposition (PLD) at an oxygen partial pressure of 10 Pa. The laser fluence was set to 2 J cm^{-2} with 2 Hz repetition rate, while the deposition temperature was varied. All films were deposited from a sintered rutile target. The advantage of PLD is the possibility of growing a low temperature phase from a high temperature phase target.

Silva et al. [81] found that on both $LaAlO_3$ (100) and $SrTiO_3$ (100), the films were anatase with a (001) orientation, i.e. the a and b axes of the unit cell grew in plane, with the c axis perpendicular to the interface of film and substrate. This was unsurprising, giving the close match of the a/b lattice parameters: 3.9050 Å for $SrTiO_3$, 3.792 Å for $LaAlO_3$, and 3.7852 Å for anatase. The ordering of the film on $LaAlO_3$ was seen to be better than the film on $SrTiO_3$, due to the better lattice match with anatase. On MgO (100), the films grew as anatase (100), i.e. the anatase c axis was in the plane of the interface between the film and the substrate. The authors suggested an epitaxial relation for the anatase film on MgO, which the author of this thesis believes to be in error (see Chap. 8). The epitaxial relation of anatase to MgO is discussed in detail in Chap. 8 of this thesis.

The same authors also grew anatase (102) films on $SrTiO_3$ (110) and rutile (110) on MgO (110). Films grown on $(1\bar{1}00)$ M-plane sapphire were rutile (001); those grown on (0001) C-plane sapphire were a mixture of anatase and rutile, and

Table 2.1 Potential substrates and temperature for growth of oriented TiO_2 films by pulsed laser deposition

Orientation	Substrate	Temperature (K)	Reference
Rutile (110)	Glass/FTO MgO (110)	>873	Hitosugi et al. [83] Silva et al. [81]
Anatase (101)	Glass/FTO		Hitosugi et al. [83]
Anatase (100)	MgO (100)	<873	Silva et al. [81]
Anatase (001)	STO (100) LAO (100)	873 873	Silva et al. [81] Hsieh et al. [84]

those grown on (1102) R-plane sapphire were rutile (011). The study of Silva et al. also tabulates the results of numerous studies on the growth of epitaxial TiO_2 films on single crystals by a range of deposition methods.

Table 2.1 suggests some suitable substrates and conditions for TiO_2 thin films that were used as a starting point to produce oriented films of anatase (101), (100), and (001), which were of primary interest. The growth of anatase (101) films is discussed further in the following section.

It is also possible to grow epitaxial TiO_2 films on Si substrates using an intermediate buffer layer of $SrTiO_3$ or TiN prior to the deposition of TiO_2 [85–87].

It is also noted that certain surfaces of films are more likely to form than others, due to having a lower surface energy; e.g. rutile (110) and anatase (101) are most likely [88]. This can be seen in studies with films deposited on amorphous substrates, on which the films were amorphous on deposition and annealed thereafter; higher annealing temperatures resulted in rutile instead of anatase. These films are discussed further in the following section.

2.12 Conductive TiO_2 Thin Films

Growth of conductive TiO_2 thin films were investigated due to the interest in; making surface measurements, and using the thin films as electrodes. In both cases, the films would need to have a measure of conductivity to allow the current-voltage response to be established. Un-doped TiO_2 films are too resistive for use as electrodes and cannot be imaged by scanning tunnelling microscopy (STM).

Several publications have demonstrated that it is possible to grow crystalline anatase films with a (101) orientation on borosilicate glass [82, 83, 89–91]. The publications all stem from a group at the Kanagawa Academy of Science and Technology, Kawasaki, Japan. The group reported the growth of a transparent conducting oxide comprised of Nb-doped anatase TiO_2 [89]. The initial films were grown on $SrTiO_3$ single crystals, and demonstrated resistivity on the order of 10^{-3} Ω cm with doping by Nb of 1 % and above [89, 90]. The same authors grew $Nb_{0.06}Ti_{0.94}O_2$ films with the (101) orientation on borosilicate glass, and demonstrated that annealing the films in H_2 resulted in lower resistivity. They then investigated the effect of growing the films at room temperature followed by annealing in H_2; films grown at room temperature where amorphous, but when annealed subsequently appeared to be more crystalline than the films grown at temperature. The films deposited at room temperature also appeared to be more uniform with a flat surface and an absence of visible grains compared to the films deposited at temperature.

The films grown on borosilicate glass appeared to be comprised of (101) oriented grains only. On this basis, borosilicate glass and fused silica was identified as potential substrates for the growth of (101) oriented films, as presented in Chap. 7 of this thesis.

The observation of conductivity in the Nb-doped anatase films was also promising, as it suggested that it would be possible to deposit conductive films that might be suitable for study by surface probe techniques, such as STM.

In this project, given the interest in Nb-doped TiO_2 for both the thin film deposition and as the substrate for the high performance plasmonic photocurrent generation reported by Nishijima et al. [34], a review of the defect chemistry of TiO_2 and Nb-doped TiO_2 was undertaken and is presented in Appendix A. This is discussed further in Chap. 7, which reports the growth of anatase films with Nb-doping. The effect of doping on the stability of anatase is also discussed. The reason for including the defect chemistry in such detail was that it was reviewed in detail to try to understand why TiO_2 films doped with Nb failed to demonstrate measurable levels of conductivity. Additionally, it was used to try to understand the differences between Nb-doped and reduced rutile, as reported in Chap. 6.

2.13 Summary of Literature Review

The preceding literature review focuses on two main areas: plasmonic Au-TiO_2 photo-anodes for solar water splitting, and to a lesser extent TiO_2 thin films and the defect chemistry of Nb-doped TiO_2.

The review of Au-TiO_2 was used to organize and contrast themes in the many studies regarding its use in photo-electrochemical cells. The purpose was to try to understand if the mechanism of plasmonic water splitting had been previously elucidated in the literature. Based on the current literature, it appears there are several mechanisms proposed, based on experimental observation, but no theoretical framework that adequately describes the process of that can be used to predict or analyse the performance of Au-TiO_2 photo-anodes. However, a paper published by Sundararaman et al. before final submission of this thesis appears to be beginning to develop a theoretical framework [57].

The review was also used to inform the choice of direction for the experimental work presented in Chaps. 5 and 6 of this thesis. Based on the work of Nishijima et al. rutile (110) single crystals doped with Nb were identified as a substrate capable, in principle, of high performance. The suggestion by Liu et al. that dopants might be partly responsible led to the decision to compare Nb-doped and un-doped rutile (110) single crystals.

The review of TiO_2 thin films was used to identify initial growth conditions and substrates for pulsed laser deposition of thin films. The aim of the experimental work was to identify the particular conditions required in the apparatus available, rather than in general. As such, an exhaustive review of TiO_2 thin films was felt to be unnecessary. The focus of the work reported in Chaps. 7 and 8 was to map out the parameter space of the equipment.

The review of the defect chemistry of TiO_2 was carried out (and included here) as it provided useful information on the behaviour of both Nb-doped and reduced TiO_2, which is relevant to the work on both the Au-TiO_2 photo-anodes and the deposition of thin films.

In the following chapter, the principles of photo-electrochemical cells as applied to TiO_2 are covered in greater detail, and lead into the experimental methods used.

References

1. De-La-Rosa, Y.: World Population to 2300, pp. 1–254. United Nations Publications, NY (2004)
2. International Energy Agency: World Energy Outlook 2010 International Energy Agency, Paris (2010)
3. Solomon, S.: Climate Change 2007—The Physical Science Basis. Cambridge University Press, Cambridge (2007)
4. Britain, G.: Climate Change Act 2008. The Stationery Office, London (2008)
5. Boyle, G.: Renewable Energy. Oxford University Press, Oxford (2012)
6. Twidell, J. Weir, A.D. Renewable Energy Resources. Taylor & Francis, London (2006)
7. Larminie, J., Dicks, A.: Fuel Cell Systems Explained. Wiley, NY (2003)
8. Hoffmann, P.: Tomorrow's Energy. MIT Press, Cambridge (2012)
9. Khaselev, O., Bansal, A.: High-efficiency integrated multijunction photovoltaic/electrolysis systems for hydrogen production. Int. J. Hydrogen Energy. 1–6 (2000)
10. Pijpers, J.J., Winkler, M.T., Surendranath, Y., Buonassisi, T., Nocera, D.G.: Light-induced water oxidation at silicon electrodes functionalized with a cobalt oxygen-evolving catalyst. Proc. Natl. Acad. Sci. U.S.A. **108**, 10056–10061 (2011)
11. Vayssieres, L.: On Solar Hydrogen and Nanotechnology. Wiley, NY (2010)
12. Grätzel, M.: Photoelectrochemical cells. Nature **414**, 338–344 (2001)
13. Kudo, A.: Z-scheme photocatalyst systems for water splitting under visible light irradiation. MRS Bull. **36**, 32–38 (2011)
14. Rajeshwar, K., McConnell, RD., Licht, S.: Solar Hydrogen Generation. Springer, New York (2008). doi:10.1007/978-0-387-72810-0
15. Dresselhaus, M., Crabtree, G., Buchanan, M.: Basic Research Needs for the Hydrogen Economy. (2003)
16. Atkins, P., de Paula, J.: Atkins' Physical Chemistry. Oxford University Press, Oxford (2010)
17. Osterloh, F.E., Parkinson, B.A.: Recent developments in solar water-splitting photocatalysis. MRS Bull. **36**, 17–22 (2011)
18. Cronemeyer, D.C.: Electrical and optical properties of rutile single crystals. Phys. Rev. **87**, 876 (1952)
19. Diebold, U.: The surface science of titanium dioxide. Surf. Sci. Rep. **48**, 53–229 (2003)

20. Liao, P., Toroker, M.C., Carter, E.A.: Electron transport in pure and doped hematite. Nano Lett. **11**, 1775–1781 (2011)
21. Tian, Y., Tatsuma, T.: Plasmon-induced photoelectrochemistry at metal nanoparticles supported on nanoporous TiO_2. Chem. Commun. 1810 (2004). doi:10.1039/b405061d
22. Thimsen, E., Le Formal, F., Grätzel, M., Warren, S.C.: Influence of plasmonic Au nanoparticles on the photoactivity of Fe_2O_3 electrodes for water splitting. Nano Lett. **11**, 35–43 (2011)
23. Xu, Y., Schoonen, M.A.: The absolute energy positions of conduction and valence bands of selected semiconducting minerals. Am. Mineral. **85**, 543–556 (2000)
24. Archer, M.D., Nozik, A.J.: Nanostructured and Photoelectrochemical Systems for Solar Photon Conversion. World Scientific Publishing, Singapore (2008)
25. Cesar, I., Kay, A., Gonzalez Martinez, J.A., Grätzel, M.: Translucent thin film Fe_2O_3 photoanodes for efficient water splitting by sunlight: nanostructure-directing effect of Si-doping. J. Am. Chem. Soc. **128**, 4582–4583 (2006)
26. Mavroides, J.G., Kafalas, J.A., Kolesar, D.F.: Photoelectrolysis of water in cells with $SrTiO_3$ anodes. Appl. Phys. Lett. **28**, 241–243 (1976)
27. Fujishima, A.: Electrochemical photolysis of water at a semiconductor electrode. Nature **238**, 37–38 (1972)
28. Boddy, P.J.: Oxygen evolution on semiconducting TiO_2. J. Electrochem. Soc. **115**, 199–203 (1968)
29. Grimes, C.A., Varghese, O.K., Ranjan, S.: Light, Water, Hydrogen. Springer Science & Business Media, Berlin (2007)
30. Sivula, K., Le Formal, F., Graetzel, M.: WO_3–Fe_2O_3 photoanodes for water scaffold splitting: a host. Guest Absorber Approach Chem. Mater. **21**, 2862–2867 (2009)
31. Maeda, K., Domen, K.: Oxynitride materials for solar water splitting. MRS Bull. **36**, 25–31 (2011)
32. Nishijima, Y., Ueno, K., Yokota, Y., Murakoshi, K., Misawa, H.: Plasmon-assisted photocurrent generation from visible to near-infrared wavelength using a Au-nanorods/TiO_2 electrode. J. Phys. Chem. Lett. **1**, 2031–2036 (2010)
33. Nishijima, Y., Nigorinuma, H., Rosa, L., Juodkazis, S.: Selective enhancement of infrared absorption with metal hole arrays. Opt. Mater. Express **2**, 1367–1377 (2012)
34. Tian, Y., Tatsuma, T.: Mechanisms and applications of plasmon-induced charge separation at TiO_2 films loaded with gold nanoparticles. J. Am. Chem. Soc. **127**, 7632–7637 (2005)
35. Sakai, N., Tatsuma, T.: Photovoltaic properties of glutathione-protected gold clusters adsorbed on TiO_2 electrodes. Adv. Mater. **22**, 3185–3188 (2010)
36. García, M.A.: Surface plasmons in metallic nanoparticles: fundamentals and applications. J. Phys. D Appl. Phys. **44**, 283001 (2011)
37. Centeno, A., Xie, F., Alford, N.: Light absorption and field enhancement in two-dimensional arrays of closely spaced silver nanoparticles. J. Opt. Soc. Am. B **28**, 325–330 (2011)
38. Catchpole, K.R., Polman, A.: Plasmonic solar cells. Opt. Express **16**, 21793–21800 (2008)
39. Beck, F.J., Verhagen, E., Mokkapati, S., Polman, A., Catchpole, K.R.: Resonant SPP modes supported by discrete metal nanoparticles on high-index substrates. Opt. Express **19**, A146–A156 (2011)
40. Maier, S.A.: Plasmonics: Fundamentals and Applications. Springer, Berlin (2007)
41. Zhang, G., Wang, D.: Colloidal lithography-the art of nanochemical patterning. Chem. Asian J. **4**, 236–245 (2009)
42. Centeno, A., Breeze, J., Ahmed, B., Reehal, H., Alford, N.: Scattering of light into silicon by spherical and hemispherical silver nanoparticles. Opt. Lett. **35**, 76–78 (2009)
43. Liu, Z., Hou, W., Pavaskar, P., Aykol, M., Cronin, S.B.: Plasmon resonant enhancement of photocatalytic water splitting under visible illumination. Nano Lett. **11**, 1111–1116 (2010)
44. Naseri, N., Amiri, M., Moshfegh, A.Z.: Visible photoenhanced current–voltage characteristics of Au : TiO2 nanocomposite thin films as photoanodes. J. Phys. D Appl. Phys. **43**, 105405 (2010)

45. Chandrasekharan, N., Kamat, P.V.: Improving the photoelectrochemical performance of nanostructured TiO_2 films by adsorption of gold nanoparticles. J. Phys. Chem. B **104**, 10851–10857 (2000)
46. Kittel, C.: Introduction to Solid State Physics. Wiley, NY (2004)
47. Petek, H., Ogawa, S.: Femtosecond time-resolved two-photon photoemission studies of electron dynamics in metals. Prog. Surf. Sci. **56**, 239–310 (1997)
48. Petek, H., Nagano, H., Ogawa, S.: Hole decoherence of d bands in copper. Phys. Rev. Lett. **83**, 832–835 (1999)
49. Sachtler, W., Dorgelo, G., Holscher, A.A.: The work function of gold. Surf. Sci. **5**, 221–229 (1966)
50. Giugni, A., et al.: Hot-electron nanoscopy using adiabatic compression of surface plasmons. Nat. Nanotechnol. **8**, 845–852 (2013)
51. Daniel, M.-C., Astruc, D.: Gold nanoparticles: assembly, supramolecular chemistry, quantum-size-related properties, and applications toward biology, catalysis, and nanotechnology. Chem. Rev. **104**, 293–346 (2004)
52. Knight, M.W., Sobhani, H., Nordlander, P., Halas, N.J.: Photodetection with active optical antennas. Science **332**, 702–704 (2011)
53. Mubeen, S., et al.: An autonomous photosynthetic device in which all charge carriers derive from surface plasmons. Nat. Nanotechnol. **8**, 247–251 (2013)
54. Mukherjee, S., et al.: Hot electrons do the impossible: plasmon-induced dissociation of H_2 on Au. Nano Lett. **13**, 240–247 (2013)
55. Cushing, S.K., et al.: Photocatalytic activity enhanced by plasmonic resonant energy transfer from metal to semiconductor. J. Am. Chem. Soc. **134**, 15033–15041 (2012)
56. Li, J., et al.: Ag@Cu_2O core-shell nanoparticles as visible-light plasmonic photocatalysts. ACS Catal. **3**, 47–51 (2013)
57. Sundararaman, R., Narang, P., Jermyn, A.S., Goddard III, W.A., Atwater, H.A.: Theoretical predictions for hot-carrier generation from surface plasmon decay. Nat. Commun. **5**, 5788 (2014)
58. Manjavacas, A., Liu, J.G., Kulkarni, V., Nordlander, P.: Plasmon-induced hot carriers in metallic nanoparticles. ACS Nano **8**, 7630–7638 (2014)
59. Govorov, A.O., Zhang, H., Gun'ko, Y.K.: Theory of photoinjection of hot plasmonic carriers from metal nanostructures into semiconductors and surface molecules. J. Phys. Chem. C. **117,** 16616–16631 (2013)
60. Forbes, R.G., Deane, J.H.B.: Reformulation of the standard theory of Fowler-Nordheim tunnelling and cold field electron emission. Proc. R. Soc. Lon. A. Math. Phys. **463**, 2907–2927 (2007)
61. Murphy, E.L., Good Jr, R.H.: Thermionic emission, field emission, and the transition region. Phys. Rev. **102**, 1464 (1956)
62. Fowler, R.H. Nordheim, L.: Electron emission in intense electric fields. Proc. R. Soc. Lond. A Math. Phys. Sci. 173–181 (1928)
63. Lau, Y.Y., Liu, Y., Parker, R.K.: Electron emission: from the Fowler-Nordheim relation to the Child-Langmuir law. Phys. Plasmas **1**, 2082–2085 (1994)
64. Lin, H.Y., Chou, Y.Y., Cheng, C.L., Chen, Y.F.: Giant enhancement of band edge emission based on ZnO/TiO_2 nanocomposites. Opt. Express **15**, 13832–13837 (2007)
65. Hernández-Martínez, P. Govorov, A.: Exciton energy transfer between nanoparticles and nanowires. Phys. Rev. B. **78**, (2008)
66. Neretina, S., et al.: Plasmon field effects on the nonradiative relaxation of hot electrons in an electronically quantized system: CdTe–Au core–shell nanowires. Nano Lett. **8**, 2410–2418 (2008)
67. Makhal, A., et al.: Dynamics of light harvesting in ZnO nanoparticles. Nanotechnology **21**, 265703 (2010)
68. Andrews, D.L.: A unified theory of radiative and radiationless molecular energy transfer. Chem. Phys. **135**, 195–201 (1989)

69. Andrews, D.L., Bradshaw, D.S.: Virtual photons, dipole fields and energy transfer: a quantum electrodynamical approach. Eur. J. Phys. **25**, 845–858 (2004)
70. Andrews, D.L., Curutchet, C., Scholes, G.D.: Resonance energy transfer: Beyond the limits. Laser Photon. Rev. **5**, 114–123 (2010)
71. Fox, M.: Optical Properties of Solids. Oxford University Press, Oxford (2010)
72. Furube, A., Du, L., Hara, K., Katoh, R., Tachiya, M.: Ultrafast plasmon-induced electron transfer from gold nanodots into TiO_2 nanoparticles. J. Am. Chem. Soc. **129**, 14852–14853 (2007)
73. Karp, G.: Cell and Molecular Biology. Wiley, NY (2009)
74. Allongue, P.: In: Bockris, J., Conway, B.E. White, R.E. (eds.) Modern Aspects of Electrochemistry. vol. 23, pp. 239–314 (1992)
75. Patel, M., Mallia, G., Liborio, L., Harrison, N.M.: Water adsorption on rutile TiO_2 (110) for applications in solar hydrogen production: a systematic hybrid-exchange density functional study. Phys. Rev. B **86**, 045302 (2012)
76. Blomquist, J., Walle, L.E., Uvdal, P., Borg, A., Sandell, A.: Water dissociation on single crystalline anatase TiO_2 (001) studied by photoelectron spectroscopy. J. Phys. Chem. C **112**, 16616–16621 (2008)
77. Sumita, M., Hu, C., Tateyama, Y.: Interface water on TiO_2 anatase (101) and (001) surfaces: first-principles study with TiO_2 slabs dipped in bulk water. J. Phys. Chem. C **114**, 18529–18537 (2010)
78. Walle, L.E., et al.: Mixed dissociative and molecular water adsorption on anatase TiO_2 (101). J. Phys. Chem. C **115**, 9545–9550 (2011)
79. Yang, H.G., et al.: Anatase TiO_2 single crystals with a large percentage of reactive facets. Nature **453**, 638–641 (2008)
80. Grinter, D.C., Nicotra, M., Thornton, G.: Acetic acid adsorption on anatase TiO_2 (101). J. Phys. Chem. C **116**, 11643–11651 (2012)
81. Silva, V.F., et al.: Substrate-controlled allotropic phases and growth orientation of TiO_2 epitaxial thin films. J. Appl. Cryst. **43**, 1502–1512 (2010). doi:10.1107/S0021889810041221
82. Hitosugi, T., et al.: Fabrication of TiO_2-based transparent conducting oxide films on glass by pulsed laser deposition. Jpn. J. Appl. Phys. **46**, L86–L88 (2007)
83. Hitosugi, T., et al.: Electronic band structure of transparent conductor: Nb-doped anatase TiO_2. Appl. Phys. Express **1**, 111203 (2008)
84. Hsieh, C., et al.: Monophasic TiO_2 films deposited on $SrTiO_3$ (100) by pulsed laser ablation. J. Appl. Phys. **92**, 2518–2523 (2002)
85. Nakamura, R., Okamura, T., Ohashi, N., Imanishi, A., Nakato, Y.: Molecular mechanisms of photoinduced oxygen evolution, PL emission, and surface roughening at atomically smooth (110) and (100) n-TiO_2 (Rutile) surfaces in aqueous acidic solutions. J. Am. Chem. Soc. **127**, 12975–12983 (2005)
86. Yamamoto, S., Sumita, T., Miyashita, A.: Preparation of TiO_2-anatase film on Si (001) substrate with TiN and $SrTiO_3$ as buffer layers. J. Phys. Condens. Matter **13**, 2875 (2001)
87. McDaniel, M.D., Posadas, A., Wang, T., Demkov, A.A., Ekerdt, J.G.: Growth and characterization of epitaxial anatase TiO_2 (001) on $SrTiO_3$-buffered Si (001) using atomic layer deposition. Thin Solid Films **520**, 6525–6530 (2012)
88. Sanches, F.F., Mallia, G., Liborio, L., Diebold, U., Harrison, N.M.: Hybrid exchange density functional study of vicinal anatase TiO_2 surfaces. Phys. Rev. B **89**(24), 5309 (2014)
89. Furubayashi, Y., et al.: A transparent metal: Nb-doped anatase TiO_2. Appl. Phys. Lett. **86**(25), 2101 (2005)
90. Furubayashi, Y., et al.: Novel transparent conducting oxide: anatase TiNbO. Thin Solid Films **496**, 157–159 (2006)
91. Hitosugi, T., et al.: Transparent conducting properties of anatase $Ti_{0.94}Nb_{0.06}O_2$ polycrystalline films on glass substrate. Thin Solid Films **516**, 5750–5753 (2008)

Chapter 3
Principles of Photo-Electrochemical Cells

This chapter presents the relevant theory of the semiconductor-liquid interface and how it relates to photoelectrochemical redox processes such as water splitting, the principle application of interest in this thesis. The principles would also hold for other photoelectrochemical processes, such as oxidation of pollutants. This chapter provides supporting information for the main literature review of Chap. 2 that focuses on particular photo-electrodes.

The theory outlined in this chapter was used to support the experimental work of this thesis, particularly in the analysis of data as presented in Chaps. 5 and 6. For example, the model of the electrochemical semiconductor-liquid interface was used to interpret impedance measurements and to fit photocurrent data.

This chapter essentially provides reference material for readers to be able to check the definition of particular terms or processes. Readers of this thesis who are familiar with the background theory may wish to skip ahead to Chap. 4, Experimental Methods.

The theory of the flat band potential presented below was adapted from a publication—to which the author of this thesis contributed—that addresses issues with reported flat band potentials on hematite [1].

First, the basic reaction of interest—water splitting—is outlined, along with a few key terms related to the electrochemistry, including quasi- and solution- Fermi levels. Second, the theory and equations that are used to describe the polarized semiconductor-liquid interface are introduced from first principles. A simple model of the photocurrent is also presented. This is necessary to provide a sufficient background to understand the terminology used in this thesis and to explain some observations in the results presented in Chaps. 5 and 6.

The theory outlined below is not exhaustive; it is only presented in as much detail as necessary to explain some experimental methods used, such as Mott-Schottky plots, and the general qualitative behaviour of TiO_2 photo-electrodes. The texts from which the following theory has been compiled treat the issues in greater detail [2–4]. In particular, while it is possible to write equations for the currents in both the conduction and valence bands of the semiconductor in general terms, the

© Springer International Publishing Switzerland 2016

J.C. Alexander, *Surface Modifications and Growth of Titanium Dioxide for Photo-Electrochemical Water Splitting*, Springer Theses,
DOI 10.1007/978-3-319-34229-0_3

detailed dark electrochemistry has not been modelled in this work. In fact, there are no examples (to the knowledge of the author) where the current-potential behaviour of water splitting photoelectrodes in the dark has been modelled (for example using the gPROMS software package). There are examples of the photocurrent [5] and the capacitive behaviour of the interface being modelled, but not the 'dark' electrochemistry. This is due to the large number of parameters required to construct an adequate model of even a simple nature. This will become apparent in the following sections.

3.1 Nomenclature

The notation used follows the conventions outlined by IUPAC (International Union of Pure and Applied Chemistry) as far as possible. However, some terms have been assigned non-standard symbols to avoid conflict. The main issue that arises is that the conventions used in solid-state physics to describe a semiconductor and those used by electrochemists have not been designed to be compatible.

The following table summarises the nomenclature used in this chapter and the rest of this thesis. Constants and symbols are given in the SI form where possible, although in the text and results they may be typically quoted with more convenient forms, such as mA cm^{-3} in place of A m^{-3}.

Some symbols are given with a subscript. Others may be used with a modified subscript at a condition of interest, such as the flat band potential. The subscript will be explained at the point that the notation is introduced in the text.

3.1.1 Constants

Symbol	Description	Units
ε_0	Permittivity of vacuum, 8.854187 × 10^{-12}	C V^{-1} m^{-1}
F	Faraday constant, 96484.56	C mol^{-1}
h	Planck constant, 6.62606957 × 10^{-34}	J s
$\hbar$	Reduced Planck constant, 1.054571726 × 10^{-34}	J s
k_B	Boltzmann constant, 1.3806488 × 10^{-23}	J K^{-1}
m_e	Electronic mass, 9.10938215 × 10^{-31}	kg
q_e	Electronic charge, 1.602177 × 10^{-19}	C
R	Molar gas constant, 8.3144621	J mol^{-1} K^{-1}

3.1.2 Symbols

Symbol	Description	Units
C	Capacitance	F
D_p	Diffusion coefficient of holes	$m^2\ s^{-1}$
D_n	Diffusion coefficient of electrons	$m^2\ s^{-1}$
E_A	Electron affinity, relative to vacuum	eV
E_{CB}, E_{VB}	Energy of conduction band minimum and valence band maximum, relative to vacuum	eV
E_G	Band gap	eV
E_D	Energy of dopant, relative to vacuum	eV
E_F	Fermi level, energy relative to vacuum	eV
i	$\sqrt{-1}$	
j	Current density	$A\ m^{-2}$
L_α	Absorption depth of semiconductor, wavelength dependent $L_\alpha(\lambda)$	m
L_D	Debye length of space charge layer	m
L_p	Diffusion length of holes	m
L_n	Diffusion length of electrons	m
m_e^*, m_h^*	Effective mass of electrons, holes	kg
n	Concentration of electrons	m^{-3}
N_C, N_V	Density of states in conduction and valence bands	m^{-3}
N_D, N_A	Concentration of donor and acceptor impurities	m^{-3}
p	Concentration of holes	m^{-3}
ρ	Charge density	$C\ m^{-3}$
$q_{p,0}(\lambda)$	Photon flux of wavelength, λ	$s^{-1}m^{-2}$
T	Absolute temperature	K
$U(RE)$	Potential relative to reference electrode RE	V
$U_i(RE)$	Potential of parameter i in bulk	V
$U_{i,S}(RE)$	Potential of parameter i at the surface/interface	V
w	Width of space charge region (depletion layer)	m
w_0	Space charge width constant	$m\ V^{-1}$
x	Distance from the solid-liquid interface ($x = 0$)	m
x_{SCR}	Edge of space charge region	m
α	Absorption coefficient of semiconductor, wavelength dependent $\alpha(\lambda)$	m^{-1}
ε	Permittivity ($\varepsilon_0\varepsilon_r$)	$C\ V^{-1}\ m^{-1}$
ε_r	Relative permittivity	–
ξ	Electric field in semiconductor	$V\ m^{-1}$
ϕ	Potential in semiconductor	V
$\Delta\phi_{SCR}$	Potential drop across the space charge layer	V
$\Delta\phi_H$	Potential drop across the Helmholtz layer	V
$\Delta\phi_{H(FB)}$	Potential drop across the Helmholtz layer at the flat band potential	

(continued)

(continued)

Symbol	Description	Units
λ	Wavelength of photon	nm
μ_p	Electrical mobility of holes	m s^{-1} V^{-1}
μ_n	Electrical mobility of electrons	m s^{-1} V^{-1}
χ	Electronegativity, relative to vacuum	eV

3.2 Hydrogen Production by Water Splitting

3.2.1 *The Water Splitting Reaction*

The general reaction of interest: solar splitting of water into its constituent parts of H_2 and O_2 is:

$$H_2O \xrightarrow{hv} H_2 + \frac{1}{2}O_2 \tag{3.1}$$

This can be written in terms of two half reactions [6]:

$$\text{(HER)} \quad 2H^+ + 2e^- \rightarrow H_2 \tag{3.2}$$

$$\text{(OER)} \quad 2h^+ + H_2O \rightarrow 2H^+ + \frac{1}{2}O_2 \tag{3.3}$$

In the hydrogen evolution reaction (HER) two protons are donated by a water molecule and are then reduced by two electrons to produce one molecule of hydrogen. This reaction defines the zero of the reversible hydrogen electrode (RHE) scale. At pH 0 this also defines the standard hydrogen electrode (SHE).

In the oxygen evolution reaction (OER) a water molecule is oxidized by two positively charged holes to produce oxygen. This is a two-electron process for ½ O_2. For each complete O_2 molecule a total of four holes oxidizing two water molecules are required. As such, water splitting is often called a four-electron process.

In a photoelectrochemical cell (PEC), the HER occurs at the cathode or photo-cathode and the OER occurs at the anode or photo-anode. For photo-electrodes, the electrons and holes are generated by the absorption of a photon of light by the semiconductor electrode.

The operation of a basic water splitting PEC, including the flow of electrons and holes is shown in Fig. 3.2 of Sect. 3.3.

TiO_2 can be used as a photo-anode to drive the OER. In relation to TiO_2, water is believed to adsorb both molecularly (a physisorption) and dissociatively (a chemisorption) depending on the configuration of atoms at the surface [7]. Factors such as temperature also affect the interaction of water with the surface [8]. As such, the OER can also be presented as for a dissociated hydroxyl group adsorbed onto the surface of the anode:

$$2h^{+} + 2OH^{-} \rightarrow 2H^{+} + O_2 \tag{3.4}$$

The work of this thesis is part of a joint computational and experimental Programme Grant on the properties of different surfaces of rutile and anatase, which includes studies into how water adsorbs and is oxidized. The thin films produced during this project and discussed in Chaps. 7 and 8 of this thesis, will be used to study these processes using appropriate techniques, such as Fourier transform infra-red (FTIR) spectroscopy, photoemission spectroscopy, and scanning tunnelling microscopy [8–10].

The minimum electrical work required to break the molecular bonds in H_2O at standard conditions of 25 °C and 1 bar pressure is given by the standard Gibbs free energy, $\Delta G°$ [11]:

$$\Delta G^{\circ} = -237.178\,\text{kJ}\,\text{mol}^{-1} \tag{3.5}$$

This consists of an enthalpy term $\Delta H°$ and an entropy term $T\Delta S°$, where $\Delta H° = -285.830$ kJ mol^{-1}, $\Delta S° = -69.91$ J mol^{-1} K^{-1} and T is the temperature, such that:

$$\Delta G^{\circ} = \Delta H^{\circ} - T\Delta S^{\circ} \tag{3.6}$$

The Gibbs free energy denotes the minimum share of electrical energy required, while the entropic term denotes the maximum share of thermal energy. In practice, the energy can be supplied in the form of the reducing/oxidizing potential difference of photo-generated charge carriers in a solar cell such as a PEC. The reversible potential is given by:

$$E^{\circ} = -\Delta G^{\circ}/nF = 1.23\,\text{V} \tag{3.7}$$

where n is the electron stoichiometry of the elementary reaction (3.2) involved (2 in this case) and F is Faraday's constant (F = 96,485 C mol^{-1}) the charge on one mole of electrons [12].

The minimum required reversible potential is $E° = 1.23$ V if additional heat is provided to the system. The minimum large entropic term would also require heat from environment. To split water without producing or consuming heat requires the application of the thermoneutral potential difference [13] of 1.481 V given by:

$$E^\circ = -\Delta H^\circ / nF = 1.48 \text{ V} \tag{3.8}$$

In practice, a greater potential difference for electrons and holes is required, to cause the reaction to depart from equilibrium due to resistive losses in the circuit and the need to drive the reaction at an economically practical rate. This is this *overpotential,* at the anode, η_a, and at the cathode, η_c. The rate coefficient k of the reaction also depends partly on overpotential.

Applying a bias to the system can also increase/decrease the overpotential, modifying both the rate of oxygen formation and hydrogen production rate. The kinetics of charge transfer and mass transport are not covered further in this report.

The overpotential concept is elaborated on below, following discussion on the main difference between semiconductor and metal electrodes, and the concept of solution- and quasi- Fermi levels.

3.2.2 Metallic Versus Semiconducting Electrodes

At this point, it is crucial to state that the current-electrode potential behaviour of a semiconductor electrode is different to that of a metallic electrode. As such, the use of the Tafel equation [14] to describe the kinetics of the electrode is not appropriate. This is due to the presence of a significant space charge layer in the semiconductor, so the electric potential is distributed spatially both in the semiconductor and in the solution adjacent to the interface, unlike for metal|solution or degenerate semiconductor|solution interface, at which only the latter is operative.

Additionally, the concept of the overpotential is typically used for metallic electrodes driving electrochemical reactions in the dark, but is less clear for photo-electrodes. For an n-type semiconductor, the electrode potential describes the energy of the electrons (majority carriers), but has nothing to say about the potential of the holes (minority carriers). For a p-type semiconductor, the electrode potential corresponds to the holes (majority carriers) and gives no information about the potential of electrons (minority carriers). This can be demonstrated by considering that an illuminated electrode will give rise to a photocurrent in addition to a dark current. For example, if a TiO_2 electrode is controlled to a potential of +1.0 V (RHE), then the potential of the Fermi level is not positive enough to oxidize water, as the potential is insufficiently positive by −0.23 V. However, illumination will result in a flow of holes to the surface, and the subsequent oxidation of water. An anodic photocurrent will be observed despite the overpotential being insufficient according to the conventional definition.

This issue can be improved with the introduction of the concept of the quasi-Fermi level, which is discussed in Sect. 3.3.1 below. However, it does not help in the practical analysis of the electrode kinetics because while the existence of a minority carrier quasi-Fermi level may be demonstrated [15], it cannot be measured in general at the same time as the majority carrier Fermi level. There have been attempts to measure the minority carrier potential, but the techniques are not in common practice as they change the nature of the electrode [16]. The above discussion is presented simply to make it clear that the Tafel equation will mislead if used to analyse a photo-electrode.

3.2.3 *Solution Fermi Level*

An aqueous solution does not contain free electrons, but the electronic levels in the species of a redox couple may equilibrate with those in the photoanode, giving rise to the concept of the *solution Fermi level*, $E_{F(O/R)}$ where O/R represents a redox couple [17]. For a multi-electron process such as water splitting, the potential of each electron step should be considered, in the ideal approach, rather than the potential of the whole reaction [18]. As described previously, the water splitting reaction can be described by two redox couples separated by 1.23 V: H^+/H_2 for the HER and H_2O/O_2 or OH^-/O_2 for the OER. The potentials of these against the Standard Hydrogen Electrode (SHE) reference are 0.0 V (SHE) for the HER and +1.23 V (SHE) for the OER at pH 0. The potential of each reaction versus the SHE becomes more negative by 59 mV pH^{-1} with increasing pH, as demonstrated in Fig. 3.1. The pH-dependent potential of the hydrogen reduction reaction can also be

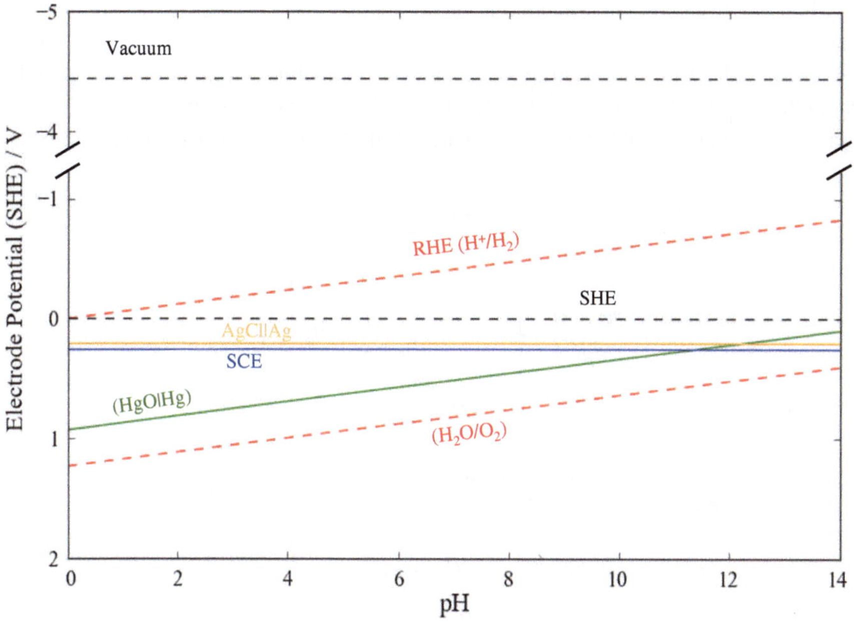

Fig. 3.1 Potential of HER and OER versus pH with a range of standard reference electrodes indicated

used to define a reference electrode termed the Reversible Hydrogen Electrode (RHE).

It is necessary to know these potentials to be able to analyse the current-potential behaviour of the electrode with respect to the reactions.

Holes that reach the surface, without recombining with an electron, are available to oxidize adsorbed species such as H_2O or OH^-. If the potential of the hole is more positive than +1.23 V (RHE) then oxidation of water can occur. Likewise electrons with a potential negative of 0.0 V (RHE) can reduce H^+ to H_2.

The equilibrium potentials change with pH, and are often presented against other reference electrodes such as the Saturated Calomel electrode (SCE) or, in this work, the HgO|Hg electrode. The HgO|Hg reference electrode is discussed further in Chap. 4, Experimental Methods.

The energy of the local vacuum level in the liquid phase has been estimated [19] to correspond to −4.44 V versus the SHE. A value of −4.5 V (SHE) is also used by some.

To understand how the electron-hole pairs are photo-generated and the behaviour of the semiconductor-liquid interface, it is necessary to outline the basic physics of the type of solar cells relevant to water splitting.

3.3 Photoelectrochemical Cells—Overview

A photoelectrochemical semiconductor-liquid junction operates on a similar principle as the Schottky barrier junction. A simplified schematic of a photoelectrochemical cell (PEC) consisting of an n-type semiconducting photoanode and a metallic counter electrode is shown in Fig. 3.2. In the most basic terms, photogenerated electron-hole pairs are separated in the local electric field, holes are

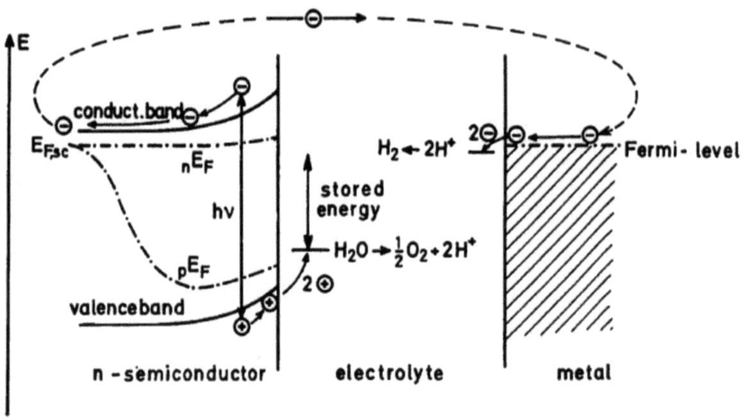

Fig. 3.2 n-type PEC for photo-electrolysis of water. Image taken from Gerischer [20]. Copyright IUPAC 1980

driven to the interface, oxidizing water, and electrons are collected at the back contact and transported to a counter electrode where they reduce protons, producing hydrogen. If the electrode potential of the anode is not sufficiently negative to reduce water, then a bias is applied to achieve the correct potential.

In particular, a Schottky barrier forms at the interface of TiO_2 (an n-type semiconductor) with water (an aqueous electrolyte with the H^+/H_2 and H_2O/O_2 redox couples present), as investigated in this thesis.

When the photo-electrode is immersed in the electrolyte, the Fermi energy levels of the electrode and solution are not in general equal (with respect to the vacuum level), and they undergo an equilibration process of electron exchange between the semiconductor and electrolyte, resulting in a *space charge region* (SCR) just inside the semiconductor. For a semiconductor in the dark, the Fermi level of the majority carriers (electrons for n-type) equilibrates with the Fermi level in solution, due to the presence of electron acceptors such as dissolved oxygen. This process is shown schematically in Fig. 3.3a, b. The space charge density field in the semiconductor is balanced in the electrolyte by the accumulation of charge at the surface in the *Helmholtz layer* (HL) in which dipoles of molecules in the electrolyte, such as H_2O, align over a few mono-layers [21]. The equations that describe these processes are presented in detail below in Sect. 3.4.

Illuminating the semiconductor with photons of energy ($E = hv$) greater than the band gap, E_G of the material, generates electron-hole pairs. The electron is promoted from the valence band (VB) to the conduction band (CB) leaving a hole (h^+) in the VB. If generated in the SCR, the electron-hole pairs separate in the electrostatic field, producing an anodic current. Holes flow towards the interface with the electrolyte and electrons flow to an ohmic back contact (usually a metal).

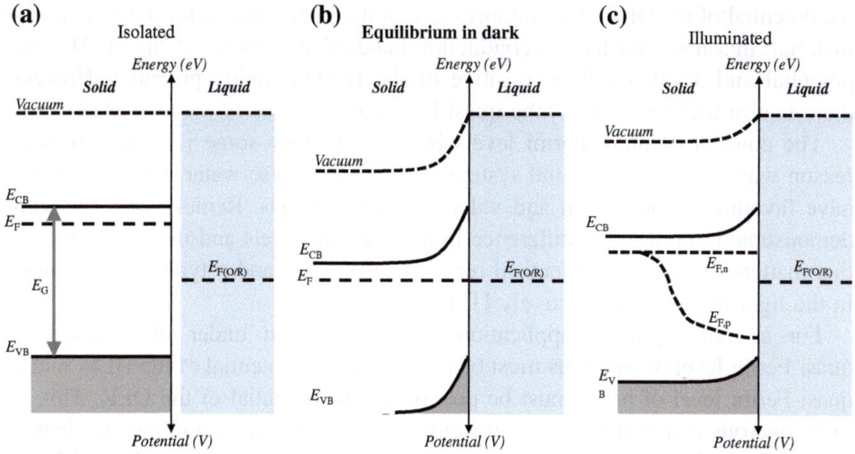

Fig. 3.3 Semiconductor-liquid interface. **a** Before contact. **b** After contact and equilibration of Fermi levels in the dark, and **c** under illumination, demonstrating the splitting of the Fermi level into quasi-Fermi levels for electrons and holes. The degree of band bending is reduced under illumination

3.3.1 Quasi-Fermi Level and Photovoltage

When a PEC junction is illuminated, generating pairs of electrons and holes, the energy of the electrons in the semiconductor is increased, raising the Fermi level relative to the Fermi level in solution. This is referred to as the quasi-Fermi level $E^*_{F,n}$ for electrons. Holes also have a quasi-Fermi level, denoted $E^*{}_{F,p}$. This is illustrated in Fig. 3.3c, in which the Fermi levels of electrons and holes are both shown. This decreases the band bending at the junction, and results in a photovoltage, V_{ph}. The photovoltage is related to the differences in the redox potential(s) and the quasi-Fermi levels as follows:

$$q_e V_{ph} = E_{F(O/R)} - E^*{}_{F,n} \tag{3.9}$$

Here q_e represents the electronic charge.

For a semiconductor-liquid system with a single redox couple such as Fe^{3+}/Fe^{2+}, the photovoltage is determined by the difference between the quasi-Fermi level of the majority carrier and the equilibrium potential of the redox couple in solution. However, there may be a number of redox couples present simultaneously in aqueous solutions. In addition to H^+/H_2 and H_2O/O_2 there may be other combinations of O-H and the reduction of dissolved O_2. Apart from complicating the analysis somewhat, the concept of the photovoltage becomes less clear. Indeed, the fact that the redox couple is not regenerative means that the PEC does not generate a potential difference across the electrodes that can be used to drive an external load. Instead, the solvent (water) is consumed and must be replenished.

For water electrolysis, this means that the potential of the electrons must be negative of the potential of the HER, and the potential of holes must be positive of the potential of the OER. This imposes one of the many criteria for a water splitting material: that it should have a conduction band edge negative of the H^+/H_2 redox potential and a valence band positive of the H_2O/O_2 redox potential. However, there is a further restriction—the quasi-Fermi level.

The concept of quasi-Fermi levels is considered by some [22] as a principal reason why some experimental systems fail to electrolyse water even though they have favourable conduction and valence band positions. Reineke and Memming demonstrated that it is the difference in quasi-Fermi levels and the redox potential that matters by achieving identical reaction rates on n- and p-type semiconductors in the light and dark, respectively [15].

For a water splitting application, this means that under illumination, the quasi-Fermi level of electrons must be negative of the potential of the HER, and the quasi-Fermi level of holes must be positive of the potential of the OER. This is a more onerous constraint than the positions of the conduction and valence bands.

However, if (as is common with metal oxide photo-anodes like TiO_2 and Fe_2O_3) the quasi-Fermi level of the electrons is insufficiently negative to reduce protons, then the electrode may be biased by the application of an external potential difference. This decreases solar-to-hydrogen energy conversion efficiencies (as discussed in the literature review, Sect. 2.3.1).

The models that are commonly used to describe and analyse the semiconductor-liquid junction are now outlined in the following sections.

3.4 Structure of the Space Charge Region

This section covers the derivation of the equations that describe the potential distribution of the semiconductor-liquid interface for an n-type semiconductor, such as TiO_2. The derivations were repeated to develop better understanding of the subtleties of the interface for analysis of the photocurrent and impedance spectroscopy data presented in Chaps. 5 and 6.

Firstly, the concept of the flat band potential is introduced for unpolarized electrodes. For polarized electrodes, Poisson's Equation can be used to derive the form of the spatial distribution of the electric field and potential in the semiconductor due to the distribution of charge concentration. Then important concepts such as the space charge width constant, the band bending, and the capacitance of the space charge region are introduced. This is necessary to be able to understand and describe the current-potential behaviour of a TiO_2 photo-electrode and to be able to determine materials properties, such as the charge carrier density, from impedance spectroscopy Finally, the potentials in the semiconductor are related to the reference electrode scale through a description of the Helmholtz layer.

The potential at any point inside the semiconductor is denoted by ϕ when referenced to another arbitrary level in the semiconductor (some authors use V in place of ϕ). A possible choice of reference is the vacuum level, which is taken as the reference for energy related terms such as the Fermi level E_F and the electron affinity, E_A. When a potential is referenced to a reference electrode potential, the notation $U_X(\mathrm{RE})$ is used where the subscript X identifies the quantity of interest and RE indicates the type of reference electrode or *vac* for a level referenced to the local vacuum level.

One option for the reference level in the semiconductor might be the flat band potential, which is described in more detail in Sect. 3.4.2. A more thorough treatment of the flat band issue is given by Hankin et al. [1].

The following treatment is presented in one dimension for simplicity.

3.4.1 Semiconductor Terminology

The main physical parameters that describe a semiconductor electrode are shown in Fig. 3.4. The unpolarized semiconductor is shown in Fig. 3.4a, while a polarized electrode with a depletion layer is shown in Fig. 3.4b, with additional features. The

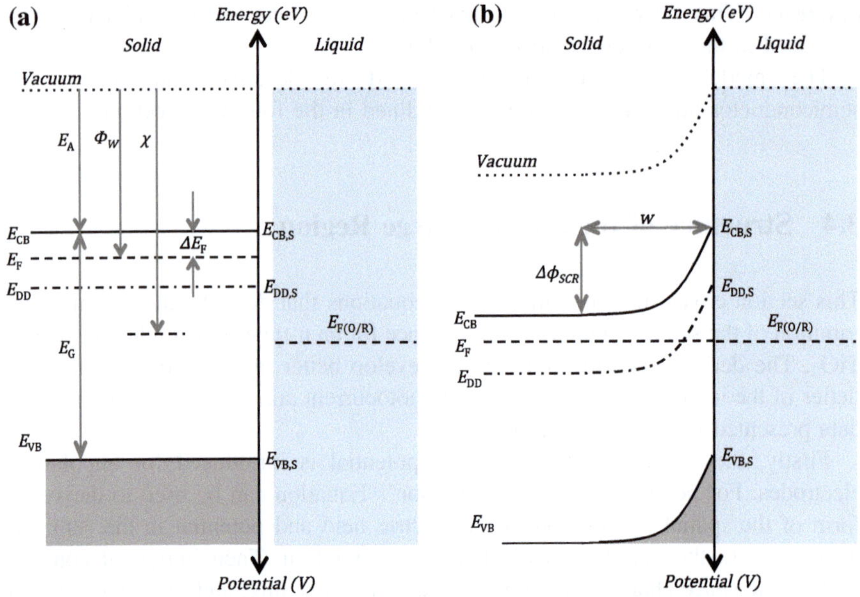

Fig. 3.4 Schematic of energy levels in n-type semiconductor. The diagram is drawn for pH = pzc, so the potential drop in the Helmholtz layer is omitted. Panel **a** shows a semiconductor that is isolated from the electrolyte. It could also be viewed as the flat band condition. Panel **b** shows the semiconductor in equilibrium with a redox couple O/R and the resultant band bending

following discussion is in general terms, followed by the derivation of the equations that describe the interface.

The band gap of the semiconductor E_G in units of eV is the minimum energy of a photon that may be absorbed by the semiconductor. The band gap of TiO_2 is indirect and has a magnitude of 3.05 eV for rutile and 3.20 eV for anatase [23]. In a material with an indirect band gap, for a photon to be absorbed it is necessary for a phonon to be created or destroyed in order to conserve both energy and momentum. In a material with a direct band gap, a photon may be absorbed without the need for a phonon. Typically, an indirect bandgap decreases the probability of a photon being absorbed, and results in a smaller absorption coefficient: a greater thickness of material is required to absorb the same number of photons as for a direct band gap material.

The band gap is bounded by the conduction band minimum energy, E_{CB} and the valence band maximum energy, E_{VB}. At absolute zero temperature, the valence band is considered full and the conduction band empty. The energy required to promote an electron from the valence to the conduction band is E_G. The energy is usually quoted using the physical scale, which takes a free electron in vacuum as the reference [19]. As such, energies are considered to be negative i.e. the electrons are bound to the solid. The energy of the conduction band minimum is also referred to as the electron affinity, $E_A(vac)$, where the fact that it is with reference to vacuum is given in parentheses. The electron affinity is also defined as the energy gained by moving an electron from outside the semiconductor to the bottom of the conduction band. Therefore the energy of the valence band edge is:

$$E_{VB}(vac) = E_{CB}(vac) - E_G \tag{3.10}$$

The electron affinity of a semiconductor is not always known a priori or easy to determine. Butler and Ginley [5] proposed a method to estimate the electron affinity from the electronegativity, χ and the band gap. The electronegativity was assumed to be a good estimate for the centre of the band gap, giving the electron affinity as:

$$E_A(vac) = -\chi + \frac{1}{2}E_G \tag{3.11}$$

The bulk electronegativity of a compound is the geometric mean of the electronegativity of its constituent atoms:

$$\chi(TiO_2) = \left[\chi(Ti)\chi(O)^2\right]^{1/3} \tag{3.12}$$

As $\chi(Ti) = 3.454$ and $\chi(O) = 7.540$, for TiO_2 this gives $\chi(TiO_2) = 5.812$ eV. This corresponds to $E_A = -4.31$ eV with a band gap of 3 eV for rutile, and $E_A = -4.21$ eV with a band gap of 3.2 eV for anatase.

It is important to note that this approach does not take account of the differences between rutile and anatase, except in the slightly different magnitude of the band gap. It is not clear that this approach for estimating the electron affinity is appropriate for metal oxides; to the knowledge of the author, it has not been shown to be formally exact. Calculation of the band edges using computational methods such as density functional theory may be more accurate [24].

If the semiconductor contains impurities such as intentional dopants, such as Nb as used in this thesis, then the dopants may result in a discrete energy level or narrow band in the band gap. These are denoted by E_{DD} (for donor impurities) and E_{DA} (for acceptor impurities). The donor impurity level is shown as E_{DD} in Fig. 3.4, because it is more usual to dope n-type semiconductors with a donor impurity. Doping in TiO_2 is discussed further in the main literature review in Chap. 2.

The Fermi level of the semiconductor E_F, lies between the conduction and valence band edges. For an n-type semiconductor (such as TiO_2), the Fermi level will be close to the conduction band, while for a p-type semiconductor, it will be close to the valence band.

When the electrode potential is controlled by a potentiostat, it is the Fermi level of the semiconductor that is held constant with respect to the reference electrode i.e. the Fermi level is also the electrode potential. Importantly, knowing the Fermi level (the electrode potential) of the majority carriers in an n-type semiconductor like TiO_2 or Fe_2O_3 does not provide any information about the potential of the minority carriers (the holes). Given that it is the holes that drive the reaction at the surface of the photo-anode, this is a major difficulty in gaining a deep understanding of the reaction kinetics.

While the energy levels are quoted as being negative relative to vacuum, the negative charge of the electron means that the potential is positive relative to vacuum. Therefore, the direction of increasingly positive (or negative) potential is the same in both the semiconductor and the electrolyte. This is often confused when the units of V and eV are used interchangeably for quantities in the semiconductor. Strictly, the sign should change: an electron in a potential well of +1 V has energy of −1 eV.

For a non-polarized electrode, the Fermi level remains at the same energy with respect to vacuum throughout the semiconductor. Likewise, the conduction and valence band edges. This is assumed to be the case for an isolated semiconductor (Fig. 3.4a).

The electrode may be polarized by immersion in the electrolyte. The Fermi level of the semiconductor, E_{F} is likely to be different to the solution-Fermi level $E_{F(O/R)}$ relative to vacuum and so charge will flow until equilibrium is established. This results in the establishment of a space charge region and hence a corresponding distribution of potential at the interface of the semiconductor and the electrolyte; the following sections provide the mathematical description. Within the space charge region, the separation between the Fermi level and the conduction band edge varies with distance from the interface, resulting in what is termed *band bending*. The conduction band and valence band edges remain parallel to each other and the vacuum level throughout the semiconductor. Overall, the vacuum level in the semiconductor appears to bend with respect to the vacuum level in the electrolyte. In fact, the vacuum level in the electrolyte will also bend, but because of the comparatively low permittivity of the electrolyte and the higher mobility of ionic charge, the bending occurs over a much shorter distance of approximately one monolayer. Generally, it is not shown on energy level diagrams. Outside the space charge region, the electronic band structure of the semiconductor is as in the isolated, unpolarized case. However, there is an offset in the vacuum level in the bulk of the semiconductor with respect to the vacuum level in the electrolyte. The electron affinity E_{A} is unchanged by the band bending. However, the potential of the conduction band minimum changes with respect to a reference electrode in the liquid phase. As such, it is important to be able to distinguish between the position of the conduction band in the bulk, E_{CB} and the position of the conduction band at the surface/interface, which is denoted $E_{\mathrm{CB,S}}$. For any position in the space charge region, this could be termed $E_{\mathrm{CB}}(x)$. Likewise, it is sensible to define the edge of the valence band at the interface as $E_{\mathrm{VB,S}}$. Here, we can also define the energy levels of dopants at the interface, $E_{\mathrm{DD,S}}$ and $E_{\mathrm{DA,S}}$. This will be important in understanding the Mott-Schottky plot of doped and reduced TiO_2 (see Results, Chaps. 5 and 6).

The principles and mathematics, as described below, are identical to those used to describe Schottky barriers and p-n junction devices. There are additional features that arise due to the pH of the liquid. The schematic shown in Fig. 3.4 and described above is for a very specific case in which there is no net absorbed charge due to the pH of the electrolyte, i.e. the pH of the electrolyte coincides with the point of zero charge of the semiconductor surface. The general, pH dependent, case is introduced below in Sect. 3.5.6.

3.4.2 Flat Band Potential

If the potential of the semiconductor electrode, in electrical contact with the electrolyte, is such that the electrode is not polarized in the interfacial region, it is termed the *flat band potential*. There is no space charge region (neither depletion nor accumulation); there is no electric field present in the semiconductor to act on charge carriers.

The equilibration of the electrode and electrolyte results in band bending. The potential of the electrode can be changed using a potentiostat to the flat band potential. As such, the flat band potential is not an equilibrium condition. Charge must flow to maintain the separation of the energy levels in the semiconductor and electrolyte. It is possible that at the flat band potential, dark redox processes can occur if there is a suitable redox couple present.

3.5 Mathematical Description of the Space Charge Layer

The following sections present a mathematical model commonly used to describe the space charge layer. This is of use in understanding the potential dependence of various aspects of electrode performance including: band bending, space charge width, photocurrent, and capacitance. The model is derived in detail to provide a comprehensive reference. The principles outlined are used in Chap. 6 of this thesis to study the differences in behaviour of reduced and doped rutile electrodes. In particular, the model is used to derive the Mott-Schottky equation, which is used to estimate the charge carrier concentration of the electrodes from impedance measurements. In turn the charge carrier concentration is used to estimate the width of the space charge layer with potential to be able to fit a model to photocurrent data.

3.5.1 Poisson's Equation

The spatial distribution of potential close to the semiconductor|liquid junction is determined by solving Poisson's equation, the general form of which in one dimension is:

$$\frac{d^2\phi}{dx^2} = -\frac{\rho}{\varepsilon} \tag{3.13}$$

where ϕ is the potential, x is distance, ρ is the charge carrier density, and ε is the electric permittivity. This can also be written in terms of the electric field strength ξ.

$$\frac{d\xi}{dx} = \frac{\rho}{\varepsilon} \tag{3.14}$$

where the electric field strength and the potential are defined by

$$\frac{d\phi}{dx} = -\xi \tag{3.15}$$

Therefore, it is possible to derive the potential ϕ from the charge density ρ.

In the general case of a semiconductor with impurity concentrations of N_D donor and N_A acceptor impurities, the charge density can be written as

$$\rho = q_e(p - n + N_D - N_A) \tag{3.16}$$

where n and p are the charge carrier concentrations of electrons and holes respectively, in units of m^{-3} (more commonly cm^{-3}). The magnitude of the electronic charge q_e is required to convert from a number concentration, m^{-3} to a charge concentration, C m^{-3}. The resulting Poisson equation is then

$$\frac{d^2\phi}{dx^2} = -\frac{d\xi}{dx} = -\frac{q_e}{\varepsilon}(p - n + N_D - N_A) \tag{3.17}$$

The intrinsic charge carrier concentrations n and p are more strictly represented with a dependence on the potential $\phi(x)$, and the temperature T. The Boltzmann constant k_B is introduced to ensure the exponent is dimensionless [2].

$$n = n_0 e^{q_e\phi(x)/k_BT} \tag{3.18}$$

$$p = p_0 e^{-q_e\phi(x)/k_BT} \tag{3.19}$$

The form of Poisson's equation given assumes that all electronic impurities present are fully ionized and potential independent. The effect of potential dependent impurity concentrations may be important in understanding the behaviour of TiO_2. For example, the donor levels in rutile TiO_2 may be deep lying by 0.5–1 eV below the conduction band edge. As a result, at room temperature they will be only partially ionized. The effect of the potential in the semiconductor on the charge carrier concentration is demonstrated for both reduced and Nb-doped rutile in Chap. 6 of this thesis. As the potential of the rutile electrode becomes positive of the flat band potential, the Mott-Schottky plot of the reciprocal of the capacitance-squared against potential shows a change in slope. This can be explained if the charge carrier concentration increases suddenly, such as by the ionization of impurities as the Fermi level of the electrode passes through an impurity energy level.

The negative and positive charge concentrations must balance to maintain charge neutrality.

$$p + N_D = n + N_A \tag{3.20}$$

A more accurate formulization accounts for the ionization potential of the donors and acceptors. Based on an impurity concentration of N_D, the concentration of ionized impurities $N_D{}^*$ can be expressed as:

$$N_D{}^* = \frac{N_D}{1 + \frac{1}{2}\exp(\Delta E_\mathrm{D}/k_B T)} \tag{3.21}$$

$$N_A{}^* = \frac{N_D}{1 + 4\exp(\Delta E_\mathrm{D}/k_B T)} \tag{3.22}$$

where $\Delta E_\mathrm{D} = E_F - E_D$ is the ionization energy of the impurity: the energy difference between the Fermi level E_F and the energy level of the impurity E_D. The factors of one half in Eq. (3.21) and four in Eq. (3.22) are due to the ground state degeneracy of the electrons. These values are given for Si; they are not known for TiO_2 (at least, this author was unable to determine them). TiO_2 is also complicated by the presence of Ti^{3+} sites.

If the condition $\Delta E_\mathrm{D} \gg k_B T$ is met then the following Boltzmann approximation is valid;

$$N_D{}^* = 2N_D\exp(-\Delta E_\mathrm{D}/k_B T) \tag{3.23}$$

$$N_A{}^* = \frac{1}{4}N_A\exp(-\Delta E_\mathrm{D}/k_B T) \tag{3.24}$$

At room temperature $k_B T$ is 0.025 eV, so the approximation is valid if the ionization potential is greater than 0.1 eV or so.

3.5.2 *Electric Field Distribution in the Semiconductor*

The potential and parameters that describe the space charge region are now derived based on a standard, simplified form of Poisson's equation as applied to the formation of a depletion layer in an n-type semiconductor.

The usual assumptions are that $N_A = 0$ and $n \gg p$. This results in a charge neutrality condition $N_D \simeq n$, and a simplified form of Poisson's equation. It is also assumed the impurities are homogeneously distributed in the semiconductor.

$$\frac{d^2\phi}{dx^2} = -\frac{d\xi}{dx} = -\frac{q_e N_D}{\varepsilon} \tag{3.25}$$

Integrating to obtain the electric field ξ gives:

$$\xi = \frac{q_e N_D}{\varepsilon} x + C \tag{3.26}$$

where the constant of integration, C is obtained by applying boundary conditions. The interface with the liquid is defined to be at $x = 0$. The other boundary is at the edge of the space charge region at $x = x_{SCR}$ where it is assumed that $\xi = 0$. This gives the constant of integration to be:

$$C = -\frac{q_e N_D}{\varepsilon} x_{SCR} \tag{3.27}$$

The full form of the electric field $\xi(x)$ is then:

$$\xi(x) = \frac{q_e N_D}{\varepsilon}(x - x_{SCR}) \tag{3.28}$$

Evaluated at $x = 0$, ξ is a maximum. So:

$$\xi(0) = -\frac{q_e N_D}{\varepsilon} x_{SCR} \tag{3.29}$$

If the convention is taken such that the distance x from the interface into the bulk is negative, $x_{SC} < 0$, then this gives a positive electric field. This would be expected to drive holes (virtual positive charge carriers) in the positive x direction, i.e. towards the interface. Therefore, electrons are driven away from the interface, into the bulk as per the observed behaviour of n-type semiconductor photo-electrodes.

Hence, to determine the electric field fully, it is necessary to calculate the width of the space charge layer, $w = |x_{SCR}|$, which itself is determined by the potential, as derived in the following section.

3.5.3 Potential Distribution at the Semiconductor-Liquid Interface

The width of the space charge region is determined by the applied potential, so it is necessary to integrate the electric field with boundary conditions to arrive at an expression for the width of the space charge layer as a function of band bending and electrode potential.

$$\phi = -\int \xi(x)dx = -\frac{q_e N_D}{\varepsilon}\int (x - x_{SCR})dx \tag{3.30}$$

If we make the substitution $s = x - x_{SCR}$ this gives $ds = dx$ and

$$\phi = -\frac{q_e N_D}{\varepsilon}\int s ds \tag{3.31}$$

So

$$\phi = -\frac{q_e N_D}{2\varepsilon}s^2 + C \tag{3.32}$$

Substituting back for $s = x - x_{SCR}$

$$\phi = -\frac{q_e N_D}{2\varepsilon}(x - x_{SCR})^2 + C \tag{3.33}$$

The constant of integration is evaluated by assigning boundary conditions. In this case, the potential at $x = x_{SCR}$ is taken to be $\phi(x_{SCR})$, giving

$$C = \phi(x_{SCR}) \tag{3.34}$$

And a completely general form of the potential

$$\phi = -\frac{q_e N_D}{2\varepsilon}(x - x_{SCR})^2 + \phi(x_{SCR}) \tag{3.35}$$

At $x = 0$, the potential is

$$\phi(0) = -\frac{q_e N_D}{2\varepsilon}(-x_{SCR})^2 + \phi(x_{SCR}) \tag{3.36}$$

The choice of a zero for the potential is essentially an arbitrary one. Taking the potential in the bulk of the semiconductor as zero, i.e. $\phi(x_{SCR}) = 0$, then the potential at the surface evaluates as negative with respect to the bulk value by $-\frac{q_e N_D}{2\varepsilon}{x_{SCR}}^2$. Likewise, taking $\phi(0) = 0$, the bulk is defined as positive with respect to the surface/interface.

$$\phi(x_{SCR}) = \frac{q_e N_D}{2\varepsilon}(x_{SCR})^2 \tag{3.37}$$

The second option is perhaps the most appropriate for use in photo-electrochemistry, where all potentials are represented on the absolute electrode potential scale. This convention makes it (relatively) simple to add the potential in the bulk to the calculated positions of the potentials of interest at the surface (such as the potential of the edges of the conduction and valence bands). In this convention, it is also clear that with positive-going electrode potential sweeps, the bulk potentials also change positively on the absolute electrode potential scale. Hence, the general form of the potential in the semiconductor as a function of distance from the interface, taking the potential at the interface as $\phi(0) = 0$, is:

$$\phi(x) = \frac{q_e N_D}{2\varepsilon}(x)^2 \tag{3.38}$$

This is valid from $x_{SCR} \leq x \leq 0$, where x is increasingly negative into the bulk of the semiconductor.

3.5.4 Width of the Space Charge Region

Now the width of the space charge region (or depletion layer) w can be calculated for a given potential drop, which over the semiconductor space charge region is referred to as the *band bending*, and may be denoted by $\Delta\phi_{SCR}$. In both of the possible conventions for referencing the potential discussed above, the total band bending can be written as:

$$\Delta\phi_{SCR} = \phi(x_{SCR}) - \phi(0) \tag{3.39}$$

Therefore, when the potential in the bulk is more positive than the surface, the band bending is positive, so the bands as drawn would curve upwards as they approach the interface. Other notations are often used, such as ϕ_{SC} and ϕ_{bb}, but using the delta form is explicit, in that it shows a potential difference rather than a relative potential. It is possible to reference the potential drop to the potential in the bulk of the semiconductor when no band bending is present at the flat band potential. Likewise, the extent of band bending could be given as the potential of the conduction band in the bulk, relative to the potential of the conduction band at the surface. However, this assumes a number of things including band-edge pinning that are not true in general. As such, here it is preferred to express the band bending as a potential difference, which can be added to a potential on a reference scale.

The width of the space charge region is often denoted by w in the photo-electrochemical literature [25], where $w = x_{interface} - x_{SCR}$. In the convention for x defined above, this gives $w = -x_{SCR}$, a positive quantity. Substitution of w and $\Delta\phi_{SCR}$ into (3.36) gives:

$$\Delta\phi_{SCR} = \frac{q_e N_D}{2\varepsilon} w^2 \tag{3.40}$$

Rearranging for w in terms of the applied band bending,

$$w = \sqrt{\frac{2\varepsilon\Delta\phi_{SCR}}{q_e N_D}} \tag{3.41}$$

This can be written as,

$$w = w_0\sqrt{\Delta\phi_{SCR}} \tag{3.42}$$

where w_0 is the space charge region width constant given by

$$w_0 = \sqrt{\frac{2\varepsilon}{q_e N_D}} \tag{3.43}$$

In physical terms, the width constant w_0 is the width of the space charge region for an applied band bending of 1 V, as the units of w_0 are m V^{-1}. Figure 3.5 shows the variation of w_0 with carrier concentration and relative permittivity. The relative permittivity depends on the orientation. For rutile TiO_2 the expected in the range is 80–200 (with $\varepsilon_r = 160$ perpendicular to the c-axis, which lies in the (110) plane

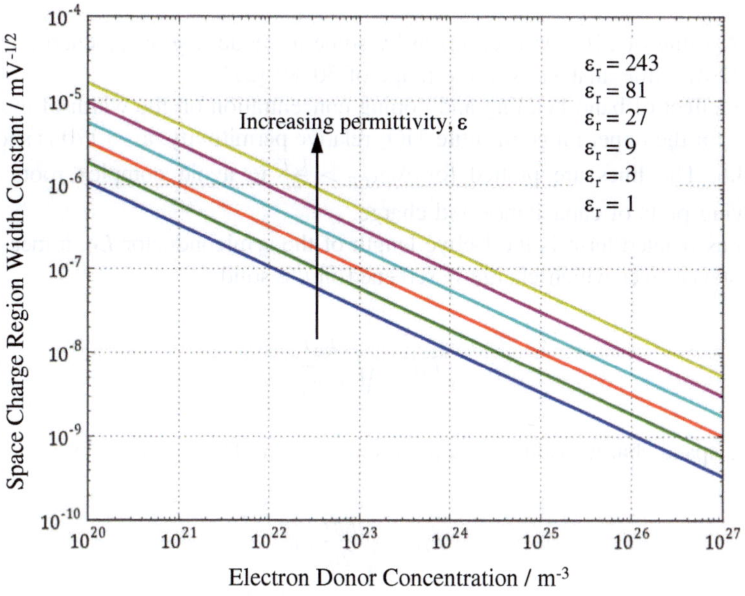

Fig. 3.5 Width constant of space charge region for range of doping concentrations and relative permittivity. The plot shows lines for a range of the relative permittivity ε_r increasing from $\varepsilon_r = 1$ (*blue line*) to from $\varepsilon_r = 243$ (*yellow line*). The top three lines encompass the range of relative permittivity expected for TiO_2

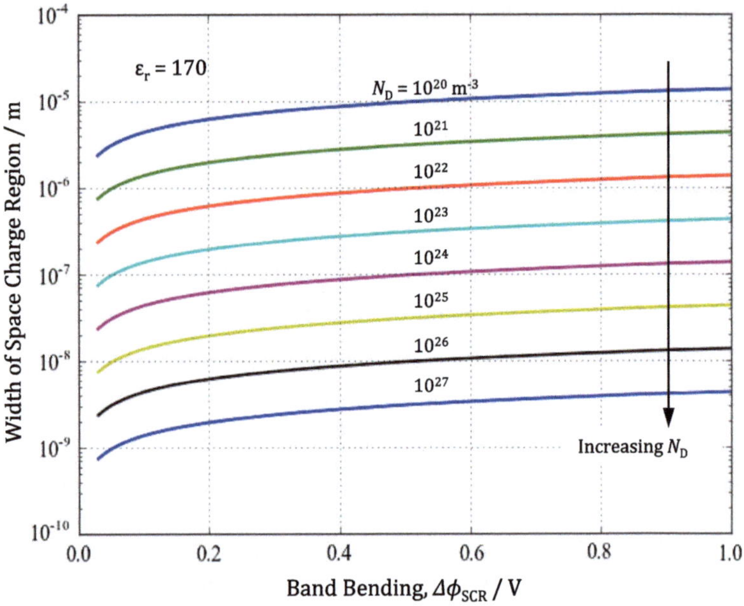

Fig. 3.6 Width of space charge region for range of doping concentrations 10^{20} to 10^{27} m^{-3} and ε_r = 170 for TiO_2

[23]. A value of 170 has been taken by some to model the space charge layer of TiO_2 [26]) while anatase is in the range of 30–80 [27].

The effect of band bending and doping concentration on the width of the space charge for the upper range of rutile TiO_2 relative permittivity, $\varepsilon_r = 170$ is shown in Fig. 3.6. The lines are plotted for $\Delta\phi_{SCR} > \frac{k_B T}{q_e}$ to avoid complex roots in the following plots of capacitance and charge.

An associated term is the Debye length of the semiconductor L_D, a measure of the distance over which charge is screened in the solid

$$L_D = \sqrt{\frac{\varepsilon k_B T}{q_e^2 N_D}} \tag{3.44}$$

The space charge width constant is related to the Debye length by

$$L_D = \sqrt{\frac{k_B T}{2q_e}} w_0 \tag{3.45}$$

The two characteristic lengths should not be confused. At 25 °C, the value of L_D is 0.11 w_0.

At this point, there is an expression for the width constant of the space charge region in terms of the doping concentration and electric permittivity of the material. With knowledge of the applied band bending (from knowledge of both the flat band and electrode potentials), it is possible to construct the electric field strength and the potential distribution inside the semiconductor. This model of the space charge region (depletion layer) is usually sufficient for modelling the photocurrent of solar cells, as discussed in Sect. 3.6 below. A more sophisticated treatment would also include diffusion of charge carriers, and modifications of the potential based on varying charge carrier concentration (say due to photo-generation). The treatment so far is sufficient for a description of the capacitance of the interface and the construction of the Mott-Schottky equation. The Mott-Schottky equation (derived in the following section) is typically employed to fit the capacitive element of an impedance measurement in order to derive material properties of the electrode.

3.5.5 Capacitance—Mott-Schottky Plots

When the potential of the electrode is not at the flat band potential, the electrode is polarized and, as shown in Fig. 3.4b and described in the preceding sections, a potential distribution is established between the interface with the electrolyte and the edge of the space charge region. For relevance to TiO_2 in water splitting, the following discussion assumes a depletion layer. An applied potential drop in the semiconductor $\Delta\phi_{SCR}$ gives a total charge in the space charge region of:

$$Q_{SCR} = q_e N_D w = q_e N_D w_0 \sqrt{\Delta\phi_{SCR}} \tag{3.46}$$

This is simply the charge density $q_e N_D$ integrated over the width of the space charge layer, w. It should be noted that this is the charge per unit area of the interface i.e. the units of charge are C m^{-2}. The capacitance is defined as the change in charge with potential.

$$C = \frac{dQ}{dV} \tag{3.47}$$

Which applied to Eq. (3.46) gives

$$C_{SCR} = \frac{1}{2} q_e N_D w_0 \frac{1}{\sqrt{\Delta\phi_{SCR}}} \tag{3.48}$$

Substituting for w_0, using Eq. (3.43) gives

$$C_{SCR} = \sqrt{\frac{\varepsilon q_e N_D}{2}} \frac{1}{\sqrt{\Delta\phi_{SCR}}} \tag{3.49}$$

This can be written in the usual form of the Mott-Schottky equation used to analyse the interfacial capacitance of the semiconductor-liquid junction

$$\frac{1}{{C_{SCR}}^2} = \left(\frac{2}{\varepsilon q_e N_D}\right)\Delta\phi_{SCR} \tag{3.50}$$

The units of capacitance here are F m^{-2}. Equation (3.50) states that: the square of the reciprocal of the capacitance per unit area increases linearly with the band bending, with a gradient of $(2/\varepsilon\, q_e\, N_D)$. The linear relationship between the inverse square of the capacitance and the band bending is seen to hold in many experimental results [2]. However, there are many examples where non-linear and/or frequency dependent Mott-Schottky plots are common, such as on nanostructured hematite [28], for which additional considerations must be taken into account. For the TiO_2 single crystals studied in this work, the linear relationship is generally sufficient. A more complete version of the Mott-Schottky Eq. (3.51) includes the effect of temperature on the capacitance. The term $k_B T/q_e$ arises if the charge and capacitance are derived from the full self-consistent Poisson Eq. (3.17), which includes a potential dependent charge carrier concentration. The full derivation is to be found in Pleskov and Gurevich [2].

$$\frac{1}{{C_{SCR}}^2} = \left(\frac{2}{\varepsilon q_e N_D}\right)\left(\Delta\phi_{SCR} - \frac{k_B T}{q_e}\right) \tag{3.51}$$

Integrating Eq. (3.51) gives a modified expression for the charge that includes the temperature dependence.

$$Q_{SCR} = \sqrt{2\varepsilon\, q_e N_D}\sqrt{\Delta\phi_{SCR} - \frac{k_B T}{q_e}} \tag{3.52}$$

The capacitance of the semiconductor depletion layer is shown in Fig. 3.7 and the stored charge in Fig. 3.8 for a range of doping concentrations and a relative permittivity of $\varepsilon_r = 170$. The temperature dependent term $\frac{k_B T}{q_e}$ of magnitude ca. 26 mV at 298 K is included, and the plot is over the potential range $\Delta\phi_{SCR} > \frac{k_B T}{q_e}$ to ensure that the root evaluates as real.

These plots show the expected range of interfacial capacitance for rutile electrodes with a range of doping concentrations.

Figure 3.9 shows the usual format of a Mott-Schottky plot in which $1/{C_{SCR}}^2$ varies linearly with the band bending. Typically, a Mott-Schottky plot takes the electrode potential as the x-axis rather than the band bending. As shown later in the discussion, this is not a strictly correct assumption, but is reasonable for the rutile electrodes investigated in this thesis. It is clear that the doping concentration has a significant effect on the gradient. The relative permittivity has a similar effect, however, it is a more constrained parameter. While the doping concentration may

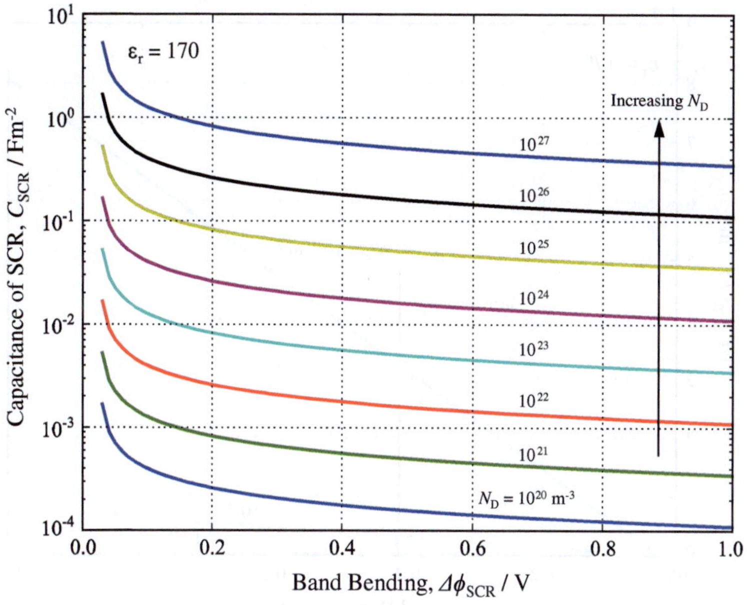

Fig. 3.7 Capacitance of space charge region for range of doping concentrations 10^{20} to 10^{27} m^{-3} and $\varepsilon_r = 170$ for TiO_2

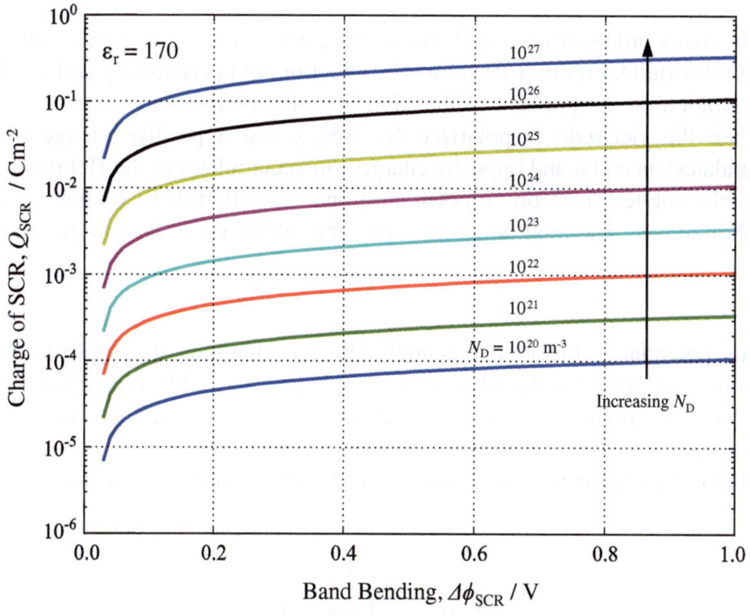

Fig. 3.8 Charged stored in space charge region during depletion for range of doping concentrations 10^{20} to 10^{27} m^{-3} and $\varepsilon_r = 170$ for TiO_2

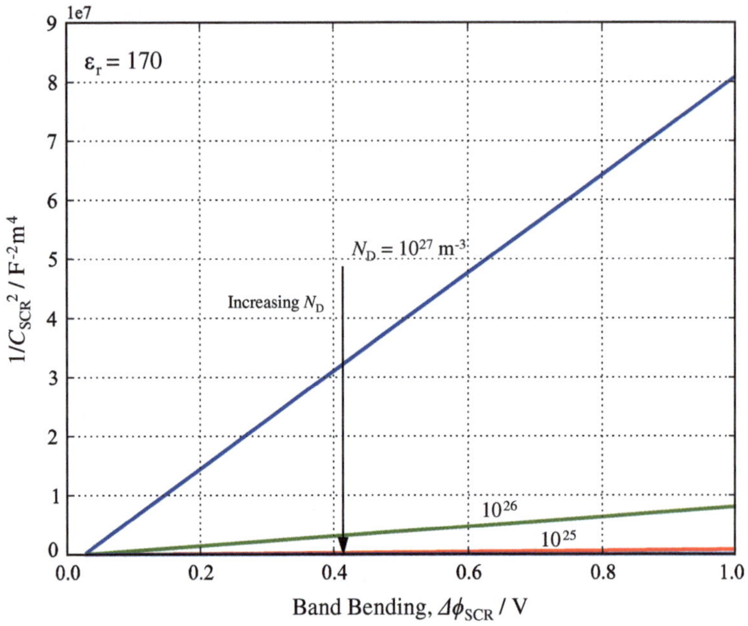

Fig. 3.9 Mott-Schottky plot of C_{SCR}^{-2} for a range of doping concentrations 10^{20} to 10^{27} m^{-3} and $\varepsilon_r = 170$ for TiO_2

vary by many orders of magnitude the relative permittivity is generally between 80 and 200 for rutile. Figure 3.10 shows the effect of both permittivity and the doping concentration on the gradient of the Mott-Schottky plot.

When the electrode is polarized by $\Delta\phi_{SCR}$ and a positive charge of Q_{SCR} accumulates; an equal and opposite charge, Q_H accumulates in the Helmholtz layer of the electrolyte. Therefore, a potential drop $\Delta\phi_H$ will also be present across the Helmholtz layer. The total potential across the interface can be written as:

$$\Delta\phi_{TOTAL} = \Delta\phi_{SCR} + \Delta\phi_H \tag{3.53}$$

This potential is the total potential difference between the applied electrode potential and the flat band potential, measured against suitable reference electrodes. The spatial distribution of this potential, due to the different capacitances of the semiconductor and the electrolyte, is now considered.

The total capacitance of the interface, with the two capacitors in series, is given by:

$$\frac{1}{C_{TOTAL}} = \frac{1}{C_{SCR}} + \frac{1}{C_H} \tag{3.54}$$

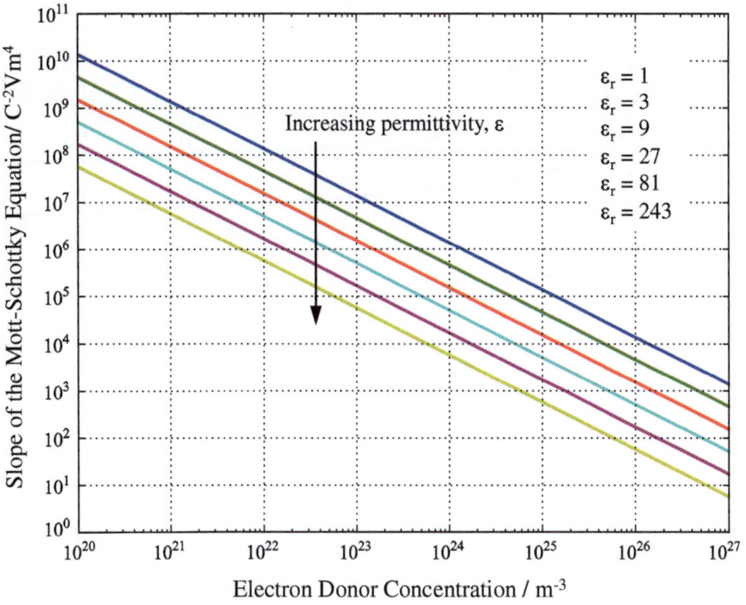

Fig. 3.10 Effect of doping concentrations 10^{20} to 10^{27} m^{-3} and relative permittivity ε_r on gradient of the Mott-Schottky plot. This plot can be used to check the range of doping and permittivity likely for a measured gradient

It is often assumed that C_H is invariant with potential and also that $C_H \gg C_{SCR}$. Typically, the value of the capacitance of the Helmholtz layer, C_H, as for a metal, is taken as 0.1 F m^{-2}. The capacitance of the interface is then given as:

$$C_{TOTAL} \approx C_{SCR} \tag{3.55}$$

As $Q_{SCR} = -Q_H$, this then implies that $\Delta\phi_{SCR} \gg \Delta\phi_H$ and the change in the potential drop in the Helmholtz layer, $\Delta\phi_H$ can be neglected. As discussed by some [2, 26, 29], this assumption may not always be accurate.

The assumption of band-edge pinning is also made, and is discussed at the end of this section.

In the following figures and text, the assumption that $C_H \gg C_{SCR}$ is not made a priori. The following figures and discussion are based on an approach outlined by Uosaki [27]; however, it is noted here that there is an error in the equation for the charge in that work. The error was to omit the square root in the dependence of the charge stored with band bending. It is shown correctly here in Eq. (3.52). The purpose is to assess whether the assumption $C_H \gg C_{SCR}$ is applicable to TiO_2 (and other high electric permittivity materials).

Figure 3.11 shows the ratio of C_{SCR} to C_H for a semiconductor with a high relative permittivity, $\varepsilon_r = 170$, often assumed for TiO_2. It is apparent that when the semiconductor is highly doped (10^{25} m^{-3}), that the assumption that $C_H \gg C_{SCR}$ is

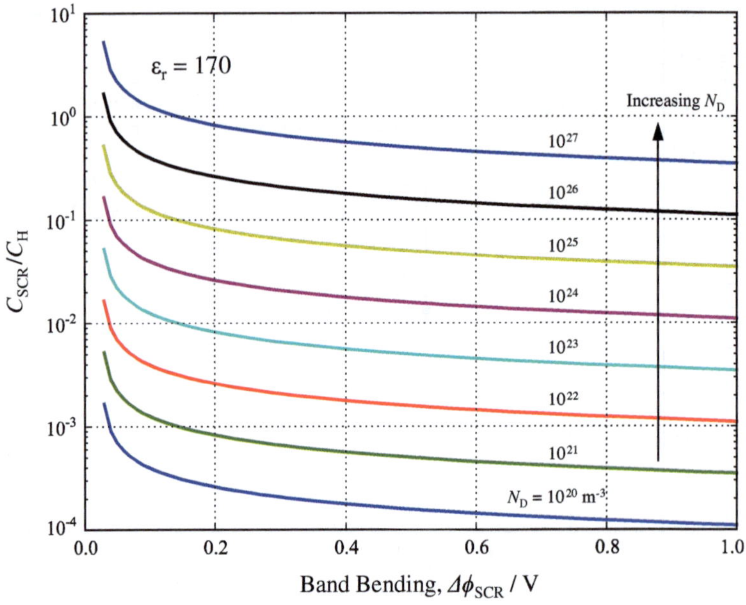

Fig. 3.11 Ratio of capacitance of space charge layer to Helmholtz layer for a range of doping concentrations 10^{20} to 10^{27} m^{-3} and relative permittivity $\varepsilon_r = 170$

poor. In particular, below 0.2 V band bending, the assumption $C_H \gg C_{SCR}$ is questionable. This suggests that in order to apply the Mott-Schottky equation for high permittivity semiconductors (like TiO_2), it is necessary to take the capacitance of the Helmholtz layer into consideration.

In a general case, both $\Delta\phi_H$ and $\Delta\phi_{SCR}$ need to be taken into account. In principle, it is possible to calculate the charge for a given potential drop in the semiconductor, and to use this charge to calculate the potential drop in the Helmholtz layer. The potential drop in the Helmholtz layer depends on the capacitance, C_H and the charge in the semiconductor space charge region, Q_{SCR}. Electroneutrality ($\sum_i z_i c_i = 0$) requires that Q_H is equal and opposite to Q_SCR.

$$\Delta\phi_H = \frac{Q_{SCR}}{C_H} = \frac{\sqrt{2\varepsilon q_e N_D}}{C_H}\sqrt{\Delta\phi_{SCR} - \frac{k_B T}{q_e}} \tag{3.56}$$

This can be used to derive an expression for the total capacitance of the interface, C_{TOT}, which can be used in turn to analyse the Mott-Schottky plot of the measured capacitance. Equation (1.52) was derived by De Gryse et al. [29] who examined the case in which the assumptions about C_H and $\Delta\phi_H$ do not hold. A full derivation is also given in by Pleskov and Gurevich [2]. The equation still assumes that C_H is independent of potential.

$$\frac{1}{C_{TOT}^2} = \frac{1}{C_H^2} + \frac{2}{\varepsilon q_e N_D}\left(\Delta\phi_{TOT} - \frac{k_B T}{q_e}\right) \tag{3.57}$$

Here, $\Delta\phi_{TOT} = U - U_{FB}$, is the applied potential relative to the flat band potential. Setting $1/C_{TOT}{}^2 = 0$ results in a relation between the electrode potential versus flat band and the capacitance of the Helmholtz layer C_H:

$$U - U_{FB} = \frac{k_B T}{q_e} - \frac{\varepsilon q_e N_D}{2C_H{}^2} \tag{3.58}$$

This demonstrates that the measured intercept of the Mott-Schottky plot with the potential axis is actually offset from the flat band potential. For highly doped semiconductors with a large permittivity, such as TiO_2, this must be taken into account. However, as seen in Eq. (3.57), the gradient of the Mott-Schottky plot is unaffected by the need to take account of C_H.

Figures 3.12 and 3.13 show plots of the Helmholtz layer potential against band bending and as a ratio of the total potential for a range of N_D and $\varepsilon_r = 170$. For high doping concentrations, it is clear that the potential drop across the Helmholtz layer (due to band bending) is tens to hundreds of mV even at low band bending, and so needs to be considered.

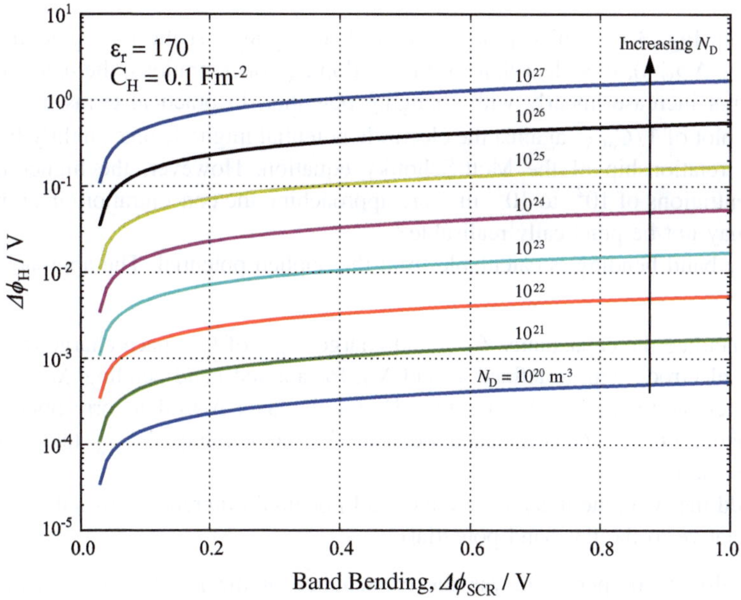

Fig. 3.12 Potential drop in Helmholtz layer for a range of doping concentrations 10^{20} to 10^{27} m^{-3}, relative permittivity $\varepsilon_r = 170$, and $C_H = 0.1$ Fm^{-2}

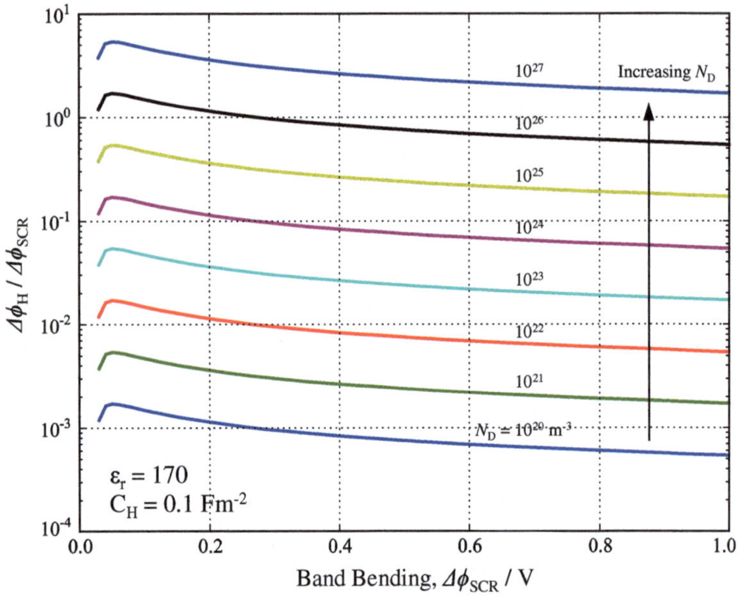

Fig. 3.13 Ratio of potential drop in Helmholtz layer to space charge region for a range of doping concentrations 10^{20} to 10^{27} m^{-3} and relative permittivity ε_r = 170 and C_H = 0.1 Fm^{-2}

From Fig. 3.14, which plots the band bending against the total potential drop $(\Delta\phi_H + \Delta\phi_{SCR})$, it is clear that for higher doping concentrations, the band bending does not increase linearly with changing potential. In practical terms, this means that a plot of $1/C_{SCR}{}^2$ against the electrode potential might deviate slightly from the linear relationship of the Mott-Schottky equation. However, the higher doping concentrations of 10^{25} to 10^{27} m^{-3} are approaching the concentration of Ti in TiO_2 and may not be practically realisable.

The band bending is calculable from the applied potential. The approach taken is to:

1. Calculate the capacitance C_{SCR} and charge Q_{SCR} of the space charge region for an electrode with a defined ε_r and N_D, for applied band bending $\Delta\phi_{SCR}$.
2. Calculate the potential drop in the Helmholtz layer $\Delta\phi_H$ that corresponds to the charge $Q_H = -Q_{SCR}$ at each degree of band bending $\Delta\phi_{SCR}$ taking C_H as constant.
3. Add the two potentials to give the total potential difference across the electrode (relative to the flat band potential).

Ideally, to properly analyse the capacitance of the interface over a range of applied electrode potentials, it may be necessary to convert the electrode potential into the band bending. Additionally, whether the assumption of C_H as a constant is applicable to TiO_2 is unknown.

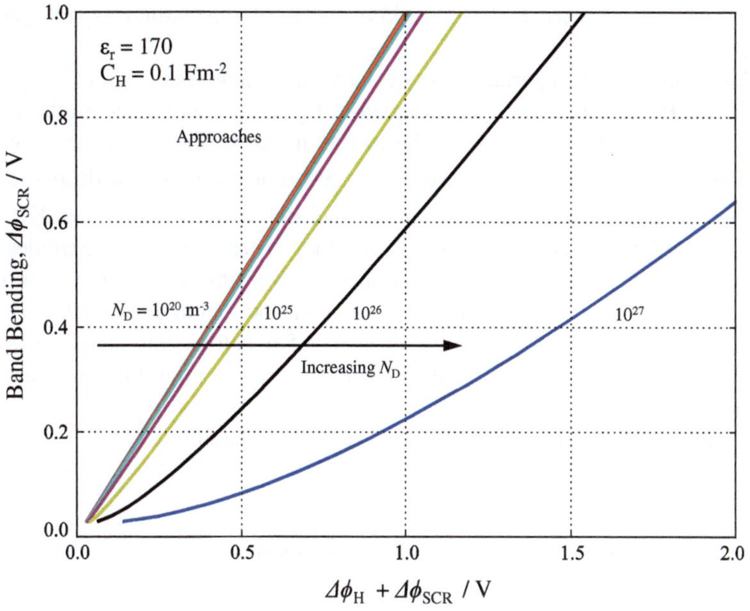

Fig. 3.14 Band bending versus applied potential (relative to the flat band potential) for a range of doping concentrations 10^{20} to 10^{27} m^{-3} and relative permittivity ε_r = 170 and C_H = 0.1 Fm^{-2}. At low doping levels, there is a one-to-one correspondence. At higher doping concentrations, the band bending does not increase linearly with changing electrode potential

Furthermore, another common assumption made (and assumed above) is termed *band-edge pinning*, which assumes that the potential of the conduction and valence band edges at the solid|liquid interface are unchanged relative to the solution potential as the electrode potential is varied. Only the potential of the band edges in the bulk changes with the electrode potential. In this assumption, the capacitance of the Helmholtz layer is invariant with potential, and all applied potential relative to the flat band potential results in band bending.

The opposite situation is *Fermi level pinning* (or more intuitively *band-edge unpinning* [2]) in which the potential of the band edges at the solid-liquid interface change with the applied potential. In this case it is assumed that most, if not all, of the full potential drop over the interface occurs in the Helmholtz layer i.e. one cannot ignore $\Delta\phi_H$, and $\Delta\phi_H \gg \Delta\phi_{SCR}$. As the electrode potential is scanned, the potential drop in the Helmholtz layer increases, while the band bending remains small and fixed.

In the general case, the semiconductor can exhibit both band-edge pinning and unpinning as the electrode potential is varied [30]. When the energy corresponding to electrode potential is scanned through the energy corresponding to a surface or defect state, then the charge carrier concentration increases, resulting in unpinning until the state is depleted of charge (or filled). A region of band bending may then occur until another surface or defect state is reached. This greatly complicates the analysis of the interface and to the knowledge of this author is not generally attempted.

What is not readily apparent from the above treatment is the sign of the potential drop in the Helmholtz layer. For a depletion layer, the charge accumulated in the space charge region is positive, so the charge in the Helmholtz layer is negative. Using the relation $\Delta\phi_H = \frac{-Q_{SCR}}{C_H}$ would suggest that the potential drop should be negative. So the potential at the edge of the Helmholtz layer (outer Helmholtz plane) should be negative relative to the interface. Therefore, the magnitudes of the potentials add as expected. For an accumulation layer, the signs of the accumulated charge and potentials are both changed, so again the magnitudes of the potentials are additive. However, as discussed in Sect. 3.5.6 below, this causes a problem when trying to account for the pH dependence of the potential difference across the Helmholtz layer.

The Helmholtz layer is now discussed in further detail.

3.5.6 Helmholtz Layer Potential Difference Due to PH

In addition to the potential drop across the Helmholtz layer arising from the compensation of the charge accumulated by band bending in the semiconductor, there is an additional potential in the Helmholtz layer that must be considered. This potential, $\Delta\phi_{H(FB)}$ is dependent only on the pH of the electrolyte and is not influenced by the band bending. The potential is the potential drop in the Helmholtz layer due to the adsorption of ions at the flat band potential: hence, the use of the subscript *FB*. The total potential drop in the Helmholtz layer is then

$$\Delta\phi_{H(TOTAL)} = \Delta\phi_{H(FB)} + \Delta\phi_H(U) \tag{3.59}$$

where $\Delta\phi_H(U)$ corresponds to the potential dependent contribution to the Helmholtz layer to due to the compensation of the space charge in the semiconductor. The pH dependent potential drop is given by

$$\Delta\phi_{H(FB)} = \log 10 \frac{RT}{nF}(pzc - \mathrm{pH}) \tag{3.60}$$

where *pzc* denotes the point of zero charge. This is the pH at which the surface of the semiconductor adsorbs equal amounts of positive and negative ionic charge. At the $\mathrm{pH} = pzc$, the pH dependent potential drop in the Helmholtz layer is zero. At $\mathrm{pH} > pzc$, $\Delta\phi_{H(FB)} < 0$ and at $\mathrm{pH} < pzc$, $\Delta\phi_{H(FB)} > 0$. For monovalent ionic adsorption ($n = 1$), the potential drop varies by 59 mV pH^{-1}. There are reports of slopes between 29 mV pH^{-1} and 73 mV pH^{-1}, although these are not necessarily for TiO_2 [31]. Typically, this is established by estimation of the flat band potential for a range of pH.

Even at the flat band potential, there is still a potential difference across the Helmholtz layer (unless pH = *pzc*). This has the effect of shifting the conduction band and valence band edges by $\Delta\phi_{H(FB)}$ relative to their position at the pzc measured against an appropriate reference electrode, such as SHE.

As a material property, the pzc vary from material to material, and for different surfaces of the same material. For example, the (100) and (001) planes of a crystal could in principle exhibit a different pzc, resulting in modified photoelectrochemical performance at a given pH.

The potential $\Delta\phi_{H(FB)}$ can be viewed as a discontinuity in the vacuum level between the semiconductor and the electrolyte, and is depicted in Fig. 3.15. The effect is shown for a range of pH at the flat band condition i.e. no band bending. The flat band potential $U_{F(FB)}$ is also shown relative to a reference electrode. This demonstrates the importance of using the electrode potential scale and the following algorithm for constructing the interface. The electrode potential is the Fermi level of the electrode. As measured with respect to the vacuum level in the semiconductor, the Fermi level is (or should be) invariant with pH. However, as measured on a reference electrode potential scale (in practice), the flat band potential shifts to more negative values, as has been observed for electrodes including TiO_2 [32]. The other energy levels in the semiconductor $E_{CB,S}$ and $E_{VB,S}$ also shift.

The total potential drop over the Helmholtz layer will depend on both the pH and the band bending. For example, at pH > pzc, $\Delta\phi_{H(FB)}$ will be negative, i.e. the vacuum energy level in the semiconductor is shifted to a more negative potential on the electrode potential scale relative to the vacuum energy level in the electrolyte. If the electrode potential is positive of the flat band, $\Delta\phi_{SCR} > 0$ then there will be an associated potential $\Delta\phi_H(V)$. Relative to the potential of the electrode at the interface, taken as the reference zero in the preceding derivations, this would

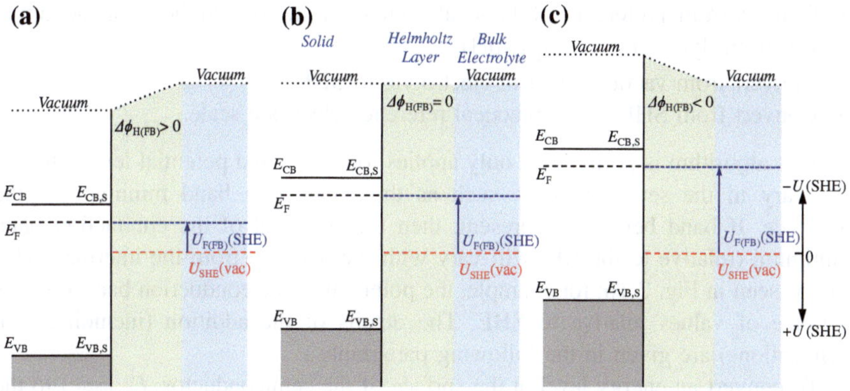

Fig. 3.15 Potential drop in the Helmholtz layer at the flat band condition for a range of pH, **a** pH < pzc, **b** pH = pzc, **c** pH > pzc

suggest that the negative potentials add. However, this is not the case and arises due to the difficulties in matching the conventions used in physics and electrochemistry. This arises because the potential $\Delta\phi_{H(FB)}$ is defined from the liquid to the semiconductor, while $\Delta\phi_H(V)$ as derived is defined as from the semiconductor to the liquid. Therefore, it is reasonable to change the sign of the potential drop $\Delta\phi_H(V)$ to be from the liquid to the solid: the same definition as the pH dependent component. As such, positive band bending due to a depletion layer results in a positive change $\Delta\phi_{H(TOTAL)}$, and negative band bending results in a negative change in $\Delta\phi_{H(TOTAL)}$.

Hence, for the example given above at pH > *pzc* and $\Delta\phi_{SCR} > 0$, the total potential drop in the Helmholtz layer $\Delta\phi_{H(TOTAL)}$ decreases from the value at the flat band condition. In theory, if the band bending is large enough, the two potential drops in the Helmholtz layer may cancel at a certain potential.

The above discussion shows the complexity of semiconductor|electrolyte interfaces.

3.5.7 Full Interfacial Model

To discuss the features of the semiconductor, such as conduction and valence band edges, flat band potential relative to a redox process, it is necessary to quote against a reference electrode, such as SHE or the HgO|Hg electrode at pH > ca. 10.

The full interfacial model is constructed by adding the various potentials that arise in the correct manner. To convert from the physical scale used in semiconductor physics to the electrode potential scale, the following algorithm can be used:

- Refer semiconductor energy levels at the interface to the vacuum level in the solid, $E(vac)$
- Calculate $\Delta\phi_{\mathrm{H(FB)}}(\mathrm{pH})$ based on the operational pH and the pzc
- Convert from a reference of the local vacuum of the solid to the local vacuum in the electrolyte using $\Delta\phi_{\mathrm{H(FB)}}(\mathrm{pH})$
- Convert from vacuum level in electrolyte to SHE
- Convert from SHE to the practical reference electrode scale.

This algorithm is partial as it only applies to energy and potential levels that do not vary in the semiconductor, such as the conduction band minimum at the interface. If band bending is present, then the potential of the conduction band minimum (relative to the RE) will vary with the distance from the interface. This can be seen in Fig. 3.16; for example, the potential of the conduction band E_{CB} has a range of values relative to SHE. The details of the addition (including sign conventions) are given in the following paragraphs.

To convert an energy level at the surface of the semiconductor, $E_{\mathrm{i,s}}(vac)$ to the reference electrode scale $U_{i,s}(RE)$, start with the flat band condition and add the pH dependent drop in the Helmholtz layer, $\Delta\phi_{\mathrm{H(FB)}}$. The quantity $E_{\mathrm{i,s}}(vac)$ will be an

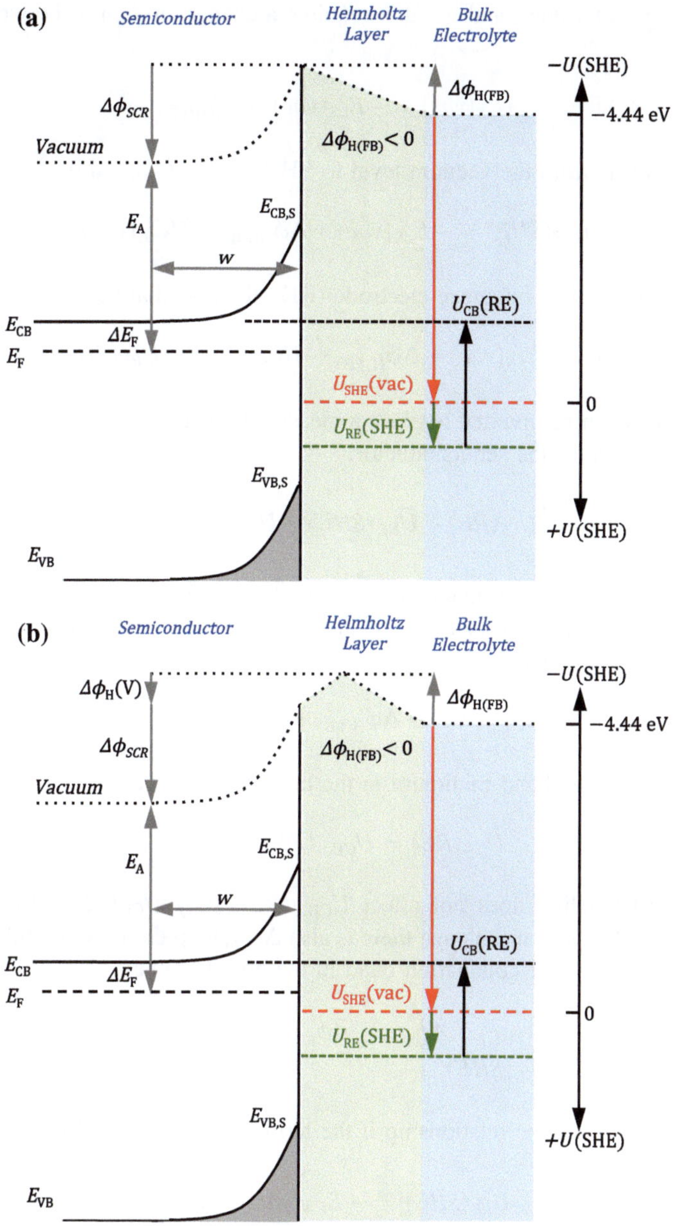

Fig. 3.16 Potentials making up the semiconductor-liquid junction at basic pH and positive band bending. In **a** the potential drop in the Helmholtz layer, $\Delta\phi_H(V)$ due to band bending is omitted. In **b** the same degree of band bending results in a small positive shift in the potential due to the change in the potential drop over the Helmholtz layer. The potential of the conduction band in the bulk $U_{CB}(RE)$ against an arbitrary reference electrode is used to demonstrate the conversion between the physical scale in the semiconductor and the absolute electrode potential scale in solution. The *arrows* indicate the addition of the potentials. The redox potential is not shown, i.e. the electrode potential is arbitrary

energy (negative value), and as such requires a change of sign to be written as a potential.

$$U_{i,S}(vac) = -E_{\mathrm{i,S}}(vac) + \Delta\phi_{\mathrm{H(FB)}} \tag{3.61}$$

Then convert from the vacuum level to SHE, where $U_{SHE}(vac) = +4.44$ V

$$U_{i,S}(SHE) = -E_{\mathrm{i,S}}(vac) + \Delta\phi_{\mathrm{H(FB)}} - U_{SHE}(vac) \tag{3.62}$$

Against a practical reference electrode (RE), the potential becomes

$$U_{i,S}(RE) = -E_{\mathrm{i,S}}(vac) + \Delta\phi_{\mathrm{H(FB)}} - U_{SHE}(vac) - U_{RE}(SHE) \tag{3.63}$$

The model can be inverted to place a measured potential of interest $U_{i,s}(RE)$ on the vacuum scale of the semiconductor.

$$E_{\mathrm{i,S}}(vac) = -\left(U_{i,S}(RE) + U_{RE}(SHE) + U_{SHE}(vac) - \Delta\phi_{\mathrm{H(FB)}}\right) \tag{3.64}$$

For a condition away from the flat band condition, it is necessary to introduce the band bending $\Delta\phi_{\mathrm{SCR}}$. For example, if the potential of the edge of the conduction band at the surface is given as

$$U_{CB,S}(RE) = -E_{\mathrm{CB,S}}(vac) + \Delta\phi_{\mathrm{H(FB)}} - U_{SHE}(vac) - U_{RE}(SHE) \tag{3.65}$$

then the conduction band minimum in the bulk will be given by

$$U_{CB}(RE) = U_{\mathrm{CB,S}}(RE) + \Delta\phi_{\mathrm{SCR}} \tag{3.66}$$

The band bending does not affect $U_{\mathrm{CB,S}}(RE)$, only $U_{\mathrm{CB}}(RE)$. However, the introduction of $\Delta\phi_{\mathrm{SCR}}$ means that there is also $\Delta\phi_H(V)$ present, so the full equation for the potential of the conduction band in the bulk becomes

$$\begin{aligned} U_{CB}(RE) = & -E_{\mathrm{CB}}(vac) + \Delta\phi_{\mathrm{SCR}} + \Delta\phi_{\mathrm{H}}(V) + \Delta\phi_{\mathrm{H(FB)}} - U_{SHE}(vac) \\ & - UL_{RE}(SHE) \end{aligned} \tag{3.67}$$

It is easier to see the relationship if the terms are grouped as follows

$$[U_{CB}(RE) + U_{SHE}(vac) + U_{RE}(SHE)] = -E_{\mathrm{CB}}(vac) + \Delta\phi_{\mathrm{SCR}} + \left[\Delta\phi_{\mathrm{H}}(V) + \Delta\phi_{\mathrm{G(FB)}}\right] \tag{3.68}$$

A simplified model of the photocurrent at a semiconductor|liquid junction is discussed in the following section based on the principals set out in the preceding sections.

The purpose of introducing the extensive theory of the junction and the photocurrent is to allow the interpretation (and fitting by model) of experimental data as presented in Chaps. 5 and 6. A rigorous reading of the theory was required to ensure that deviations of experimental data from models were real rather than being due to misinterpretation of either the data or the model.

3.6 Photocurrent

When the photo-electrode is illuminated, electron-hole pairs are generated. If there is band bending, then the electric field in the space charge region separates any electrons and holes generated therein. For a TiO_2 photo-anode with positive band bending forming a depletion layer, the holes are driven to the interface with the electrolyte, where they can oxidize water. The electrons are driven into the bulk, to be collected by the counter electrode. This results in a photocurrent, which can be predicted based on knowledge of the structure of the space charge layer. The equations developed in the preceding sections can be used to estimate the width of the space charge layer for a given electrode potential, and in turn to calculate the expected photocurrent. The model outlined below is used in Chap. 6 to try to fit photocurrent data for a Nb-doped TiO_2 electrode.

A simple model of the photocurrent in a Schottky barrier junction was proposed by Gärtner [33] and is re-derived below. This model was successfully applied to the semiconductor-liquid interface by Butler [25] and developed further by others [34, 35]. Typically, only the simplest form (excluding diffusion) is used to analyse junctions. It is also common to use numerical solutions (e.g. COMSOL) of the transport equation if including processes such as recombination [36].

The model of the photocurrent accounts for both the current arising from photo-generated charge carriers in the space charge region, as well as the diffusion towards the interface. The derivation of the photocurrent from the depletion layer is outlined below, while the derivation for the diffusive photocurrent is included in Sect. 3.6.2. A common use of this analytical model is in estimating the flat band potential of the photo-electrode. This is discussed in Sect. 3.6.5 below.

A sketch of the interface is shown in Fig. 3.17. The model is derived for the general case of a photovoltaic junction and ignores the effect of recombination and thermal generation of carriers, assuming that all electron-holes pairs excited in the semiconductor are available.

The total photocurrent density j_{TOT} in a semiconductor junction is the sum of the current densities due to carriers generated in the depletion layer j_{DL}, and those generated outside it, that diffuse into the depletion layer and are transported to the current collecting interface, j_{DIFF}.

$$j_{TOT} = j_{DL} + j_{DIFF} \tag{3.69}$$

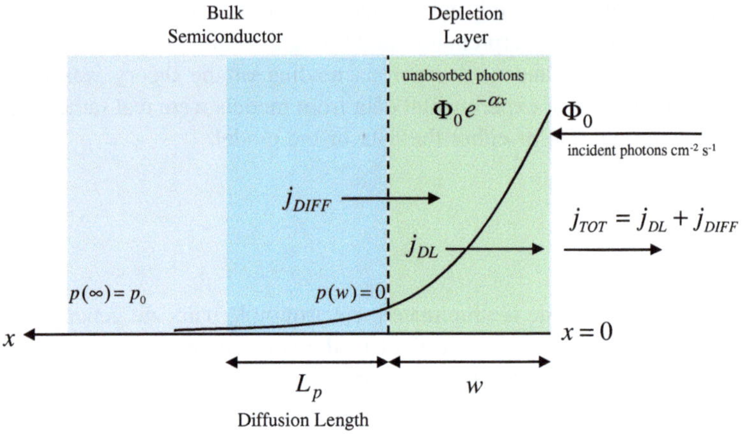

Fig. 3.17 Photocurrent model according to Gärtner [33]. The width of the depletion layer and the diffusion length of holes are shown in addition to the exponential absorption of incident photons. The boundary conditions for the photocurrent due diffusion are also shown

3.6.1 Depletion Layer Current

The depletion region current is calculated by assuming that all electron-hole pairs generated in the space charge region result in anodic current, i.e. the holes reach the interface and oxidize water. A model of the depletion layer photocurrent is used to fit photocurrent data in Chap. 6.

It is assumed that there is an incident normal photon flux of $q_{p,0}(\lambda)$ (photons of wavelength λ per unit area per second), and on a semiconductor with absorption coefficient of $\alpha(\lambda)$. As the photons pass through the semiconductor, they are absorbed such that after a distance x from the interface there are $q_{p,0}(\lambda)e^{-\alpha x}$ photons remaining (see Lambert's law, Experimental Methods chapter). The absorption coefficient is α, so the rate of electron-hole pair generation $g(x)$ at x is given by:

$$g(x) = \alpha\, q_{p,0}(\lambda)\mathrm{e}^{-\alpha x} \tag{3.70}$$

It is noted that for broad spectrum rather than monochromatic illumination, it is necessary to integrate over the spectrum.

To calculate the photocurrent generated in the depletion layer, it is necessary to integrate over the whole of the width of the space charge layer, w, where the width is given by Eq. (3.42). This gives:

$$j_{DL} = q \int_0^w g(x)dx = -q_e q_{p,0}(\lambda)[e^{-\alpha w} - 1] \tag{3.71}$$

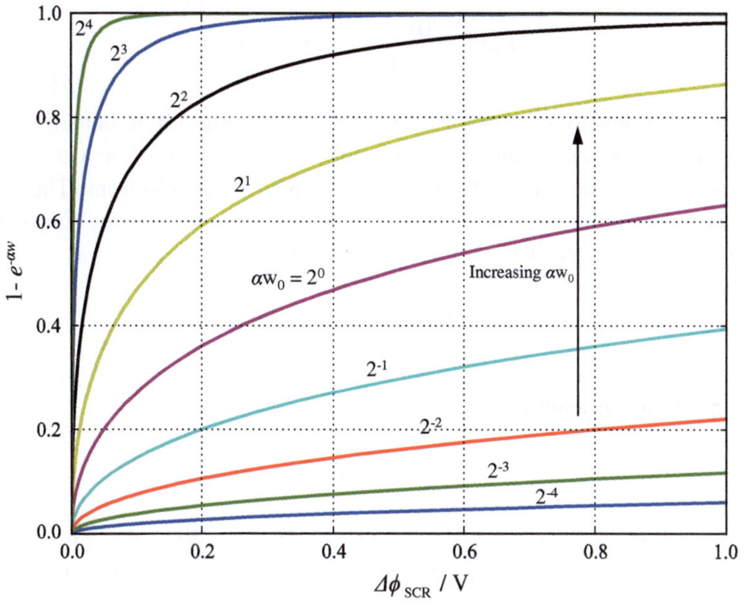

Fig. 3.18 Shape of the photocurrent arising in the depletion layer (space charge region) against band bending. The shape is plotted over the range of $2^{-4} \leq \alpha w_0 \leq 2^4$

The current is implicitly dependent on the band bending through w. Showing this more clearly:

$$j_{DL} = q\Phi_0\left[1 - \mathrm{e}^{-\alpha w_0\sqrt{\Delta\phi_{SCR}}}\right] \tag{3.72}$$

The shape of the photocurrent is given by $(1 - \mathrm{e}^{-\alpha w})$ and is shown in Fig. 3.18 for a range of values, $2^{-4} \leq \alpha w_0 \leq 2^4$. This demonstrates that the ratio of the space charge width constant w_0 to the absorption depth α^{-1} defines the shape of the photocurrent with band bending. If the absorption depth is much larger than the space charge width constant, then the photocurrent increases slowly. Conversely, a small absorption depth compared to the width constant results in a rapid rise in the photocurrent with band bending.

3.6.2 *Diffusion Photocurrent*

The diffusion current density is determined by solving the diffusion equation with applied boundary conditions for the concentration of holes in the bulk at $x < x_{SCR}$. Gartner writes the diffusion equation for holes, including a generation term $g(x)$ as:

$$D_p p'' - \frac{(p - p_0)}{\tau_p} + g(x) = 0 \tag{3.73}$$

where p is the density of holes at x, p_0 is the equilibrium hole density, τ_p is the lifetime of excess holes, and D_p is the diffusion coefficient. For a p-type semiconductor, the equation would be written in terms of excess electrons. The assumed boundary conditions are $p = 0$, at $x = w$ and $p = p_0$ at $x = \infty$. The diffusion current density is given here (without derivation) as:

$$j_{DIFF} = q_e q_{p,0}(\lambda) \frac{\alpha L_p}{(1 + \alpha L_p)} e^{-\alpha w} + q_e p_0 D_p / L_p \tag{3.74}$$

where the diffusion length of holes L_p is

$$L_p = (D_p \tau_p)^{1/2} \tag{3.75}$$

Typically, in metal oxides such as TiO_2, Fe_2O_3, and WO_3, the diffusion length L_p is small in comparison to the absorption depth, so $\alpha L_p \ll 1$. Furthermore, it is clear that the diffusion current decreases exponentially as the band bending increases, i.e. the width of the depletion layer increases, resulting in fewer photons reaching the diffusion region. The shape of the photocurrent due to diffusion is given by $e^{-\alpha w}$ and is shown in Fig. 3.19 for a range of values, $2^{-4} \leq \alpha L_p \leq 2^4$ and a fixed ratio of the space charge region width constant to the absorption depth, $\alpha w_0 = 1$. This demonstrates that the ratio of the space charge width constant w_0 to the absorption depth α^{-1} defines the shape of the photocurrent with band bending.

The diffusive photocurrent is shown in Fig. 3.20 for a fixed product of absorption coefficient (α) and diffusion length (L_p), $\alpha L_p = 0.1$.

3.6.3 *Total Photocurrent*

The total photocurrent is obtained by adding the equations for j_{DL} and j_{DIFF}:

$$j_{TOT} = q_e q_{p,0}(\lambda) \left[1 - \frac{e^{-\alpha w}}{(1 + \alpha L_p)} \right] + q_e p_0 D_p / L_p \tag{3.76}$$

Figure 3.21 shows the total photocurrent for $\alpha L_p = 0.1$. Compared to the photocurrent from just the depletion region, the rate of carrier generation at the flat band potential increases by $\alpha L_p / (1 + \alpha L_p)$. Although, the photocurrent due to diffusion decreases exponentially with band bending, due to this factor (which is always less than unity), it does so at a slower rate than the current from the depletion region increases. The net effect is that the photocurrent always increases with increased band bending, and the inclusion of the diffusive current results in a more rapid rise

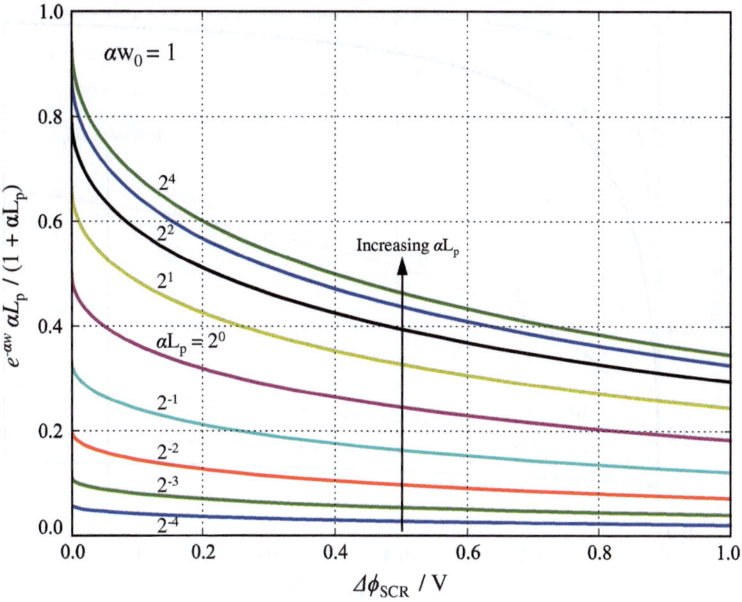

Fig. 3.19 Shape of the photocurrent arising from diffusion into the space charge region as a function of the band bending. The shape is plotted over the range of $2^{-4} \leq \alpha L_P \leq 2^4$ and a fixed value of the space charge width constant to the absorption depth $\alpha w_0 = 1$. The current density is obtained by multiplying by $q_e q_{p,0}(\lambda)$

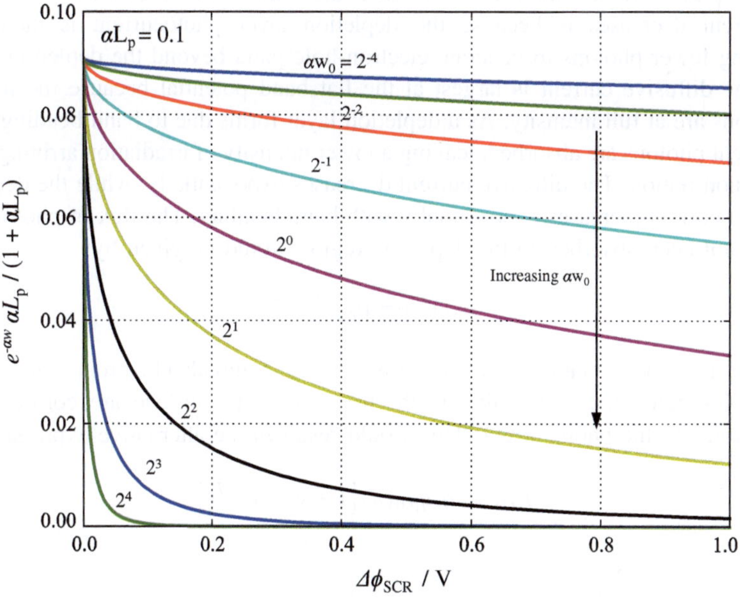

Fig. 3.20 Shape of the photocurrent arising from diffusion into the space charge region, as a function of the band bending, for a fixed value $\alpha L_P = 0.1$. The shape is plotted over the range of $2^{-4} \leq \alpha w_0 \leq 2^4$

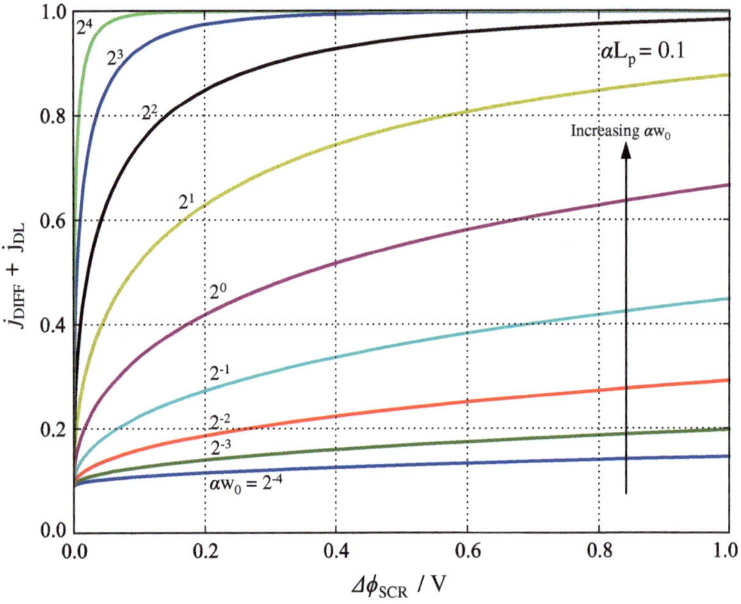

Fig. 3.21 Effect of band bending and the product αw_0 on shape of the total photocurrent arising from both diffusion and depletion layer, for a fixed value $\alpha L_P = 0.1$. The shape is plotted over the range of $2^{-4} \leq \alpha w_0 \leq 2^4$

in the photocurrent than without. In fact, the only reason that the diffusive photocurrent decreases is because the depletion layer photocurrent is increasing, leaving fewer photons to generate electron-hole pairs beyond the depletion layer.

The diffusive current is largest at the flat band potential because the incident photons are at full intensity. As a depletion layer forms due to band bending, more incident photons are absorbed, leaving a lower intensity of irradiation arriving in the diffusion region. The diffusive current decreases exponentially, while the depletion region current increases exponentially with band bending. The flux of photons that have not been absorbed in the depletion region would be given by:

$$q_{p,0}(\lambda) = q_{p,1}(\lambda)\mathrm{e}^{-\alpha w} \tag{3.77}$$

It is noted here that a much simpler expression could also be proposed, whereby all holes generated within the depth $(w + L_p)$ of the surface are considered to contribute to the photocurrent. This would result in the alternative expression

$$j_{TOT} = q_e q_{p,0}(\lambda)\left[1 - \mathrm{e}^{-\alpha w}\mathrm{e}^{-\alpha L_p}\right] \tag{3.78}$$

3.6.4 Determining the Characteristics of the Electrode

Ignoring the (typically small) term $q_e p_0 D_p / L_p$ in Eq. (3.76), the expression for the total photocurrent is of the form

$$j_{TOT} = A\left[1 - Be^{-C\sqrt{\Delta\phi_{SCR}}}\right] \tag{3.79}$$

And as such, the photocurrent can be fitted by three parameters. Combinations of three physical parameters α, w_0, and L_p determine the shape of the photocurrent, and the magnitude is determined by $q_e q_{p,0}(\lambda)$. It should be possible to determine the parameters by fitting the shape of the photocurrent using Eq. (3.79). Butler and Ginley outlined this approach for a WO_3 photoanode [25].

An issue that needs to be considered in practical terms is that the measured photocurrent is likely to be less than that given in (3.76) due to losses, such as reflection (from the electrode and cell window) and recombination. This is most easily accounted for by the introduction of an efficiency term Φ. While this is a gross oversimplification of the loss processes, it should assist in using the shape of the photocurrent to establish the value of αw_0. The efficiency factor can be introduced into a simplified expression for the photocurrent as follows.

$$j_{TOT} = \Phi q_e q_{p,0}(\lambda)\left[1 - \frac{e^{-\alpha w}}{(1 + \alpha L_p)}\right] \tag{3.80}$$

This approach does not account for the effect of scan rate during a practical measurement using cyclic voltammetry, i.e. transient effects, so it is important to make sure the measurement is as close to steady-state if the data is to be fit.

3.6.5 Flat Band Potential Estimation from the Photocurrent

A means of estimating the flat band potential from the photocurrent was outlined by Butler [25] based on the Gärtner model. This is an alternative to using the Mott-Schottky equation with a measurement of the capacitance. Both approaches may also be used together to check agreement.

Butler suggests that the flat band potential can be obtained in terms of the photocurrent by expanding of Eq. (3.76) in the limit $\alpha L_p \ll 1$. Additionally, the terms in $q p_0 D_p / L_p$ for TiO_2 can be neglected, as p_0 of the order of 10^{-12} m^{-3}. Then the current is given by:

$$j_{TOT} = q_e q_{p,0}(\lambda)\left[1 - e^{-\alpha w_0 (\phi_{fb} - \phi_{ap})^{1/2}}\right] \tag{3.81}$$

For arguments of the exponent $\alpha w \ll 1$, the approximation $e^x \approx 1 - x$ results in the photocurrent being given by:

$$j_{TOT} = q_e q_{p,0}(\lambda)\left[\alpha w_0 \left(U_{fb} - U\right)^{1/2}\right] \tag{3.82}$$

Squaring both sides leads to

$$U_{fb} - U = \left(\frac{j_{TOT}}{q_e q_{p,0}(\lambda)\alpha w_0}\right)^2 \tag{3.83}$$

Therefore, a plot of $j_{TOT}{}^2$ against potential should cross the x-axis at the flat band potential. It should be noted that this is based on the photocurrent alone. Using the total current may result in a shift of the estimated flat band potential, for example, if there is cathodic dark current flowing at the same time as anodic photocurrent. The photocurrent alone can be obtained experimentally by using a lock-in amplifier and a chopped light source to remove the dark current (see Experimental Methods).

Butler suggested that the approximations leading to (3.83) are valid only close to the flat band condition. In contrast, it is argued here that the approximations are somewhat contradictory and are in fact better at larger band bending (provided $\alpha w \ll 1$ still holds).

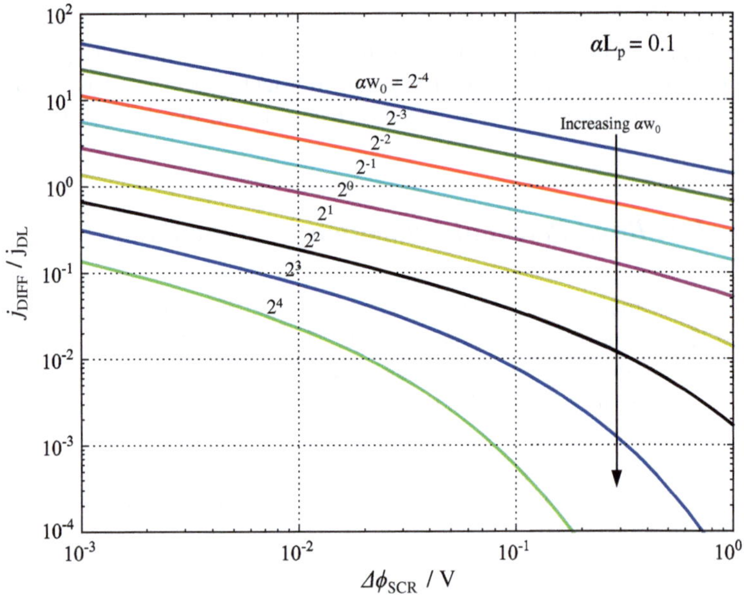

Fig. 3.22 Ratio of photocurrent arising from diffusion to that from the depletion layer, against band bending. The shape is plotted over the range of $2^{-4} \leq \alpha w_0 \leq 2^4$ and a fixed value of the diffusion length to the absorption depth $\alpha L_P = 0.1$

It is noted that the first approximations made by Butler reduce the total photocurrent to an equation equal to the photocurrent from the depletion layer alone, i.e. Equation (3.81) is the same as (3.71). Essentially, this is the same as assuming that $j_{DL} \gg j_{DIFF}$, which is reasonable for a material like TiO_2 that has a very short diffusion length. The next assumption is that the width of the space charge layer is much smaller than the absorption depth. This is typically true if the band bending is small, i.e. close to the flat band potential. However, close to the flat band potential, the depletion photocurrent is not necessarily larger than the diffusive photocurrent. Hence, there is inherently some contradiction between the assumptions made. This may result in an error in the estimation of the flat band potential. The ratio of the diffusive photocurrent to the depletion layer photocurrent is shown in Fig. 3.22.

References

1. Hankin, A., Alexander, J.C., Kelsall, G.H.: Constraints to the flat band potential of hematite photo-electrodes. Phys. Chem. Chem. Phys. **16**, 16176 (2014)
2. Pleskov, I.V., Gurevich, I. I.: *Semiconductor Photoelectrochemistry*. Consultant Bureau, New York (1986)
3. Morrison, S.R.: *The Chemical Physics of Surfaces*. Plenum Publishing Corporation, New York (1990)
4. Bard, A.J., Stratmann, M., Licht, S.: *Encyclopedia of Electrochemistry, Semiconductor Electrodes and Photoelectrochemistry. Encyclopedia of electrochemistry*. Wiley-VCH, Weinhiem (2002)
5. Butler, M.A., Ginley, D.S.: Correlation of photosensitive electrode properties with electronegativity. Chem. Phys. Lett. **47**, 319–321 (1977)
6. Grätzel, M.: Photoelectrochemical cells. Nature **414**, 338–344 (2001)
7. Patel, M., Mallia, G., Liborio, L., Harrison, N.M.: Water adsorption on rutile TiO_2 (110) for applications in solar hydrogen production: a systematic hybrid-exchange density functional study. Phys. Rev. B **86**, 045302 (2012)
8. Blomquist, J., Walle, L.E., Uvdal, P., Borg, A., Sandell, A.: Water dissociation on single crystalline Anatase TiO_2 (001) studied by photoelectron spectroscopy. J. Phys. Chem. C **112**, 16616–16621 (2008)
9. Sumita, M., Hu, C., Tateyama, Y.: Interface water on TiO_2 Anatase (101) and (001) surfaces: first-principles study with TiO_2 slabs dipped in bulk water. J. Phys. Chem. C **114**, 18529–18537 (2010)
10. Walle, L.E., et al.: Mixed dissociative and molecular water adsorption on anatase TiO_2 (101). J. Phys. Chem. C **115**, 9545–9550 (2011)
11. Bard, A.J., Parsons, R., Jordan, J.: *Standard Potentials in Aqueous Solution*. CRC Press, New York (1985)
12. Atkins, P., de Paula, J.: *Atkins' Physical Chemistry*. Oxford University Press, Oxford (2010)
13. Bockris, J., Dandapani, B., Cocke, D., Ghoroghchian, J.: On the splitting of water. Int. J. Hydrogen Energy **10**, 179–201 (1985)
14. Larminie, J., Dicks, A.: *Fuel Cell Systems Explained*. Wiley, London (2003)
15. Reineke, R., Memming, R.: Comparability of redox reactions at n-and p-type semiconductor electrodes. 1. The quasi-fermi level concept. J. Phys. Chem. **96**, 1310–1317 (1992)
16. Pinson, W.E.: Measuring the quasi-fermi levels and flat band potential of an illuminated Au/n-$GaAs_{0.6}P_{0.4}$ anode. Surf. Sci. **101**, 251–260 (1980)

17. Archer, M.D., Nozik, A.J.: *Nanostructured and Photoelectrochemical Systems for Solar Photon Conversion*. World Scientific Publishing, Singapore (2008)
18. Xu, Y., Schoonen, M.A.: The absolute energy positions of conduction and valence bands of selected semiconducting minerals. Am. Mineral. **85**, 543–556 (2000)
19. Trasatti, S.: The absolute electrode potential: an explanatory note (Recommendations 1986). Pure Appl. Chem. **58**, 955–966 (1986)
20. Gerischer, H.: Heterogeneous systems for solar energy conversion. Pure Appl. Chem, 1–19 (1980)
21. Nelson, J.: *The Physics of Solar Cells*. Imperial College Press, London (2003)
22. Vayssieres, L.: *On Solar Hydrogen and Nanotechnology*. Wiley, New York (2010)
23. Diebold, U.: The surface science of titanium dioxide. Surf. Sci. Rep. **48**, 53–229 (2003)
24. Toroker, M.C., et al.: First principles scheme to evaluate band edge positions in potential transition metal oxide photocatalysts and photoelectrodes. Phys. Chem. Chem. Phys. **13**, 16644 (2011)
25. Butler, M.A.: Photoelectrolysis and physical properties of the semiconducting electrode WO_2. J. Appl. Phys. **48**, 1914 (1977)
26. Uosaki, K.: Effects of the Helmholtz Layer capacitance on the potential distribution at semiconductor/electrolyte interface and the linearity of the Mott-Schottky Plot. J. Electrochem. Soc. **130**, 895 (1983)
27. Kim, J.Y., Kim, D.-W., Jung, H.S., Hong, K.S.: Influence of anatase-rutile phase transformation on dielectric properties of sol-gel derived TiO_2 thin films. Jpn. J. Appl. Phys. **44**, 6148–6151 (2005)
28. Cesar, I., Sivula, K., Kay, A., Zboril, R., Grätzel, M.: Influence of feature size, film thickness, and silicon doping on the performance of nanostructured hematite photoanodes for solar water splitting. J. Phys. Chem. C **113**, 772–782 (2009)
29. De Gryse, R., Gomes, W.P., Cardon, F., Vennik, J.: On the Interpretation of Mott-Schottky plots determined at semiconductor/electrolyte systems. J. Electrochem. Soc. **122**, 711–712 (1975)
30. Sato, N.: *Electrochemistry at Metal and Semiconductor Electrodes*. Elsevier, Boston (1998)
31. Kennedy, J.H., Frese, K.W.: Photooxidation of water at α-Fe_2O_3 electrodes. J. Electrochem. Soc. **125**, 709–714 (1978)
32. Cardon, F., Gomes, W.P.: On the determination of the flat-band potential of a semiconductor in contact with a metal or an electrolyte from the Mott-Schottky plot. J. Phys. D Appl. Phys. **11**, L63 (1978)
33. Gärtner, W.W.: Depletion-layer photoeffects in semiconductors. Phys. Rev. **116**, 84 (1959)
34. Laser, D., Bard, A.J.: Semiconductor electrodes VII. Digital simulation of charge injection and the establishment of the space charge region in the absence and presence of surface states. J. Electrochem. Soc. **123**, 1828–1832 (1976)
35. Reiss, H.: Photocharacteristics for electrolyte-semiconductor junctions. J. Electrochem. Soc. **125**, 937–949 (1978)
36. Ong, C.K.: Design and performance of photo-electrochemical reactors with Fe_2O_3 photo-anodes for water splitting. (2013)

Chapter 4
Experimental Methods

The experiments reported in this thesis cover two main areas of investigation:

1. Investigation of plasmonic water splitting by gold nanoparticles on TiO_2 (Chap. 5), and analysis of the electrochemistry of rutile single crystals (Chap. 6).
2. Growth of TiO_2 thin films by pulsed laser deposition on amorphous substrates (Chap. 7) and single crystal substrates (Chap. 8) as the basis of further studies of the surface science of TiO_2.

A brief overview of the two strands of work is given first, followed by the experimental methods used, which discuss the fabrication methods, standard materials characterization techniques, and electrochemical techniques employed. The depth of discussion for each technique is based on how often the technique was used, and how integral it is to the results. Where standard settings have been used for all, or a large proportion of, measurements (such as scan resolutions for XRD), these are given. In the case the equipment settings are the main subject of the experiment (such as the temperature during growth of thin films), these are discussed in the appropriate chapter of results.

The fabrication methods presented include solid-state ceramic processing and pulsed laser deposition (PLD) for the TiO_2 thin films, and DC magnetron sputtering to form metallic thin films for back contacts and as a precursor to forming nanoparticles. The characterization techniques cover X-ray diffraction (XRD), including texture analysis, and Raman spectroscopy for structural analysis, scanning electron microscopy (SEM) and atomic force microscopy (AFM) for imaging the surfaces of samples, profilometry for measuring thin film thickness, and UV-Visible spectroscopy (UV-Vis) for determining optical properties. Electrochemical techniques discussed include cyclic voltammetry, chronoamperometry, and impedance spectroscopy. The issues associated with the measurement of the response of photoelectrodes to illumination are also discussed. The fabrication techniques used to form gold nanoparticle arrays are also outlined.

© Springer International Publishing Switzerland 2016

J.C. Alexander, *Surface Modifications and Growth of Titanium Dioxide for Photo-Electrochemical Water Splitting*, Springer Theses,
DOI 10.1007/978-3-319-34229-0_4

4.1 Outline of Experiments

4.1.1 Plasmonic Au NPs on TiO_2

Experiments were done with the aim of reproducing the visible light water splitting photocurrent reported for gold nanoparticles on TiO_2 [1]. Single crystals of (110) rutile TiO_2 doped with 0.05 %wt of Nb^V were used in the study. To assess the role of Nb^V doping on the photocurrent, the experiment was also repeated with un-doped single crystal rutile of the same orientation. The undoped crystals were also examined after reduction in a hydrogen atmosphere. The crystal surfaces were patterned with gold NPs deposited by several methods including annealing gold films, micellar nanolithography (MNL), and nano-sphere lithography (NSL). These methods gave a range of particle sizes and separations. Ohmic back contacts were deposited by sputtering of Ti–Au thin films.

The response of the gold nanoparticle coated TiO_2 and the bare single crystals were examined using the materials characterization and electrochemical techniques described below.

4.1.2 Thin Film Deposition and Characterization

Experiments were done with the aim of establishing PLD growth conditions for thin films of anatase TiO_2 with a range of orientations. These films were to be used subsequently for a variety of surface science investigations. Based on the existing literature, a range of substrates was selected; each should have given a different film orientation. The thin films were grown over ranges of temperature and pressure and characterized using the techniques described below, to try to understand the effect of those conditions on the growth, the structure, and the film properties.

In the long term, the study of plasmonic photocurrent by gold nanoparticles on TiO_2 could be extended to study the adsorption and reaction processes on the various orientations of thin films fabricated.

4.2 Single Crystal Preparation

4.2.1 Nb-Doped Rutile (110)

Rutile (110) TiO_2 single crystal substrates were purchased (from Crystal GmbH, Germany). The single crystals were all $10 \times 10 \times 0.5$ mm^3. Undoped crystals and 0.05 %m Nb-doped crystals were bought throughout the project in batches of five. The first batches of crystals were single side epi-polished only. The single-side polishing made it difficult to make optical transmittance measurements, due to scattering by the rough side. They were polished subsequently, using progressively finer diamond spray until forward and backward transmittance curves were equal. The second batches of crystals were bought with both sides epi-polished by the supplier. Measurements described in Chaps. 5 and 6 would later show differences in the properties of the different batches of Nb-doped rutile crystals.

According to correspondence with the supplier, the crystals were fabricated using the Verneuil process and the dopant was introduced at the initial synthesis step by mixing of TiO_2 and Nb_2O_5 powder, which should have ensured homogenous distribution (as compared to doping by diffusion driven by concentration gradients).

Before use, all as-received crystals were cleaned in an ultrasonic bath using acetone, isopropanol, and de-ionized water in turn.

The Nb^V-doped crystals were used as electrodes without further treatment of the substrates other than cleaning. In particular, there was no annealing of the substrates (for either oxidation or reduction).

4.2.2 *Reduced Undoped Rutile (110)*

The undoped rutile (110) TiO_2 single crystal substrates proved to be too resistive for voltammetric measurements. As such, several samples were reduced in hydrogen to impart conductivity. The method used to reduce the crystals was determined by the work by Diebold [2].

Kai Wang of the Department of Materials, Imperial College London, performed the reduction. The reduction was done in a tube furnace with the following conditions:

- 1050 °C, temperature ramp 50 °C min^{-1}.
- 5 h dwell at 1050 °C.
- Pressure 0.7 mbar.
- H_2 flow rate of 40 atm cm^3 min^{-1}.
- Cool-down to room temperature at 10 °C min^{-1}.

The resulting rutile crystals were blue in colour, indicating they were at least partially reduced, and exhibited similar UV-Vis spectra to the as-received Nb-doped rutile crystals.

4.2.3 *Fabrication of Electrodes*

Following the required treatment of the single crystals and deposition of nanoparticles (or not for reference samples), a back contact was sputtered onto the reverse side of the single crystal. A 5 nm layer of Ti was used to form an ohmic contact with TiO_2 [3]. Then a 50 nm layer of gold was deposited on top to prevent oxidation of the Ti and provide greater in-plane conductivity, allowing the current to be collected via a silver wire bonded to the back by conductive silver electrode paint (Electrolube, RS Components). The metal films were deposited using DC magnetron sputtering, as outlined in Sect. 4.4.3.

The electrode assembly is shown and discussed in further detail in Sect. 4.6.2.

4.3 Nanoparticle Fabrication

Following a literature review of many available techniques to form nanoparticle arrays, and consideration of the existing expertise at Department of Materials, Imperial College London (ICL), a number of feasible techniques were identified:

- Annealed metal thin films (AMTF).
- Micellar Nanolithography (MNL).
- Nano-sphere Lithography (NSL).

The techniques are outlined below. Other techniques, including electron-beam lithography (EBL) and nano-imprint lithography (NIL), were considered due to the ability to form well ordered arrays with small deviations in sizing and spacing. However, these techniques were not sufficiently established in the Department at the time (the focus of this study was not to develop new nanoparticle deposition techniques) and are beyond the scope of this thesis.

4.3.1 Annealing Metal Thin Films

A fast and large-scale method of forming nanoparticles is to deposit a thin metal film on the substrate, then anneal it at an appropriate temperature [4]. The formation of the nanoparticles depends on the surface energy (surface tension) of the substrate-air interface $\gamma_{sub-\mathrm{air}}$, the metal-air interface $\gamma_{met-\mathrm{air}}$, and the substrate-metal interface $\gamma_{sub-\mathrm{met}}$. Whether the film forms individual islands or a continuous film depends on the balance of these forces [Eq. (4.1)]. The contact angle θ_c indicates whether the interface between the metal and substrate is energetically favourable. For example, a large angle indicates a poor wetting of the substrate by the metal film:

$$\gamma_{sub|air} - \gamma_{met|air} - \gamma_{sub|met} \cos \theta_c = 0 \tag{4.1}$$

During heating (below the melting temperature of the metals), energy is given to the sample, allowing the metal film to minimise surface energy to best accommodate thermal expansion. The film slowly de-wets, forming nano-spheres, and leaving an array of nanoparticles on the surface after cooling. Control over the

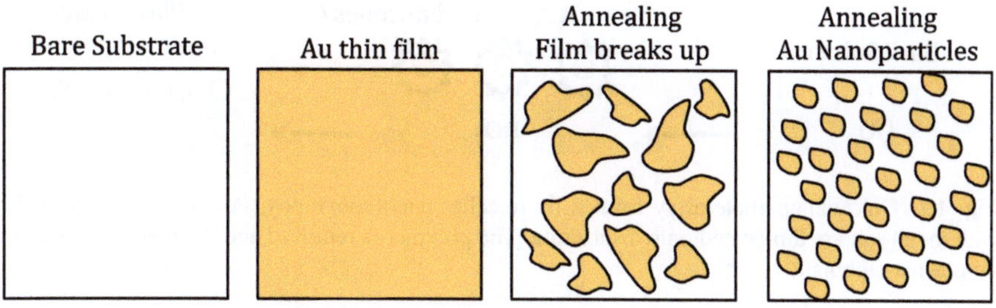

Fig. 4.1 Gold nanoparticle array formed by annealing a thin film. The film breaks up and forms individual nanoparticles, the size of which depend on the film thickness and annealing conditions

distribution and separation of particle size is relatively difficult, as formation is governed by the local properties of the surface such as roughness and contact angle of the materials in question, rather than the precise annealing temperature and time [5].

To establish the conditions to form a homogeneous (mono-disperse) layer of gold nanoparticles by annealing, a study of initial film thickness and annealing temperature was done.

Thin films of Au of 2, 4, 8, and 16 nm thicknesses were sputtered onto indium-tin oxide (ITO). The films were annealed at a range of temperatures: 300, 400, 500, and 600 °C giving 16 combinations in all. The aim was to establish the uniformity of size and distribution of nanoparticles formed under different conditions before applying this technique to an expensive single crystal of TiO_2. The results of the study are presented in Chap. 5. A sketch of the process is shown in Fig. 4.1. A film thickness of 2 nm and annealing conditions of 600 °C for 1 h was chosen for the deposition on the rutile (110) single crystals.

4.3.2 Micellar Nanolithography

Micellar nanolithography—developed by Spatz [6]—offers more control over size and separation, with particle sizes from 1 to 15 nm and spacing between 30 and 140 nm. The micelles consist of block copolymers of poly(styrene)-block- poly (2-vinylpyridine) (PS-b-P2VP) loaded with a metal salt such as $HAuCl_4$ in solution. A substrate is dip-coated in the solution and the micelles self-assemble into a hexagonal closed packed array, resulting in uniform spacing of well-defined spheroid Au particles (Fig. 4.2). The sample is treated in an oxygen-plasma, removing the polymer, and leaving the array of Au nanoparticles. A sketch of the basic process is shown in Fig. 4.2.

Derfogail Delcassian (Department of Materials, ICL) performed the MNL on undoped and Nb-doped rutile (110) single crystal substrates. One set of depositions resulted in a nanoparticle spacing of 100–150 nm. Another set of depositions,

Fig. 4.2 Gold nanoparticle array formed by micellar nanolithography. Polymer micelles loaded with metal salt are dip-coated onto a substrate. The polymer is removed and the metal salt reduced by plasma etching

aiming for a smaller spacing, was unsuccessful i.e. the nanoparticles did not form. This was attributed to the fact that the process used was optimized for deposition on glass wafers rather than TiO_2 [7]. In particular, the plasma treatment optimized for glass substrates may have been too aggressive, and removed the micelles. The successful set of depositions was considered to be of sufficient quality for an initial comparison to the method of annealing gold thin films.

4.3.3 Nano-sphere Lithography

In typical colloidal nanolithography, a sample is coated with a monolayer of polystyrene spheres (PS) of diameter <1 μm that self-assemble into a hexagonal close packed lattice. This is used as a mask for physical deposition methods such as thermal evaporation or sputtering [8]. The material deposited in the interstices between the nano-spheres forms triangular particles after the PS is etched away. These nano-triangles can be annealed to form nano-dots (Fig. 4.3).

An alternative approach was used to form ca.150 nm diameter gold nano-islands with a layer of Ti below, to improve adhesion of the gold to the TiO_2. In this approach, the polystyrene spheres are used as a mask, and the uncovered areas of a

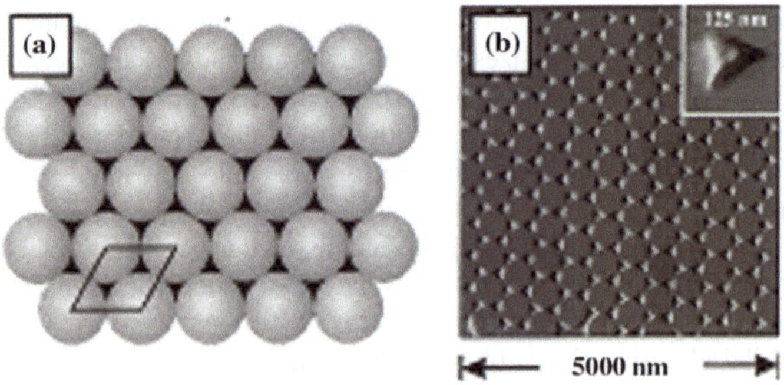

Fig. 4.3 **a** Monolayer of polystyrene nanospheres **b** triangular Au nanoparticles following plasma treatment [8]

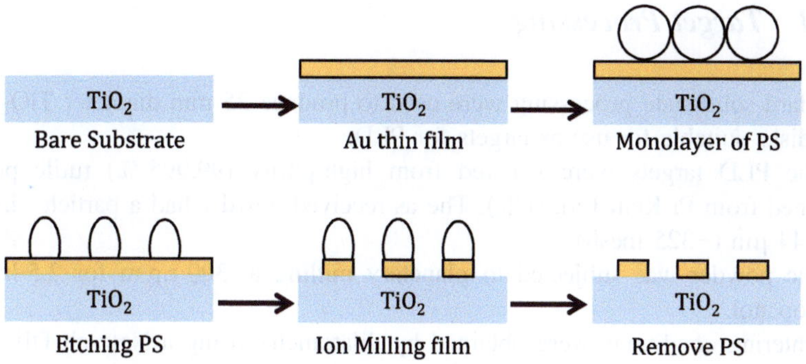

Fig. 4.4 Nano-sphere lithography process used to form gold nanoparticle arrays on rutile single crystals. The process is valid for films with multiple layers such as Au-Ti

thin film are etched away, leaving circular islands of Au. The basic steps in the process are shown in Fig. 4.4.

The technique was developed and done on several TiO_2 rutile (110) crystals by Jing Pang (Department of Materials, ICL) as part of a parallel project, researching fabrication methods and behaviour of plasmonic arrays. The method and parameters used were:

- Deposition of gold (or other metal) thin film by sputtering or thermal evaporation.
- Deposition of a monolayer of 300 nm diameter polystyrene spheres (PS) onto the thin film by the "fishing" technique: dipping sample into a layer of PS on surface of liquid containing a surfactant.
- Etching in an oxygen plasma for 30–45 s to reduce size of PS.
- Argon ion milling for 12 min to remove Au/Ti between the PS (45° angle for rotation).
- Removal of remaining PS by ultra-sonication in toluene.

The results are presented and discussed further in Chap. 5, Results: Plasmonic Au–TiO_2 Systems.

4.4 Thin Film Deposition

The following techniques and apparatus were used in the production of thin films.

- Solid-state processing of ceramic targets for pulsed laser deposition (PLD).
- Pulsed Laser Deposition to produce thin films of TiO_2.
- Sputtering of gold and titanium thin films to form plasmonic nanoparticles.

4.4.1 Target Processing

Standard solid-state processing were used to produce 25 mm diameter TiO_2 ceramic disks suitable for use as targets for PLD.

The PLD targets were sintered from high-purity (99.995 %) rutile powder obtained from Pi-Kem Ltd. (UK). The as-received powder had a particle size less than 44 μm (−325 mesh).

The powder was subjected to planetary milling at 300 r.p.m for 2.5 h, with isopropanol.

Sintering conditions were obtained by dilatometry using a Netzsch DIL 402C instrument with 8 mm diameter pellets of 3–5 mm thickness. A sintering temperature of 1400 °C was chosen and revised to 1500 °C following initial target processing tests.

The 25 mm target was pressed from milled powder at 70 MPa and sintered at 1500 °C with a 5 °C min^{-1} heating and cooling rate and 8 h dwell, resulting in a density of 83 %. While this was lower than the 90 % density aimed for, the target was sufficiently hard to allow polishing and remained intact during laser ablation.

The TiO_2 targets doped with 1 and 10 % Nb were pressed at 70 MPa and sintered at 1500 °C for 8 h (10 °C min^{-1} heating and cooling rates.) The Nb-doping was introduced by mixing high-purity Nb_2O_5 powder (99.999 %, Pi-Kem Ltd.) with the rutile powder. The powders were first mixed dry then mixed with isopropanol as a binder, followed by the milling, pressing, and sintering procedure outlined above. XRD scans of the sintered pellets doped with Nb suggested the pellets were single-phase rutile i.e. there was no Nb_2O_5 detected. A pellet doped with a higher concentration of Nb of ca.30 % appeared to be $TiNb_2O_7$.

4.4.2 Pulsed Laser Deposition

Pulsed laser deposition was used to synthesis thin films of TiO_2 with thicknesses ranging from 50 to 300 nm.

The PLD system used was that of the Thin Film Technology Laboratory located in the London Centre for Nanotechnology (LCN) at Imperial College London. In particular, all films were deposited in Chamber 1. The main components of the PLD system (Neocera, USA) are:

- 248 nm KrF excimer pulsed laser, used to ablate a ceramic target inside a vacuum chamber.
- Mirror to direct the laser beam into the chamber.
- Lens to focus laser beam in front of target (considered fixed).
- 1028 nm Nd:YAG laser heater for substrate heating.
- Steel vacuum chamber and associated turbo-pumps.
- Mass-flow controllers for oxygen and nitrogen and associated pressure monitoring gauges.

- Substrate holder and carousel.
- Target carousel.

In PLD, the target absorbs the intense energy of the laser pulses, causing local heating and explosive evaporation (Coulomb explosion) of material near the surface of the target. This forms a plume of ionized material ejected perpendicular to the surface of the target. The material ejected in the plume deposits onto a substrate mounted opposite the target (Fig. 4.5). The incident material then diffuses on the surface of the substrate (or previous layers of the film) to nucleation sites such as defects, step edges, and islands and is incorporated into the crystal structure of the thin film.

PLD offers a great deal of control over deposition conditions including: chamber pressure, gas composition, substrate temperature, beam energy, and repetition rate. Additional targets can also be used to deposit multilayer films. Samples can be annealed in situ or subsequently ex situ at different conditions [5].

The rate at which material is ejected from the target is controlled by parameters such as the fluence (energy intensity of pulse in mJ cm^{-2} per pulse) of the incident laser pulse and the repetition rate. For example, a faster repetition rate reduces the amount of time that material has to diffuse before additional material is added and may affect the growth mechanism or the concentration of defects. The plume also interacts with the gases, typically oxygen, in the chamber. This can reduce the energy of the material reaching the substrate and hence the growth of films is sensitive to the gas pressure in the chamber. The substrate-target distance therefore also affects the interaction of the plume with the gases in the chamber.

The growth of the film is further determined by the choice of substrate and the temperature during deposition or post-annealing. If a single crystal substrate is used, and there is a match between the lattices of the substrate and the material to be deposited, it is possible that an epitaxial film will grow. In this case, the orientation

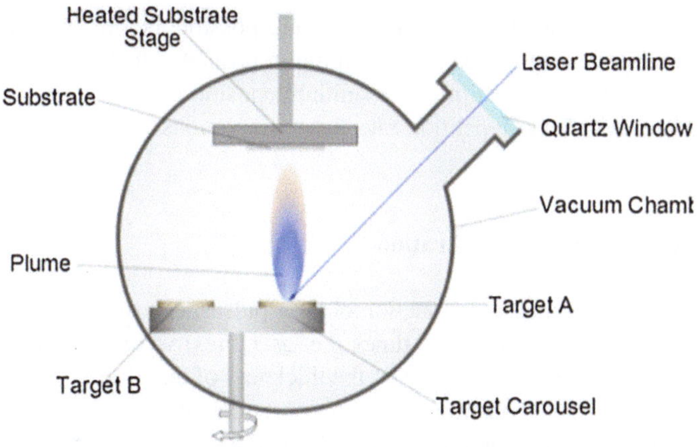

Fig. 4.5 Schematic diagram of typical PLD chamber. Image courtesy of: S.N. Cook (Department of Materials, ICL)

of the film is determined by the substrate. When there is no match between the materials of the film and substrate, it is likely that either an amorphous or a polycrystalline film will grow.

The temperature during deposition can affect: whether the film crystallizes or not, what phase is achieved, for example anatase or rutile, and what orientation is achieved. Furthermore, whether the film is annealed during deposition or post-annealed can also affect growth, as shown in Chap. 7.

The precise details of how all the many parameters combine to affect the growth is complex and can be determined only by mapping out the parameter space by trial and error with a particular material and system. In particular, the precise effect of certain parameters may be valid only for the deposition system on which the original study was performed.

The following deposition parameters were investigated, the results of which are presented in Chaps. 7 and 8:

- In situ deposition temperature.
- In situ deposition oxygen pressure.
- Thickness (number of pulses).
- Ex situ post-annealing temperature of films deposited at room temperature.

In some cases, the effect of the substrate-target (working) distance and the repetition rate were also investigated.

The deposition pressures used were in the range of 10–100 mTorr. In SI units, 1 Torr is equal to 133.3224 Pa, so a chamber oxygen pressure of 10 mTorr is 1.33 Pa.

The following conditions were typically fixed during a deposition experiment:

- 520 mJ per laser pulse (220 mJ measured in chamber).
    - Spot area of 0.35 cm^2.
    - Fluence of 628 mJ cm^{-2} or 0.63 J cm^{-2}.
- 90 mm substrate-target separation (minimum possible before July 2012).
- 50 mm substrate-target separation (minimum possible after July 2012).
- 5000 pulses at 2 Hz repetition rate (initial depositions).
- 10,000 pulses at 8 Hz repetition rate (later depositions).

4.4.2.1 Film Thickness Calibration

For evaluation of thickness as a function of laser pulses, depositions with a range of numbers of pulses (the number of times the laser fires) were done, with all other conditions held constant. By measuring the thickness of the films at each number of pulses, it is possible to construct a calibration curve of thickness against pulses. Figure 4.6 shows the measured thickness of films deposited with 5, 10, 20, and 30

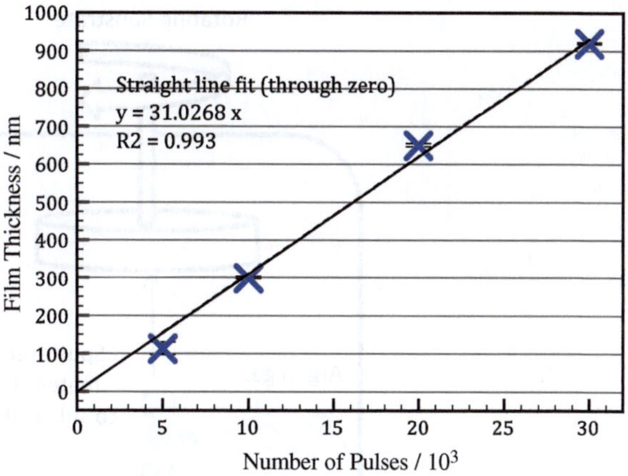

Fig. 4.6 Thickness calibration curve for PLD films. Films deposited with 5, 10, 20, and 30 thousand pulses. A straight line fit through zero gives a rate of 31 nm per thousand pulses

thousand pulses, and can be used to identify the number of pulses required to grow a film of a given thickness.

The thickness calibration was done with films deposited from the 1 % Nb-doped target on fused silica substrates. The calibration films were deposited with the substrate heated to 400 °C, at a chamber oxygen pressure of 30 mTorr, a repetition rate of 8 Hz, and a substrate-target distance of 50 mm. It is noted that a slightly different calibration curve might be expected for different choices of substrate, temperature, and oxygen pressure. The calibration curve gives the thickness as 0.031 nm pulse^{-1} i.e. 10 thousand pulses gives a film of 310 nm. The effect of the thickness on both the XRD pattern and the surface roughness (AFM measurements) is discussed in Chap. 7.

4.4.3 Sputtering

To produce gold films for the annealing method of nanoparticles and Ti films for ohmic back contacts for TiO_2 single crystals, DC magnetron sputtering (MANTIS deposition system) was used. In sputtering, a gas such as argon is ionized close to a target by applying a strong electric field. The ionized gas knocks particles out of a target composed of the material to be deposited. These are ejected in a plume towards a rotating substrate (Fig. 4.7).

Deposition rates depend principally on: applied voltage, current, and argon gas flow. The rate of deposition was monitored with a quartz-crystal-microbalance (QCM). The typical deposition conditions used for the Au and Ti thin films were (Table 4.1):

The types of sputtered films grown were:

- 5 nm Ti films for adhesive layers.
- 50 nm Au films for back contacts and NSL.
- 2, 4, 8, 16 nm Au films to study nanoparticle formation by annealing.

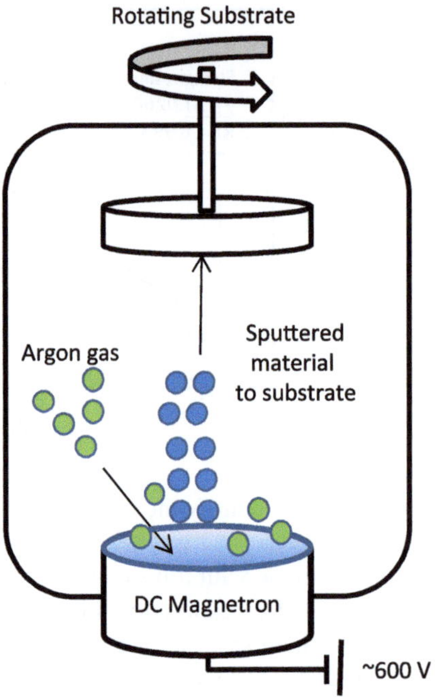

Fig. 4.7 Schematic diagram of a typical sputtering chamber

Table 4.1 Sputtering deposition conditions

Parameter	Titanium	Gold
Voltage (V)	620	620
Current (mA)	800	120
Argon flow rate (atm cm^3 min^{-1})	150	150
Pressure (mBar)	10^{-4}	10^{-4}
Deposition rate (Å s^{-1})	0.7–0.8	0.5–0.6

4.5 Materials Characterization

The standard techniques used to characterize material properties in this thesis are described below, including:

- X-ray diffraction (XRD) including texture analysis.
- Raman spectroscopy.
- Scanning electron microscopy (SEM).
- Atomic force microscopy (AFM).
- Profilometry.
- UV-visible spectroscopy (UV-Vis).

The techniques that have been used more frequently are covered in greater detail.

In addition, Scanning Tunnelling Microscopy (STM) was attempted, as it is a technique that can be used to map the atomic configuration of the surfaces of the TiO_2 thin films. This would allow studies of the reconfiguration of the different crystal facets, comparison to computation models of the surfaces (an aim of the overall Programme Grant under which this work was funded) and the investigation of the adsorption of molecules such as H_2O. However, the thin films (even those that were highly doped) were insufficiently conductive to be imaged by STM. As such, STM could not be done and no results are presented. The reasons for the lack of conductivity are proposed in the discussion of Chap. 7.

4.5.1 X-Ray Diffraction

X-ray Diffraction (XRD) uses monochromatic X-rays to analyse the crystallographic structure of materials. It is typically the first analytical technique applied following the growth of thin films, and is used to identify the phase and orientation of the films. The wavelengths of the Cu $K\alpha_1$ and $K\alpha_2$ X-rays (1.5405 and 1.5444 Å respectively) are of similar magnitude to a typical unit cell or spacing of planes of atoms: resulting in incident X-rays being diffracted by the lattice. X-rays incident on a material at angle θ to the plane of the sample are diffracted from parallel crystallographic planes of separation d, and interfere. If the path length of interfering X-rays differs by an integer number of wavelengths, $n\lambda$, then constructive interference occurs resulting in a strong peak in the diffracted signal [9].

The relationship is given by Bragg's Law:

$$n\lambda = 2d\sin\theta \tag{4.2}$$

The geometry of diffraction by successive crystallographic planes is shown below in Fig. 4.8.

For this thesis, the primary objective was to identify the phase and orientation of the films, and the cell parameters. This was done using XRD 2θ-scans made with two diffractometers:

- Bruker D2 Phaser.
- PANalytical X'Pert MPD.

The majority of 2θ scans were made with the Bruker diffractometer. Some initial scans (of films deposited on MgO) were made with the PANalytical machine.

Scans made on the Bruker machine were done with identical settings to allow the comparison of samples. The settings used for most TiO_2 thin films were:

- Scan from 2θ = 10 to 2θ = 110°.
- Interval of 0.03°.
- Time step of 0.2 s/interval giving ~10 min per sample.
- 30 kV/10 mA.

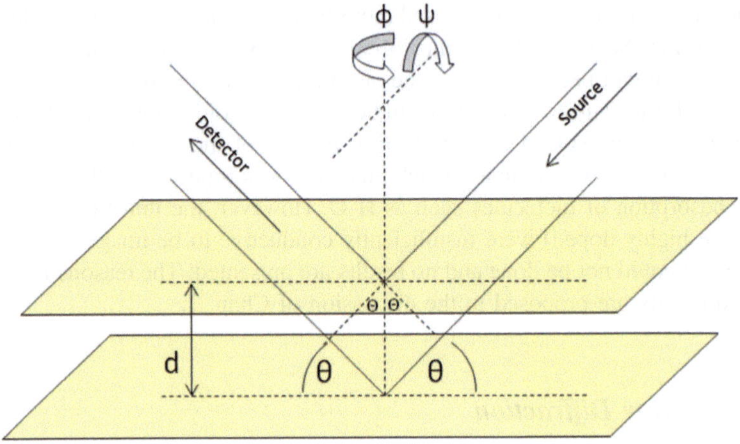

Fig. 4.8 Schematic of X-ray diffraction according to Bragg's Law. The ray diffracting from the lower plane has travelled further by a distance $2d\sin\theta$

Scans are performed by incrementing both the source and detector angle to the sample in step. As the scan progresses, peaks are seen in the patterns, corresponding to strong constructive inference. The patterns are then matched to reference patterns to identify the phases and crystal orientations present. This type of scan will detect only planes that are parallel to the plane of the sample. In a powder sample, it is usual to have randomly oriented grains, and hence planes corresponding to most Miller indices are seen. For a thin film sample, this may not be the case if the film grows with a preferential orientation due, for example, to epitaxial growth on an oriented substrate or due to the effect of surface energy.

4.5.1.1 Data Correction

The XRD patterns presented in the results chapters of this thesis have been corrected by shifting the raw data of each spectrum horizontally on the 2θ-axis. This is done in order to ensure that the substrate peaks for all samples coincide. As measured, the substrate peaks did not necessarily coincide from one sample to the next. This could be due to a number of factors such as: sample tilt, a vertical offset of the sample resulting is the movement of the focal plane, or misalignment of the optics. All measurements were made with sample rotation, which should account for sample tilt.

The correction was made by shifting the dominant substrate peak to coincide with the position given in the high-quality ICDD reference pattern. For example, for films deposited on (100) crystals of strontium titanate (STO) the (200) reflection of STO should occur at $2\theta = 46.486°$, according to ICDD reference pattern 00-035-0734. This causes the visible substrate peaks (100), (200), and (300) to line up for each scan and allow a comparison between the film peaks to be made. It is noted that in lining up the (200) peak, that the (100) and the (300) peaks do not line up perfectly with the peaks of the reference pattern. This suggests that the lattice parameter of the single crystal (3.9050 Å) is not equal to the reference pattern. This could be attributed to each substrate having a slightly different lattice constant. Alternatively, this might indicate a slight error in the setup of the apparatus, or the sample mounting.

4.5.1.2 Peak Fitting

The High Score Plus software produced by PANalytical was used to fit the peaks of the XRD spectra.

The use of fitting algorithms specific to XRD data allows both the background of the data and the presence of peaks corresponding to both Cu $K_{\alpha 1}$ and $K_{\alpha 2}$ X-rays to be fully accounted for. Firstly, the background was fit then the peak of interest was fit using a Voigt function (mixed Gaussian and Lorentzian) with a shape of 0.6 (determined by the fitting routine). The fitting function produces a pair of curves for each peak, the first corresponding to $K_{\alpha 1}$ and a second to $K_{\alpha 2}$. The second fitted peak had half the intensity of the first (amplitude).

4.5.1.3 XRD Texture Analysis (Pole Figures)

Texture analysis is an XRD technique in which diffraction is measured from planes that are not parallel to the substrate-film interface. This is used to determine a number of factors including the out-of-plane lattice parameters, preferential orientation of crystal grains, and the epitaxial relation between a film and the substrate. In this thesis, texture analysis has been used primarily to determine the epitaxial relations between the thin films and substrates for TiO_2 films grown on MgO (100) and STO (100), as described in the results of Chap. 8.

In texture analysis, the X-ray source and detector are set at a fixed value of 2θ and sample is rotated in ϕ for a range of ψ (ϕ and ψ are defined in Fig. 4.8). Typically the value of 2θ is set to a value expected for a crystallographic plane of interest that is not seen in the plane of the substrate. A pole figure is obtained by plotting the intensity of the XRD spectra on a polar contour plot in ϕ and ψ. A diffraction peak is called a pole. For example, for a cubic crystal structure such as MgO, a film that is seen to be oriented in the (100) direction in a conventional 2θ-scan would not have the (110) planes parallel to the substrate. The $\{110\}$ planes are oriented at 45° to the (100) planes in ψ. Due to the four-fold symmetry of MgO the film could be rotated to four equally spaced angles of ϕ that would each bring a different plane into a diffraction condition with the X-ray beam.

Texture analysis was made with a PANalytical X'Pert MRD diffractometer. The general approach used to generate pole figures was:

- Set 2θ to initial value of interest (based on a reference pattern).
- Run in ϕ scans over 360 ° for values of ψ from 0 to 80°.
- Identify poles of interest (if present).
- Set ϕ and ψ values for a pole of interest.
- Repeat 2θ scan to obtain accurate value.
- Run single line ϕ and ψ-scans through poles to estimate peak width.

The pole figures were typically made with a resolution of 5° in both ϕ and ψ. When greater detail was required a scan of a reduced range of ψ was done at a resolution of 2.5°. This resolution is appropriate in estimating the texture and epitaxial relations of the thin films, but is not sufficient to obtain the out-of-plane lattice parameter.

Scans of the thin film poles were carried out at 40 kV and 40 mA, while single crystal substrate peaks were analysed at 30 kV and 10 mA to avoid damage to the detector.

4.5.2 Raman Spectroscopy

Raman spectroscopy is a complementary technique to XRD, in that it allows the crystallographic structure to be probed.

In Raman spectroscopy, a sample is illuminated with laser light. The photons are absorbed and then re-emitted at a slighter lower frequency (Stokes scattering) or a slightly higher one (anti-Stokes). The difference in frequency, termed the shift, is usually measured in cm^{-1} and indicates a different initial and final vibrational or rotational state of the absorbing molecule or lattice. Different phases of materials have characteristic spectra from which they can be identified. These can be checked against reference patterns to identify phases [10].

Measurements were carried out with a Renishaw Ramascope. The spectrometer was first calibrated using a Si single crystal reference to give a peak centre of 521 nm. The laser light was focussed on the surface of the thin film (or other sample) and the Raman shift from 0 to 1000 cm^{-1} recorded for an acquisition time of typically 10–60 s, depending on the noise present and the response of the film-substrate system. Longer acquisition times were avoided if possible to prevent spot heating (that can change the properties of the sample, such as causing crystallization or a phase change) or saturation of the detector.

Raman data can be analysed by fitting the peaks present in the spectrum. Shifts in the peaks can indicate structural changes. The Raman spectra were used here simply as a means of confirming the phases present in the film by comparison to reference patterns.

4.5.3 Scanning Electron Microscopy (SEM)

Scanning Electron Microscopy (SEM) was done with a LEO Gemini 1525 FEG-SEM (Field Emission Gun SEM) to obtain micrographs of nanoparticle arrays. The electron beam is produced by application of a high tension accelerating voltage, typically 5–30 kV. This produces electrons with a wavelength of <0.02 nm (at 5 kV) and a resolution limit of several nanometres, which is suitable for imaging nanostructures <100 nm, compared to UV-light with a resolution limit of about 200 nm [11]. A typical SEM column is shown in Fig. 4.9. The SEM used for images is this report also employs an "in-lens" secondary electron detector for use with the small working distances of 7–10 mm used for high magnification.

4.5.4 Atomic Force Microscopy (AFM)

Atomic force microscopy is a scanning probe technique used to map the surface topology of a sample. This allows the roughness of the surface to be identified in

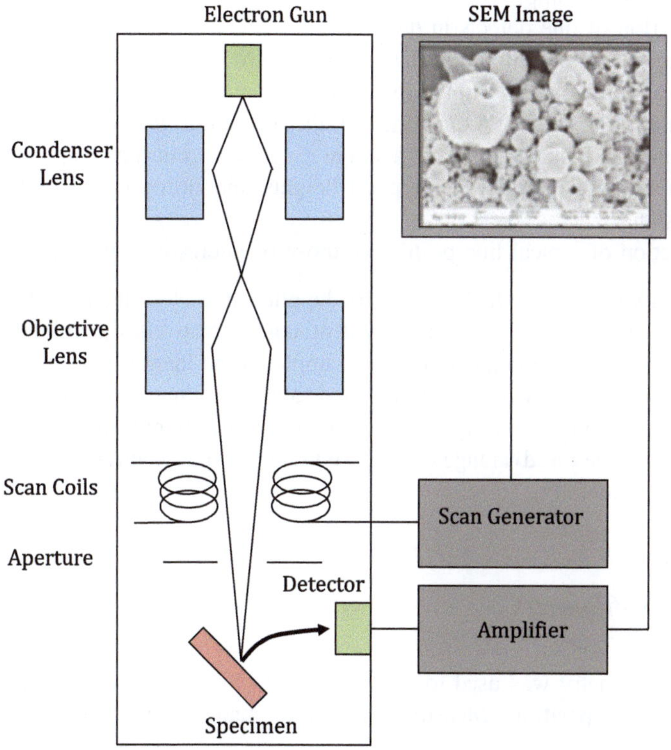

Fig. 4.9 Schematic of a typical SEM system. The electron beam is condensed and focused by the lenses, and then rastered across the sample by the scan coils. Figure taken from reference [11]

addition to quantitative estimates of the height of surface features (as opposed to SEM which does not give this information). This can provide information about the growth mechanism of a thin film for example. A tip is scanned across an area of the sample on the order of 1 μm^2: the deflection of the tip due to variations in the height of the sample is recorded, giving a three dimensional picture of the surface. The deflection of the tip is measured (on the apparatus used here) by the reflection of a laser beam from the back of the cantilever on which the tip is fabricated. Changes in the tip position move the reflected spot and change the intensity of the light incident on a detector. The detected signal is proportional to the deflection.

The surfaces of the TiO_2 films deposited with 1 % Nb-doping were characterized using Atomic Force Microscopy (AFM). The measurements were made using a Veeco Proscan AFM in tapping (non-contact) mode. The tip used was a Micro Masch NSC15/AIBS with a spring constant of 46 Nm^{-1} and a resonant frequency of 325 kHz.

Ms. Manisha Chhikara of the University of Nova Gorica operated the AFM, while the author of this thesis did the data analysis.

Each set of topographical data was imported into the software package Gwyddion [12], which was used to analyse statistically the AFM micrographs. The following procedures were applied using the tools built into the software:

- Correction of sample tilt.
- Correction of line defects in data.
- Setting the minimum height data value to zero (zeroing).
- Setting the colour-map range to 0–50 nm to aid comparison of samples.
- Calculation of statistical quantities for the whole film including; average height; median; maximum; minimum; average roughness; root-mean-square roughness.
- Calculation of distribution of pixel heights and fitting of distribution with a Gaussian.
- Extraction of typical line profiles or those of interesting features.

While most samples had a range of height values less than the 50 nm range applied to the images, using a consistent colour map allows differences in the roughness and feature heights between samples to be identified visually. Using an individual range for each sample, matched to the maximum and minimum of each, would make smooth and rough films more or less indistinguishable when presented side by side. The disadvantage of the fixed range is that features on smoother films are harder to identify.

4.5.5 Profilometry

Stylus profilometry was used to measure the thickness of thin films fabricated by pulsed laser deposition. Measurements were made with a Veeco Dektak 150 profilometer.

The bare substrates on which the thin film is deposited by PLD are mounted in a frame that covers each of the four corners. This leaves the corners of the substrate free of film and provides a step, the height of which can be measured and taken as the film thickness. In practice, the curvature and roughness of the substrate makes it difficult to get an unambiguous thickness measurement. For each sample, a profile was taken for between two and four of the corners, and the clearest taken to be the thickness.

There are a number of problems with this technique that became apparent during measurement:

- Curvature over the length of the 10 mm by 10 mm substrates due to polishing.
- Roughness of corners—this may be caused by the application of pressure to hold the substrate in place while at high temperatures of 400–800 °C.
- Varying thickness of the film across the substrate due to the concentration gradient across the plume (differential rates of deposition).

The results of several films measured using this technique are presented in the results in Chaps. 7 and 8.

4.5.6 *UV-Vis Spectroscopy*

UV-Visible (UV-Vis) spectroscopy is used to analyse the optical properties of materials, including: transmittance, T; reflectance, R; absorbance, A; and the bandgap, E_G. UV-Vis was done with a Bentham system consisting of: a quartz-halogen light source (IL1) with current stabilized power source (B 605), a Monochromator (TM300), and a silicon photodiode detector. A background was first measured in air (without a sample) to give the reference intensity I_0 over the range 300–1100 nm in 1 nm increments. The measurement was then repeated with the sample in place between the source and the detector, giving a new measured intensity, I. To measure the reflectance the sample is tilted slightly (to 10° from normal incidence) and the detector is repositioned to receive the reflected beam. Ideally, the reflectance would be measured at normal incidence, but the configuration of the spectrometer did not allow this. Ignoring reflection, the transmittance of the sample T is equal to the ratio I/I_0 and is related to the absorption by the Beer-Lambert's law [13]:

$$I = I_0 e^{-\alpha t} \tag{4.3}$$

where α is the linear absorption coefficient of the material (related to the imaginary part of the dielectric function), and t is the thickness or optical path length. A quantity called the absorbance A is defined as by the logarithm of the transmittance:

$$A = -\log_{10}(I/I_0) = -\ln T/2.303 \tag{4.4}$$

In practice, the measured transmittance data inherently includes losses from both reflection and absorption. The measured reflectance data should be used to correct for reflection in the transmission measurement, leaving—in principle—only losses from absorption. For an incident intensity of I_0, the measured intensity should be given by:

$$I = I_0(1 - R)\mathrm{e}^{-\alpha t} \tag{4.5}$$

where R is the measured reflectance. This assumes that the substrate is completely transparent over the range of the measurement, so that all absorption can be attributed to the film. The transmittance T is therefore defined as:

$$T = \frac{I}{I_0} = (1 - R)\mathrm{e}^{-\alpha x} \tag{4.6}$$

So, the absorption coefficient is given by:

$$\alpha = -\frac{1}{x}\ln\left(\frac{T}{1 - R}\right) \tag{4.7}$$

And the absorption depth by:

$$L_\alpha = \frac{1}{\alpha} \tag{4.8}$$

When thin film interference is present, there will be fringes in both the transmittance and reflectance spectra. As the reflectance measurement is made at 10° from normal incidence, and the transmittance measurement is made at normal incidence, this induces slight offsets in the fringes with wavelength. In this case, it would be better to make both the measurements at the same angle of incidence.

The interference fringes in the transmittance (and reflectance) spectra can be used to determine the refractive index and thickness of the films. The most common method to do this—and the one used here—was outlined by Swanepoel [14].

4.5.6.1 Bandgap Estimation—Tauc Plots

The bandgap can be estimated by plotting a Tauc plot [15]. In a Tauc plot, the quantity $(\alpha h\nu)^n$, which depends on the absorption coefficient and the photon energy, is plotted against photon energy $h\nu$. For indirect bandgap materials, the exponent is set to $n = \frac{1}{2}$, while for a direct bandgap material $n = 2$. The plot of $(\alpha h\nu)^n$ should form a straight line crossing the x-axis at the value of the bandgap. By constructing both direct and indirect Tauc plots for the material it is possible to see which model best fits the data. It is likely that a straight line will only fit a portion of the data. Tauc plots assuming both direct and indirect transitions were produced for the TiO_2

thin films deposited on fused silica, and compared to establish the nature of the bandgap (see Chap. 7, Results: TiO_2 Thin Films on Fused Silica).

4.6 Electrochemical Techniques

The apparatus and techniques used in this thesis to characterize samples by electrochemical kinetics in both the dark and under illumination (photoelectrochemistry) are now described. Firstly, the physical arrangement of the photoelectrochemical reactor and system is discussed. Then the following electrochemical techniques used to characterize the solid-liquid interface and its performance in a water splitting context are presented.

- Cyclic Voltammetry (CV).
- Chronoamperometry (CA).
- Electrochemical Impedance Spectroscopy (EIS).

The only samples investigated by electrochemical techniques (to date) were the single crystals of reduced and Nb-doped rutile (110), in both their bare and gold nanoparticle covered states. The results are presented in Chaps. 5 and 6. These techniques would also be applied to analysis of the TiO_2 thin films, given time.

4.6.1 Electrochemical Measurement Setup

The photoelectrochemical system setup is shown below in Figs. 4.11 and Fig. 4.12. The main components discussed below are:

- Photoelectrochemical Cell.
- Potentiostat and PC (not pictured).
- Xenon Lamp and Monochromator.
- Enclosure (Faraday Cage).
- Chopper.
- Lock-in amplifier (not pictured).

The working, counter, and reference electrodes were housed in a single compartment photoelectrochemical reactor built in-house (Electrochemical Engineering Group, Department of Chemical Engineering, ICL). The body of the reactor was milled from polyvinyl chloride (PVC) and held approximately 100 ml of electrolyte. The working electrode was illuminated through a Pyrex® window sealed using a Viton® O-ring.

The lid of the reactor had three ports through which the connections to the working, counter, and reference electrodes pass. There was a fourth port that allows

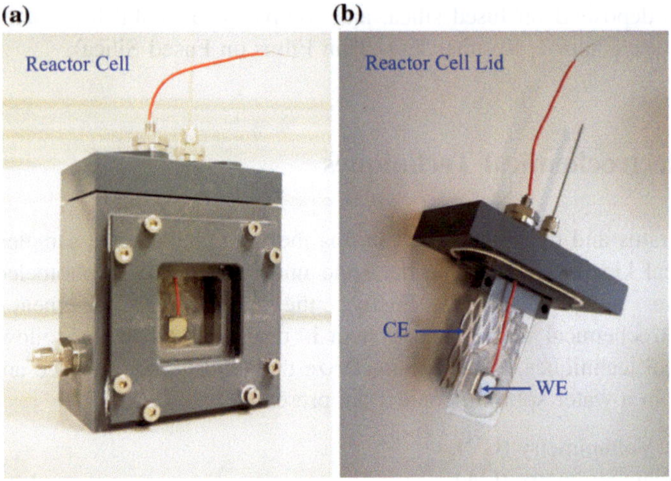

Fig. 4.10 Photoelectrochemical reactor cell and electrode configuration. Showing the assembled cell in panel (**a**) and the working and counter electrodes in panel (**b**)

an additional electrode, a gas feed such as nitrogen, or the outlet of gases. The lid also provided clamping points for both the working and counter electrodes.

Initially, a nitrogen gas feed was bubbled through the electrolyte for 10–15 min to purge the electrolyte of oxygen before measurements were made. It was decided that the bubbling was not required, considering that the action of the photoanode was to oxidize water, producing oxygen in the process.

The counter electrode consisted of a platinized titanium mesh with a wire soldered onto it that passed out through a port in the lid (Fig. 4.10). The working and reference electrodes are discussed in greater detail in the following sections.

The cell was enclosed in an earthed Faraday cage (Figs. 4.11 and 4.12). The enclosure could be covered to ensure that no light entered during measurements of the electrochemical kinetics of the samples. Both the cell and the chopper were placed on adjustable-height platforms to ensure that the illumination was incident on the working electrode.

Illumination was provided by a 300 W Xenon arc light in series with a Omni-λ 150 monochromator (LOT-Oriel) in the asymmetrical in-plane Czerny-Turner configuration. The output of the monochromator passed through sets of lenses that could be used to vary the intensity incident on the sample.

The output of the lamp-monochromator system was measured with an integrating sphere (StellarNet, Tampa, Florida). The output spectra of the lamp before and after being monochromated are presented in Chap. 5. It was discovered that higher order wavelengths were also generated by the monochromator, e.g. generation of 350 nm when monochromated to 700 nm. These harmonics were removed (when necessary) by including a UV-filter (Hoya Pro-1 digital, Japan) after the monochromator. This is discussed further in Chap. 5, in relation to measuring

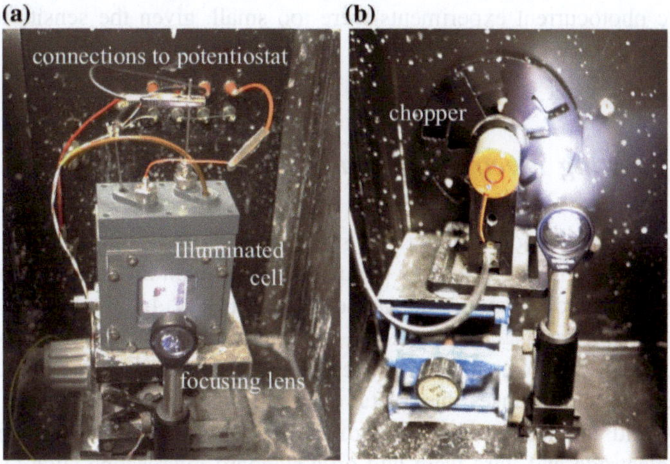

Fig. 4.11 Photoelectrochemical reactor cell in situ. Showing the connected and illuminated cell in panel (**a**) and the mechanical chopper in panel (**b**)

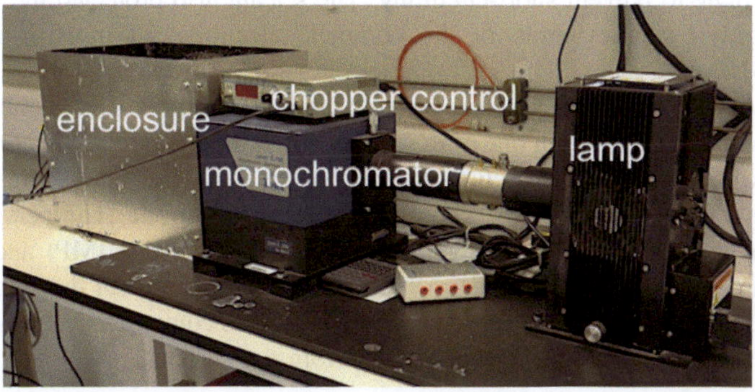

Fig. 4.12 Photoelectrochemical equipment, comprising the lamp, monochromator, enclosure and chopper control

plasmonic photocurrent. The transmittance spectra of the UV filters used are also given in Chap. 5.

The photoelectrochemical and electrochemical measurements were made using a potentiostat (Autolab PGSTAT302, MetrOhm, UK) with a frequency response analyser (FRA). This was used to control the potential of the working electrode relative to a reference electrode. The details of the working and reference electrodes are covered in the following sections. The detailed operation of the electrochemical measurement system and the potentiostat is covered in detail in Sects. 4.6.4–4.6.6 that describe each type of measurement performed.

Ideally, gas chromatography (GC) would be used to identify the products of the voltammetry and chronoamperometry. However, the currents produced during the plasmonic photocurrent experiments were too small, given the sensitivity of GC, and as such it is was not done.

4.6.2 Working Electrode Preparation

The electrodes consisted of rutile (110) single crystals of 10 mm by 10 mm with or without gold nanoparticles. A back contact of Au with an intermediate Ti adhesive layer was deposited by sputtering (as described in Sect. 4.4.3). A silver wire was attached to the back contact using conductive silver electrode paint (3 g Electrolube, RS Components). The wire could not be soldered to the back contact as the thermal shock causes the crystal to shatter in spectacular fashion. The wire was also held in place with a strip of carbon tape to prevent movement. The working electrode and wire were laid on a standard microscope slide (VWR supplies). The crystal and wire were permanently fixed by the application of a silicone sealant (RS Components). The sealant was applied such that it overlapped the active area of the working electrode, electrically insulating the back contact from the electrolyte to prevent short-circuiting. The sealant also protected the wire from the electrolyte. An example of the complete electrode assembly is shown in Fig. 4.13.

The working electrode was clamped into the lid of the photoelectrochemical cell and the wire passed through one of the access ports, from where it was connected to the working electrode connection of the potentiostat. The active area of each electrode was measured from a photograph of each electrode. The perspective of each image was corrected using image-processing software (GIMP) and the area of the un-coated electrode was estimated from the known area of the single crystal of 1 cm^2. The active area of each electrode was used to compute quantities on a unit area basis, for example the current density from the current.

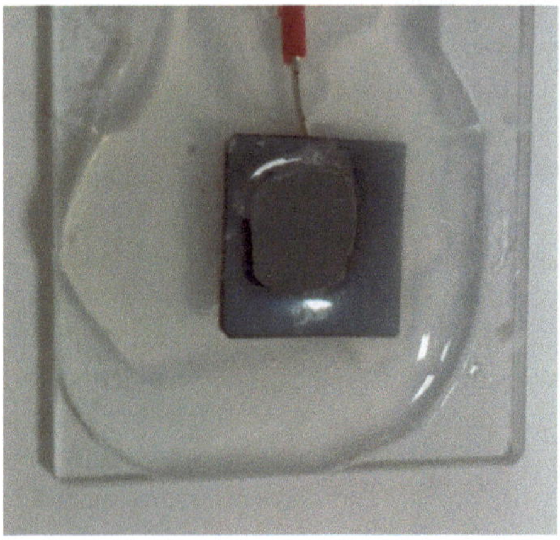

Fig. 4.13 Working electrode assembly and sealing. The wire was connected to the Au-Ti back contact by silver electrode paint. Only the central part of the TiO_2 crystal was in contact with the electrolyte

Were the work to be continued, the cell would be redesigned so that the use of sealant was unnecessary. This would remove all uncertainty as to whether or not the sealant is participating in the electrochemical reactions occurring in the cell, such as oxidation of the sealant by photo-generated holes. A cell in which the working electrode is mounted externally would be preferable.

4.6.3 Reference Electrodes and Electrolyte

All experiments were carried out in an electrolyte of 1 M NaOH (1×10^{-3} mol m^{-3}) solution. This was chosen to provide high conductivity and avoid mass transport limitations. The high concentration should also result in the lack of a Gouy-Chapman or diffuse double layer, simplifying any analysis of the solid|liquid interface.

The reference electrode used was the HgO|Hg electrode, built in-house. The electrode was constructed by inserting a copper wire into a glass tube with an upturned end shaped like a cane. A small length of platinum wire was soldered to the copper wire. The platinum wire was partly immersed in a slug of liquid Hg, which occupied the bottom of the bend in the glass tube. The electrode was completed with the addition of (red) HgO powder (Fig. 4.14). The electrode was kept immersed in 1 M NaOH electrolyte .

This was considered to be an appropriate choice of reference electrode for basic aqueous conditions pH 13.5–13.7, based on the stability of the electrode. The potential-pH (Pourbaix) diagram for the Hg–O system is shown below in Fig. 4.15. The lines are constructed from the Gibbs free energy of formation of the species at standard conditions and unity activity of Hg [16]. Some studies [17, 18] use the $AgCl|Ag_{sat}$ electrode in basic solution. However, that electrode is unstable in these conditions, potentially forming Ag_2O and AgOH. Likewise the Saturated Calomel Electrode (SCE) is also unstable at high pH; both should be avoided. Despite this, they are often used for measurements at high pH.

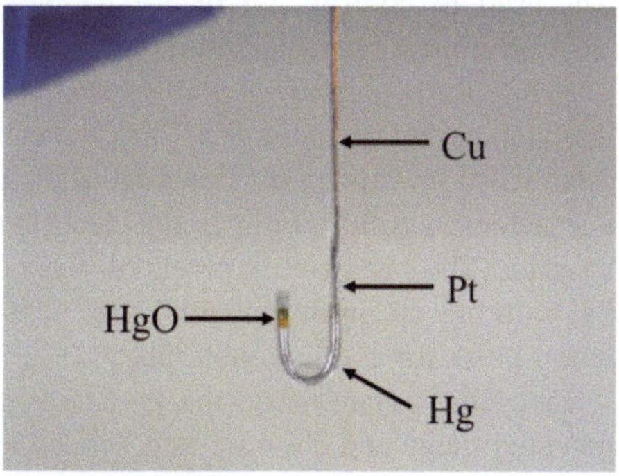

Fig. 4.14 HgO|Hg reference electrode

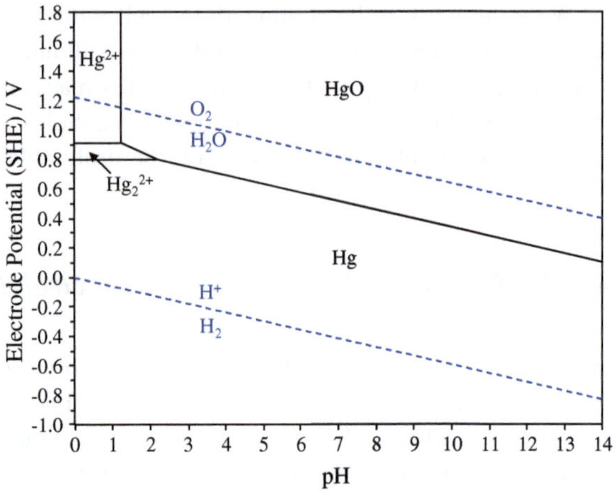

Fig. 4.15 Hg–O Pourbaix diagram. The regions of stability of HgO and Hg are shown as a function of pH and electrode potential (versus SHE)

The potential of the HgO|Hg reference electrode in basic solution ($a_{OH^-} = 1$) and at standard conditions is reported to be +99 mV (SHE). Thacker [19] reports a potential of 928 mV (at 23 °C) versus a hydrogen electrode in the same solution. Given that at pH 14 the potential of the reversible hydrogen electrode is −0.826 V (SHE), this corresponds to a HgO|Hg potential of +0.102 V (SHE), which is in approximate agreement.

The potential versus HgO|Hg electrode is pH dependent, varying with 59 mV per pH unit. Therefore, at pH 13.65, this should give a reference electrode potential of +0.124 V (SHE).

4.6.4 Voltammetry

Cyclic voltammetry was used to study both the dark electrochemical kinetics and the photoelectrochemical kinetics of the TiO_2 electrodes and the effect of gold nanoparticles. The potentiostat was used to control the potential difference between the working electrode and the reference electrode. The associated NOVA software was used to control and record the voltammetric measurements (and all other electrochemical characterizations.)

While the scan may start at an arbitrary potential, it is typical to start at or near the open circuit potential (OCP) of the electrode. The open circuit potential can be established by leaving the potential of the working electrode uncontrolled, allowing it to achieve equilibrium with the electrolyte. This can be done in both the light and the dark. The open circuit potential is obtained by monitoring the rate of change of the working electrode potential until it is negligible. This can then be set as the initial point for cyclic voltammetry.

In cyclic voltammetry, the potential of the working electrode is swept over a potential range at a constant rate, while the current passed by the counter electrode is recorded. The scan starts at the initial potential and scans to an upper limit. The scan is then reversed in direction, sweeping the potential back through the initial potential to a lower limit. The scan then reverses direction again and returns to the start potential. This constitutes one cycle. The scan may be run for several cycles to establish the reproducibility of the obtained voltammogram (a plot of current density against potential). The direction in which the scan starts may be either positive-going or negative-going depending on the requirements of the user. In the results presented in this thesis, the scans were typically begun in the positive-going direction and were repeated for several cycles to ensure reproducibility. The applied potential can be written as,

$$U(t) = U_0 \pm rt \tag{4.9}$$

where U_0 is the start potential, t is the time in seconds, and r is the scan rate in mV s^{-1}. To be precise, the equation above is for linear sweep voltammetry; however, cyclic voltammetry is simply a series of linear sweeps done in succession over a defined range of potential. This is demonstrated schematically in Fig. 4.16 for two scan rates $r_1 > r_2$.

As most practical devices, such as a solar cell, would be operated at steady-state conditions, i.e. constant potential, it is usual to use a slow scan rate that approaches steady-state behaviour to give the most accurate picture of the performance of an electrode. However, the scan rate may be varied to investigate the effect of particular processes, such as the establishment of concentration gradients in the electrolyte. Typically, a higher scan rate means that the current is sampled (and averaged) over a much shorter time interval. As such, a faster scan can capture short-lived transient processes, such as the oxidation or reduction of a previously generated species, before it can diffuse away from the interface.

In this work, a scan rate of 50 mV s^{-1} was used for the initial experiments. It was decided that this was too fast (due to the observation of transients) and so a slower rate of 10 mV s^{-1} was used thereafter; the differences between scans at 10 and 1 mV s^{-1} were small, so a slower rate than 10 mV s^{-1} was not deemed

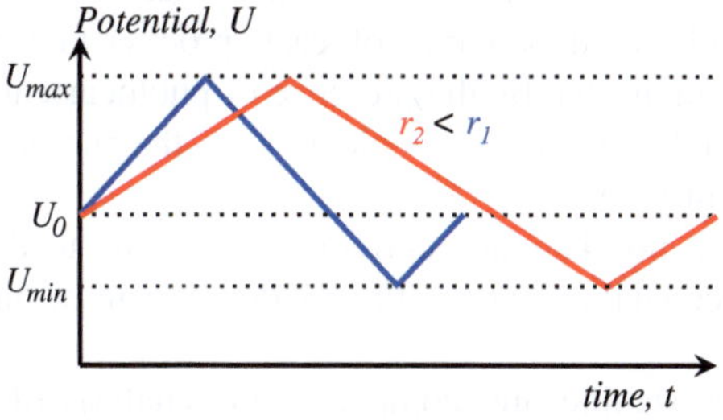

Fig. 4.16 Time dependence of potential during cyclic voltammetry. From an arbitrary starting potential U_0 the potential sweeps at rates $r_1 > r_2$, to a maximum, then changing direction and sweeping back through the start potential to a minimum, and finally changing direction again and returning to the start/stop potential. This cycle can be repeated multiple times to check repeatability

beneficial. In some cases, several different scan rates were investigated from 1 to 500 mV s^{-1}.

The potential range that the scan is carried out over also affects the processes that may be detected. For example, a positive-going potential scan may produce species in solution that are subsequently reduced during the negative-going scan. If the positive-going scan doesn't reach a potential that produces the reducible species, then it will not be available for reduction. Additionally, the presence of photo-generated holes can also result in the production of reducible species that would not be present in voltammetry in the dark. Series of scans were produced in which the upper and lower potential limits of the scans were incremented, to determine the effect of the range of potential on the processes seen in the scan.

The total current density j_{total} seen when the electrode is illuminated is composed of both the dark j_{dark} current density and the photocurrent density j_{photo}.

$$j_{total} = j_{dark} + j_{photo} \tag{4.10}$$

The analysis of the dark current density for a semiconductor electrode is somewhat more complex than for a metallic electrode. The current in the dark is made up of four contributions:

- Anodic current via valence band (electrons into valence band, holes to solution).
- Cathodic current via valence band (electrons into solution, holes to valence band).
- Anodic current via conduction band (electrons into conduction band, holes to solution).
- Cathodic current via conduction band (electrons into solution, holes to conduction band).

Typically, in analysing an n-type semiconductor such as TiO_2, the contribution from the valence band would be ignored due to the negligible minority carrier (holes) concentration expected in the valence band in dark conditions.

Analysis of the cyclic voltammetry was aided by comparison of current onset potentials to potential-pH diagrams for the O–H system and the Ti–O–H system (included in Appendix B). In principle, this allows redox processes to be identified by comparing the observed potential of each process to a range of theoretical estimates. This is complicated by the presence of photocurrent; the potential of the minority carriers is unknown. For example, in TiO_2 the photocurrent is an oxidative current deriving from holes.

There can also be non-Faradaic currents arising due to the charging and discharging of the space charge layer as the bands are bent or unbent with the swept potential.

The above issues significantly complicate the analysis of the voltammetry of semiconductor|liquid junctions. This may explain why few (if any) studies of metal oxide semiconductor electrodes in aqueous solutions go as far as assigning reactions to explain the detailed structure of the voltammetry. The theory and accompanying equations that describe the semiconductor|liquid interface were outlined in greater detail in the preceding chapter (Chap. 3, Principles of Photo-Electrochemical Cells). The discussion above puts this into the context of the experimental techniques used here.

The illuminated and dark current measurements have also been used to calculate the photocurrent, by subtracting the dark from the illuminated current. In turn, this can be used to estimate the position of the flat band potential, as described in the preceding chapter.

An alternative means of measuring only the photocurrent that was used in this work is now discussed.

4.6.5 Chopped Light Voltammetry

This section discusses the use of a mechanical chopper with lock-in amplifier to study small plasmonic photocurrents. The principle of the chopped light measurement is introduced followed by discussion of practical issues encountered while making the measurements presented in this thesis.

As the current measured under illumination is composed of both the dark and the photocurrent, it is difficult to independently assess the potential dependence of the photocurrent. It is possible to use a lock-in amplifier with a chopped light source to separate out the components. This works on the principal of a band pass filter, provided in practice by a phase sensitive detection (multiplier) and a low pass filter. The chopped light source causes the photocurrent to fluctuate at the same frequency. In simple terms, the lock-in amplifier filters out all signals that are not oscillating with the same frequency as the reference signal; this includes the dark current, which is unaffected by the varying illumination. This leaves a measurement of the photocurrent. However, steady-state behaviour may no longer apply. The equipment and the practical aspects are discussed below.

The technique can also be used to study small signals that might otherwise be dominated by other processes; in this case plasmonic photocurrent.

Plasmonic photocurrent measurements made by chronoamperometry (discussed below) on the AuNP coated TiO_2 electrodes showed the total current to be on the order of 100 nA cm^{-2} or less. This is a similar order of magnitude to the dark current. As such, it was necessary to try to distinguish between photocurrent associated with the plasmonic resonance of the gold nanoparticles, and the dark processes as a function of potential.

A mechanical chopper (Stanford Research Instruments, SR 540) was used to make to chopped-light voltammetric measurements by interrupting the output of the lamp-monochromator system. The frequency of the light chopping was variable between about 4 Hz and 3.7 kHz, although typically a frequency of between 80 and 340 Hz was used.

The frequency of the mechanical chopper was fed from the chopper controller into the reference port of the lock-in amplifier (Stanford Research Systems, SR 830). The photocurrent signal was fed from the I_{OUT} port of the potentiostat to the signal input port of the lock-in amplifier. The amplifier multiplies the reference and the signal together, resulting in a phase sensitive DC voltage signal. This result of the amplification is: one DC signal, and an AC signal with a frequency twice that of the chopping. A low band pass filter removes the AC signal, leaving a DC voltage output. The voltage output of the lock-in amplifier was fed into the auxiliary input of the potentiostat and monitored. To convert the voltage into a current it was necessary to scale the voltage by the sensitivity of the lock-in amplifier and the scale of the potentiostat. This is shown below in Eq. (4.14).

The multiplying action of the lock-in amplifier is given below

$$output = \left(\frac{signal}{sensitivity} - offset\right)(Expand)(10\,\mathrm{V}) \tag{4.11}$$

This can be re-arranged to estimate the signal (the photocurrent) from the output.

$$signal = \left(\frac{output}{(Expand)(10\mathrm{V})} + offset\right)(sensitivity) \tag{4.12}$$

The units of the *signal* and *output* and *sensitivity* are in units of volts. The *offset* and *Expand* are dimensionless.

To convert the output voltage from the lock-in amplifier into a current, it is necessary to multiply by the scale used for the potentiostat (typically 100 μA V^{-1} in the experiments done here).

$$current = (signal)(scale) \tag{4.13}$$

The *Expand* parameter was set to unity, resulting in a conversion of:

$$current = \left(\frac{output}{(10\mathrm{V})} + offset\right)(sensitivity)(scale) \tag{4.14}$$

The user manual of the lock-in amplifier notes that the output is based on multiplying the input signal by a pure sine wave at the input reference frequency. If the chopped illumination is actually a square wave (or near enough) as expected here, then there is an error introduced, and the scaled photocurrent cannot be taken as being precisely the photocurrent. From Fourier's Theorem [20], a square wave can be represented by an infinite sum of sine waves at frequencies of: f, 3f, 5f, and so on. The DC output voltage of the lock-in amplifier will give the amplitude of the first sinusoidal component. To calculate the actual current, it would be necessary to sum the Fourier series for a square wave. Additionally, the output voltage is an R. M.S value; an output of 1 V corresponds to a signal of $\sqrt{2}$ V.

The chopper and lock-in amplifier provide a number of parameters that needed to be set to obtain a reliable measurement. These were set as follows:

- Chopping frequency (Hz)—87 Hz initially, 330 Hz latterly, to reduce signal from sources of noise. A higher chopping frequency gave a better signal-to-noise ration.
- Phase difference (degrees)—90°, to provide a positive current.
- Negative/Positive edge (similar effect as phase difference)—Negative edge, to provide a positive current.
- Sensitivity (mV/V)—100 mV, to scale the output of the lock-in amplifier to maximize the signal to the potentiostat.
- Time constant (seconds)—10 ms, to ensure the positive-going and negative-going scans were the same, i.e. no hysteresis.

In practice, the original voltammetry of the chopped photocurrent against potential was made with certain default parameters on the uncoated and gold nanoparticle coated samples. A subsequent study of the effect of varying the parameters on the measured photocurrent was done following the main measurements reported in results. The parametric sensitivity study was done on uncoated Nb-doped TiO_2. The parameters used were a reasonable choice i.e. the chosen parameters resulted measurements with the lowest noise, and the same current on the positive and negative scans. The effect of increasing the chopping frequency was primarily to decrease the noise. Increasing the time constant too much results in hysteresis; the measured current of the positive-going and negative-going potential scan were not the same. Decreasing the time constant results in noise. As such, a balance between minimal noise, and capturing changes in the signal had to be found. The scale of the potentiostat (Ma V^{-1}) and the sensitivity of the lock-in amplifier (V) both act to multiply the output. They do not alter the shape of the voltammogram, just the amplitude of the signal. It is only the combination of the two that matters: the combined multiplication had to be such that the signal from the lock-in did not exceed the range of the potentiostat (which results in saturation of the measurement).

4.6.6 Chronoamperometry

Chronoamperometry was also used to determine the response of the photoanodes in both the dark and to illumination by a range of wavelengths from 300 to 1000 nm. This was used to assess the plasmonic photocurrent of AuNP–TiO_2 electrodes, as presented in Chap. 5.

In chronoamperometry, the electrode is held at a constant potential by the potentiostat, and the current is recorded over a period of time. This can be repeated for a range of different potentials. The potential can also be stepped (potential step chronoamperometry) or the illumination can be modified to see the response of the electrode. For the dark measurements made in this work, the current was simply recorded for up to 1000 s. For measurements in the light, the current was also recorded for up to 1000 s, while the wavelength of the illuminating light was changed every 20–30 s. This resulted in steps in the measured current as the photocurrent responded to the different wavelengths. The changing photocurrent was due to either: a different absorption coefficient for each wavelength of light, a change in the intensity of the illumination, or a combination of the two.

The potential of the electrode was set and the electrode was allowed to settle in the dark for a short period, to allow the initial transient currents associated with the potential step to die away. Once the initial current had decayed, the electrode was illuminated and the measurement begun.

The illumination was stepped from 350 to 1000 nm in increments of 25 nm. Each increment resulted in a step in the measured current data. The wavelength was incremented every 30 s (or so).

A UV-filter was employed between the monochromator and the photoelectrode to prevent stray UV photons from directly exciting the TiO_2. The UV light was generated by the monochromator. For example, a beam monochromated to 700 nm will generate some photons at 350 nm, which are readily absorbed by TiO_2. This was necessary as the photocurrent response of the TiO_2 to the low intensity of UV generated by the monochromator was in fact greater than the response of the plasmonic nanoparticles to visible light, complicating measurement.

4.6.7 Impedance Spectroscopy

Electrochemical impedance spectroscopy (EIS) was used to characterize the TiO_2-electrolyte interface. In addition to giving the resistance, capacitance and inductance of the electrode, EIS can yield information on the charge carrier density and electric permittivity of the electrode, and the processes occurring at the surface. The general principles of EIS are outlined below based on the reference text by Barsukov and MacDonald [21]. The practical aspects of the experiments considered as part of the research described in this document are also discussed.

As with the voltammetry and chronoamperometry, the impedance spectroscopy was controlled and recorded using the NOVA software. The data was also analysed using the tools provided in the NOVA software for fitting equivalent circuits and checking the goodness (causality) of the data.

4.6.7.1 EIS Theory

In solid-state EIS, it is typical to apply a potential difference across the electrode of interest, and investigate the response for a range of frequencies. For example, EIS is commonly used to investigate anodes and cathodes of fuel cells and electrolysers. In this case, the operating potential difference is usually fixed. The measurements may also be repeated at a range of temperatures. In the case of aqueous photoelectrochemistry, the performance of the electrode over a range of potential is also of

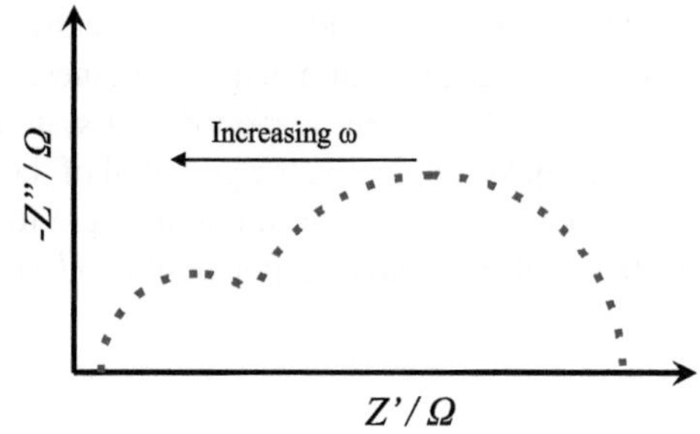

Fig. 4.17 Example of a Nyquist plot. The imaginary part of the impedance, Z″ is plotted against the real part, Z′ for a range of frequencies, ω = 2πf. The semi-circles represent different processes occurring at the interface and can be represented by elements in an equivalent circuit

interest. Given that different redox processes can occur at different potentials, this complicates the analysis of the data significantly.

The two main types of analysis are based on producing a Nyquist plot, or more usual in photoelectrochemistry, a Mott-Schottky plot. A Nyquist plot shows the imaginary part of the impedance versus the real part for a range of frequencies; an example Nyquist plot is shown below in Fig. 4.17. This can be used to fit an equivalent circuit to the data, which can be used to understand the processes occurring at the electrode. A Mott-Schottky plot is a plot of the reciprocal of the squared capacitance against potential; it is plotted at a single frequency. This is used to derive information about the nature of the space charge region of the electrode. The theory and governing equations behind the Mott-Schottky plot were presented in the previous chapter on the theory of the semiconductor|liquid junction. The following equations cover the general principle of EIS.

In general terms, the complex frequency dependent impedance, $Z(\omega)$ in units of ohms (Ω) is defined as:

$$Z(\omega) = V(\omega)/I(\omega) \tag{4.15}$$

where V is the potential difference in volts (V) and I is the current in amps (A). The impedance is a complex quantity with real and imaginary parts, Z' and Z''.

$$Z = Z' + iZ'' \tag{4.16}$$

The real part of the impedance is the resistance $Z' = R$, and the imaginary part is the reactance $Z'' = X$, where the reactance includes the capacitance, C and inductance L. The reactance of a capacitance C is, $X = -i/\omega C$, and the reactance of an inductance L is, $X = i\omega L$. The general form for the impedance of a resistor, capacitor, and inductor connected in parallel is therefore:

$$Z = R + \; - i/\omega C + i\omega L \tag{4.17}$$

The notation i is used to denote $\sqrt{-1}$ and j is used for current density. However, the conventions are sometimes reversed in electrical engineering literature.

In impedance spectroscopy, the current response of a system to a time varying potential difference is determined. Typically, the potential of the electrode is set to a value of interest U_0 and a potential $U'(t)$ varying at a single frequency $\omega = 2\pi f$ is super-imposed, giving the total time varying potential of $U(t)$.

$$U(t) = U_0 + U'(t) \tag{4.18}$$

Typically the form of the time varying potential $U'(t)$ is given by

$$U'(t) = U_m \sin(\omega t) \tag{4.19}$$

The impedance is calculated by measuring the phase shift and amplitude of the current based on the applied time varying voltage. The symbol U is used to denote a potential versus an arbitrary reference electrode (rather than a potential drop across the electrode).

A major assumption of impedance spectroscopy is that the response to the stimulus is linear. The assumption is generally considered to hold when the following conditions are met:

$$U_m < V_T = \frac{RT}{F} = \frac{kT}{e} \tag{4.20}$$

The value of V_T at 25 °C is 25 mV. An amplitude of V_m = 10 mV (r.m.s) was used for the experiments described in this thesis.

4.6.7.2 EIS Practical Aspects

Impedance measurements were made using the same electrodes and electrolyte as used in the voltammetry and chronoamperometry. Both Nyquist and Mott-Schottky plots were produced as necessary. The individual potential and frequency ranges used are discussed in the relevant results section of Chap. 6 [Results: Electrochemistry of Rutile (110)].

A measurement of the impedance at one potential and a range of frequencies can be and was used to construct a Nyquist plot of Z″ against Z′. The same data can also be used to plot the both modulus |Z| and phase Arg(Z) of the impedance against frequency to obtain a Bode modulus and Bode phase plots respectively. A Mott-Schottky plot is obtained by a measurement at a single frequency over a range of potential. Typically, frequencies of 100 kHz down to 0.1 Hz were investigated for Nyquist plots. Several Mott-Schottky plots were attempted at 1 MHz, but the measurement was not stable, resulting in the termination of the procedure. The Mott-Schottky equation was presented in Chap. 3.

When the measurement was repeated for a range of potentials, say from −1.0 V (HgO|Hg) to +1.0 V (HgO|Hg), the measurement could take up to 10 h. Measurements made over an extended range of potentials and frequencies were made in the dark. This was to avoid wearing out the lamp used to illuminate electrodes. Several Mott-Schottky measurements were made with illumination. The measurements appeared to be almost identical to those made in the dark, suggesting that the carrier concentration was not changed significantly by the addition of photo-generated charge carriers.

Nyquist plots are produced from the determined impedance at each frequency without the need for any intermediate manipulation of the data. However, to generate a Mott-Schottky plot, it is necessary to calculate the capacitance from the impedance, which was done by assuming a simple model for the solid|liquid interface. The NOVA software can be programmed to produce the Mott-Schottky plot based on two simple models, the series resistance (R_S) and capacitance (C_S) model (Fig. 4.18a) and the parallel resistance (R_P) and capacitance (C_P) model (Fig. 4.18b).

In the case of the series equivalent circuit, C_S is used to generate the Mott-Schottky, and in the case of the parallel circuit, C_P is used.

Ideally, the equivalent circuits used to generate the Mott-Schottky plots would be determined for each electrode from the Nyquist plots. Non-ideal elements in the model of the interface result in the capacitance calculated from the impedance being different to the actual capacitance and could result in misleading plots. As such, use of the standard equivalent circuit models may need to be reconsidered if the Mott-Schottky plot displays unexpected behaviour. For example, non-linear plots are often obtained for nanostructured photoanodes, such as those made from hematite (Fe_2O_3) [21, 22].

The fitting of impedance data using an equivalent circuit is a non-trivial matter, particularly for photoelectrochemical systems. The NOVA software (or others such as Zview) may be used to specify a model consisting of resistance (R) and capacitance (C) components (and others discussed below). The fitting procedure (Complex Non-linear Least Squares, CNLS) then varies the values of the components until the best fit is obtained. A χ^2 value is calculated to indicate the goodness of the fit to the data points. A value of less than $\chi^2 < 0.1$ was taken as indicating a reasonable fit.

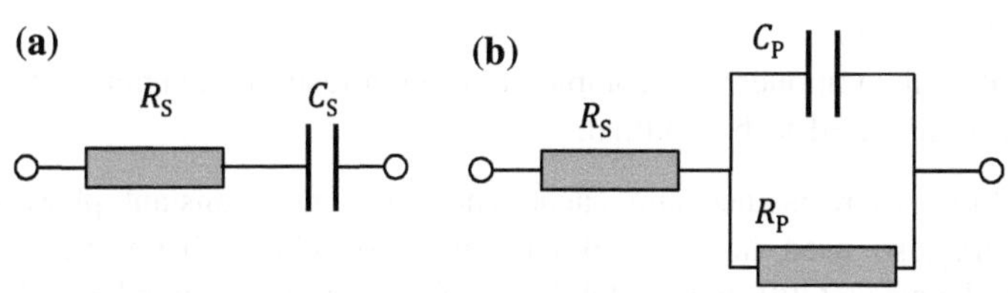

Fig. 4.18 Circuits used to extract resistance and capacitance from measured impedance to construct Mott-Schottky plots

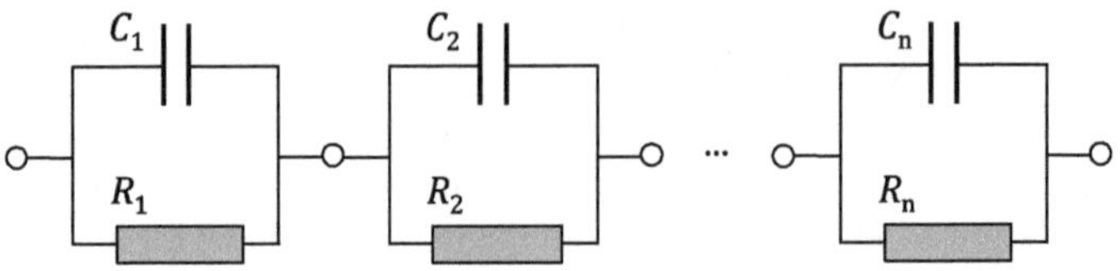

Fig. 4.19 Voigt circuits used to test causal relation between real and imaginary parts of the impedance

The NOVA software also uses a procedure outlined by Boukamp [23] to determine whether the impedance data obeys the Kramers-Kronig relations [24], which demonstrate the causal relation between the real and imaginary parts of a complex function. The test was applied before fitting each set of data. The principle of the test is that any impedance data that obeys the relations can be modelled with an equivalent circuit consisting of an infinite number of parallel resistor-capacitor pairs—a Voigt circuit—in which every other resistance is negative (Fig. 4.19).

Of course, the Voigt circuit is not a representation of a set of physical processes that may be occurring at a photoelectrode. It is also possible to fit any particular impedance data with an infinite number of equivalent circuits (hence the use of *equivalent* in the name [21]. Instead, it is necessary to try to fit the data with simple equivalent circuits that represent real processes as far as possible. These are presented in the results chapters.

An additional error has been observed in some published results, where only the shape of the impedance data in the Nyquist plot is fit, without concern for the frequency at which each value of the impedance occurs. This is incorrect. Frequency agreement of points—as well as shape—is crucial.

In the analysis of impedance spectroscopy data carried out in this work, the approach taken was as follows:

1. Run the Kramers-Kronig test to check the goodness of the data.
2. Fit a simple Voigt circuit with no more than two or three R-C pairs as an initial fit.
3. Fit more complex circuits such as ladder circuits or nested R-C pairs.
4. Replace C elements with Q elements to improve fit (see below).
5. Addition of other components such as inductance and Warburg impedance as necessary.
6. Remove terms that give odd results, such as infinite or negative resistances, to simplify circuits.
7. If additional components or complexity did not result in a better fit, the existing fit was considered to be optimal.

In addition to resistance and capacitance elements, constant phase elements (CPE) were also used in place of capacitors to achieve better fits. The letter Q usually denotes a CPE in a model. Use of CPEs in fitting impedance data can be contentious [25] as it is considered by some to be unphysical and a way to fit data. Constant phase elements have been used in the analysis in this thesis as a means of indicating how far from ideal R-C behaviour the data is. In fact, many solid electrodes at which Faradaic processes occur demonstrate this departure from ideal behaviour [26]. The behaviour may be due to a frequency dispersion of the electric permittivity. A CPE is a type of *distributed element:* the circuit is composed of a distribution of time constants, whereas an *R-C* circuit has a single time constant. They may arise due to effects such as surface roughness and inhomogeneous chemical composition of the surface, for example the distribution of dopants. The characteristic signature of a CPE in impedance data is the presence of a depressed semi-circle in the Nyquist plot. A capacitor gives a vertical line (90° to real axis) on the Nyquist plot. A CPE gives a straight line at less than 90°. A single *R-C* element gives a semi-circle in the Nyquist plot, while an *R–Q* element gives a depressed semi-circle i.e. a tilted semi-circle. The form of a CPE is given in (4.21).

$$Z_{\mathrm{CPE}} = \frac{1}{T(j\omega)^{\phi}} \tag{4.21}$$

where T is a constant in F cm^{-2} $\mathrm{s}^{\phi-1}$ and ϕ is related to the slope of the capacitance in the complex plane. For $\phi = 1$, the CPE is purely capacitive; for $\phi = 0.5$ the CPE is equivalent to the Warburg impedance; for $\phi = 0$, the CPE is purely resistive; and for $\phi = -1$, the CPE is purely inductive.

In the results for the single crystal rutile (110) electrodes [see chapter—Results: Electrochemistry of Rutile (110)], it was necessary to use CPEs in the equivalent circuits. One physical interpretation is that the electrodes are behaving as "leaky" capacitors. It is suggested here that the reasons for this is the defect chemistry of TiO_2. While a single crystal of TiO_2 might be considered to be as close to an ideal electrode as possible (compared to a sintered ceramic), the nature of the defects allows for distributed elements (see Chap. 2). If it were assumed that electronic defects that contribute to conductivity are homogeneously distributed, then a CPE would not be supportable. This should be the case for the distribution of Nb doped into the TiO_2. For reduced TiO_2, the dominant electronic defect is assumed to be interstitial Ti^{3+} and there is evidence that the defects form conductive filaments or planes through the crystal rather than being evenly distributed [27]. This could result in a "leaky" pathway, with some areas of the electrode that are dielectric and others that are metallic.

References

1. Nishijima, Y., Ueno, K., Yokota, Y., Murakoshi, K., Misawa, H.: Plasmon-assisted photocurrent generation from visible to near-infrared wavelength using a Au-Nanorods/TiO_2 electrode. J. Phys. Chem. Lett. **1**, 2031–2036 (2010)
2. Diebold, U.: The surface science of titanium dioxide. Surf. Sci. Rep. **48**, 53–229 (2003)
3. Fuke, N., et al.: Influence of TiO_2/electrode interface on electron transport properties in back contact dye-sensitized solar cells. Sol. Energ. Mat. Sol. C. **93**, 720–724 (2009)

4. Temple, T.: Optical properties of metal nanoparticles and their influence on silicon solar cells (2009)
5. Smith, D.: Thin-Film Deposition: Principles and Practice. McGraw Hill Professional (1995)
6. Spatz, J.P., et al.: Ordered deposition of inorganic clusters from micellar block copolymer films. Langmuir **16**, 407–415 (2000)
7. Delcassian, D., et al.: Nanoscale ligand spacing influences receptor triggering in T cells and NK cells. Nano Lett. **13**, 5608–5614 (2013)
8. Zhang, G., Wang, D.: Colloidal lithography-the art of nanochemical patterning. Chem. Asian J. **4**, 236–245 (2009)
9. Callister, W.D., Rethwisch, D.G.: Materials Science and Engineering: An Introduction, 9th edn. Wiley Global Education (2013)
10. Ferraro, J.R.: Introductory Raman Spectroscopy. Academic Press (2003)
11. Kelsall, R., Hamley, I.W., Geoghegan, M.: Nanoscale Science and Technology. John Wiley & Sons (2005)
12. Nečas, D., Klapetek, P.: Gwyddion: an open-source software for SPM data analysis. Cent. Eur. J. Phys. **10**, 181–188 (2011)
13. Fox, M.: Optical Properties of Solids. OUP Oxford (2010)
14. Swanepoel, R.: Determination of the thickness and optical-constants of amorphous-silicon. J. Phys. E: Sci. Instrum. **16**, 1214–1222 (1983)
15. Kasap, S., Capper, P:. Springer Handbook of Electronic and Photonic Materials. Springer (2007)
16. Bard, A.J., Parsons, R., Jordan, J.: Standard Potentials in Aqueous Solution. CRC Press (1985)
17. Dare-Edwards, M.P., Goodenough, J.B., Hamnett, A., Trevellick, P.R.: Electrochemistry and photoelectrochemistry of iron(III) oxide. *J. Chem. Soc. Faraday Trans. 1:* **79,** 2027 (1983)
18. Klahr, B., Gimenez, S., Fabregat-Santiago, F., Hamann, T., Bisquert, J.: water oxidation at hematite photoelectrodes: the role of surface states. J. Am. Chem. Soc. **134**, 4294–4302 (2012)
19. Thacker, R.: A simple procedure for making Hg HgO reference electrodes. J. Chem. Educ. **45**, 180 (1968)
20. Hecht, E.: Optics. Pearson Education India (2008)
21. Barsoukov, E., Macdonald, J.R.: Impedance Spectroscopy. Wiley-Interscience (2005)
22. Cesar, I., Sivula, K., Kay, A., Zboril, R., Grätzel, M.: Influence of feature size, film thickness, and silicon doping on the performance of nanostructured hematite photoanodes for solar water splitting. J. Phys. Chem. C **113**, 772–782 (2009)
23. Boukamp, B.A.: A linear Kronig-Kramers transform test for immittance data validation. J. Electrochem. Soc. **142**, 1885–1894 (1995)
24. Kittel, C.: Introduction to Solid State Physics. Wiley (2004)
25. Zoltowski, P.: On the electrical capacitance of interfaces exhibiting constant phase element behaviour. J. Electroanal. Chem. **443**, 149–154 (1998)
26. Brug, G.J., Vandeneeden, A., Sluytersrehbach, M., Sluyters, J.H.: The analysis of electrode impedances complicated by the presence of a constant phase element. J. Electroanal. Chem. **176**, 275–295 (1984)
27. Smyth, D. M.: The Defect Chemistry of Metal Oxides. Oxford University Press (2000)

Chapter 5
Results: Plasmonic Photocurrent with AuNP-TiO_2

This chapter presents the results of a series of experiments on doped and reduced rutile (110) TiO_2 single crystals coated with gold nanoparticle (AuNP) arrays. The structure of the chapter is outlined immediately below, followed by summaries of the experimental aims and experimental results.

Firstly, the results of the investigation into forming gold nanoparticles are presented, including scanning electron microscopy (SEM) images of the resulting nanoparticles, and UV-Visible spectroscopy (UV-Vis) of the optical characteristics of the nanoparticles.

Secondly, photoelectrochemical measurements that investigate the performance of the AuNP-TiO_2 electrodes as water splitting photo-anodes are reported. The discussion includes further consideration of the experimental methodology and the implications for the results, additional to that given in the chapter dedicated to Experimental Methods (Chap. 4).

The characteristics of the bare single crystal substrates are introduced briefly, and are then followed by a dedicated chapter on the detailed electrochemistry of rutile TiO_2. The focus of the current chapter is on the performance of the Au-TiO_2 system under illumination with visible light, i.e. the plasmonic aspects. The following chapter (Chap. 6) focuses on differences between the batches of rutile (110) single crystals, which may have lead to differences in the performance of the AuNP-TiO_2 electrodes as discussed in this chapter.

The photocurrent that is attributed the effect of surface plasmons on the gold nanoparticle will be referred to as "plasmonic photocurrent". In essence, this means a photocurrent that occurs at wavelengths that coincide with a plasmon resonance of the nanoparticles, and at a wavelength at which the underlying substrate—in this case TiO_2—is not active (does not absorb photons by electron-hole pair generation). In particular, it is not referred to as a "plasmon-enhanced" photocurrent (as is commonly used in the literature: see the Literature Review in Chap. 2). The term "plasmon-enhanced" would suggest that there is normally a photocurrent at a particular wavelength, and that the effect of the surface plasmon is to enhance the

© Springer International Publishing Switzerland 2016

J.C. Alexander, *Surface Modifications and Growth of Titanium Dioxide for Photo-Electrochemical Water Splitting*, Springer Theses,
DOI 10.1007/978-3-319-34229-0_5

current. This cannot occur with Au and TiO_2, as the band gap of TiO_2 is around 400 nm, while the plasmon resonance is typically above 500 nm.

5.1 Experimental Aims

Based on an early reading of the literature (summarised in detail in Chap. 2 of this thesis), a number of studies were identified suggesting a route of experimental investigation. The studies by Nishijima and co-workers [1, 2] demonstrated that an array of AuNPs formed on 0.05 wt% (weight percentage) Nb-doped rutile (110) could apparently drive water splitting under visible (VIS) and near infra-red (NIR) light, with quantum efficiency approaching 10 %, at wavelengths corresponding to a plasmon resonance of the nanoparticles. The efficiency was significantly higher than any other reported for water splitting; high efficiency for other redox couples has, however, been reported [3]. The study did not consider whether the doping was crucial to the performance. Likewise, the effect of a Ti interlayer on performance was not considered. Another study by Liu and co-workers [4] concluded that doping was important, based on comparing a control sample of AuNP covered high purity anatase, which did not produce visible light photocurrent, to a N and F doped anatase nanotube film that did exhibit some enhancement (the doping by N and F was as a result of the anodising process). While there are issues with the Liu study (discussed in Literature Review), the use of a control sample influenced the choice of experiments for this thesis.

The original intention of the experiments, for which results are reported here, was to assess the impact of the doping of the rutile (110) substrate on performance for water splitting. In particular, doping is known to lead to absorption of visible light (but not necessarily by inter-band transitions). Hence, the original question was whether absorption of visible light by the TiO_2 substrate was necessary for plasmonic enhancement of photocurrent.

The objectives of the experiment were to deposit nanoparticle arrays onto 0.05 wt% Nb-doped TiO_2 (shown to work well by others) and compare the performance with high purity un-doped TiO_2. The single crystal substrates were selected to have the same orientation, rutile (110), with the only difference being doped or not. In addition, it was decided to employ nanoparticle array fabrication techniques with the potential for use on a large scale (1 cm^2 and above), thus precluding e-beam lithography (EBL) from the initial work. EBL was also discounted due to cost and not being an established technique in the laboratory in which the work was done. However, initial studies with the un-doped TiO_2 found it to be un-responsive (too resistive), precluding photo-electrochemical measurements. Hence, the decision was taken to reduce the un-doped TiO_2 in hydrogen, in order to impart conductivity. The effects of this reduction on the plasmonic photocurrent are discussed in Sect. 5.7.4 (in short, no effect was observed). The effect of the reduction on the performance of the bare reduced electrode relative to the bare Nb-doped electrode is discussed in Chapter 6.

5.2 Summary of Experimental Results and Conclusions

A number of techniques for the fabrication of gold nanoparticle arrays on TiO_2 were investigated including annealing metal thin films (AMTF), micellar nano-lithography (MNL), and nano-sphere lithography (NSL). The optical characteristics were determined by UV-Visible spectroscopy (UV-Vis), and the photoelectro-chemical performance of Au-TiO_2 photoanodes under UV and visible light was measured using a range of electrochemical techniques. Finally, the results were discussed in the context of the proposed mechanisms of plasmonic water splitting, and several alternative mechanisms were proposed briefly.

The research identified a number of novel features and conclusions:

- For thin films of gold deposited by sputtering onto indium-tin oxide (ITO), using thinner films of 2–4 nm and higher annealing temperatures (up to 600 °C) resulted in the formation of an array of roughly spheroidal nanoparticles with random size and distribution on the order of 50–100 nm. The nanoparticles demonstrated localized surface plasmon resonance (LSPR) under illumination by visible light (400–700 nm). The arrays formed by annealing the 2 nm films at 600 °C demonstrated the highest extinction of visible light photons.

This suggested that annealing metal thin films was a viable approach for forming arrays of plasmonically active gold nanoparticles. In particular, the technique is ideal for forming arrays over a large area, and at relatively low cost compared to techniques like electron beam lithography. While the nanoparticles formed are all different sizes, the range of sizes are consistent over areas of at least 1 cm^2. The technique has the potential to form nanoparticles over much larger areas and is only limited by the initial thin film deposition process.

- Annealing a 2 nm thick gold thin film deposited on rutile (110) resulted in the formation of much smaller gold nanoparticles (AuNPs) of 20 ± 10 nm. These AuNPs also demonstrated LSPR under illumination by visible light, in the range 600–650 nm. Following electrochemical measurements, the LSPR shifted to around 580 nm.

This showed that while it was possible to use ITO as a substrate to establish that the technique of annealing metal thin films results in nanoparticle arrays, that the size and shape of the AuNPs is considerably different on TiO_2, i.e. the AuNPs were smaller, further apart, and appeared more uniform in cross-section. The initial study should have been done on rutile (110). However, at that time during the experimental work, the rutile (110) single crystals had not been received, and the thin film deposition of rutile (110) was not established.

- The AuNPs formed on Nb-doped rutile (110) single crystals demonstrated a small plasmonic photocurrent under illumination by visible light. The incident photon to current efficiency (IPCE) of 0.003 % was vanishingly small but measurable. The variation of the IPCE with wavelength coincided with the reflectance and absorption of the LSPR, with the maximum occurring around 575–600 nm.

This demonstrated that a photocurrent attributable to the interaction of the AuNPs with visible light could occur. However, the low efficiency of the process precluded any further analysis of the process by techniques such as transient absorption spectroscopy (TAS) to study the charge transfer dynamics. The low efficiency also precluded any analysis of the reaction products, and as such it was not possible to verify that the photocurrent is indeed due to water splitting and not another reaction, such as the oxidation or dissolution of the nanoparticles.

- The gold nanoparticle arrays fabricated by micellar nanolithography did not demonstrate plasmon resonance or a photocurrent under visible light.

This suggests that the larger nanoparticles deposited by annealing are better than the smaller particles deposited by MNL. The arrays fabricated by nano-sphere lithography did not survive long enough in solution to be measured.

- The plasmonic photocurrent was not detected on reduced rutile (110) and some Nb-doped crystals. It is suggested here that the Nb-doped crystals that do not display the phenomena were partly reduced. This is based on electrochemical measurements made on bare electrodes fabricated from each batch of crystals. This is discussed further in Chap. 6, in which the differences in the batches of doped and reduced rutile (110) crystals are presented.

This observation suggested that any reduction of the rutile (110) surface inhibits plasmonic photocurrent and should be avoided in any further studies. In particular, the electrochemistry of the reduced rutile (110) suggests that the effect of the reduction is not simply to increase conductivity, but to fundamentally alter the behaviour of the TiO_2, and in such a way that inhibits the plasmonic photocurrent. That some Nb-doped crystals appeared to have been received reduced as well as being doped, suggests that the crystal growth process may result in inadvertent reduction of the crystal. Hence, a process to ensure that as-received crystals are not reduced needs to be developed. It is also possible that the difference is simply down to the increased charge carrier concentration (as reported in Chap. 6), which reduces the width of the depletion region of TiO_2. In this case, the ideal level of doping may need to be controlled carefully. While annealing in oxygen is one possible process for ensuring the crystals are not reduced, this may also have the effect of decreasing the conductivity of the material (see discussions in Chap. 7 and in Appendix B) and preventing its use as an electrode.

- The potential dependence of the plasmonic photocurrent using chopped light voltammetry with a lock-in amplifier was found to be similar to the shape of the photocurrent for bare TiO_2 in UV light. There were also two features to the current, which may have indicated two separate processes occurring over different ranges of potential (although the processes have not been identified explicitly beyond doubt).

The fact that the shape of the potential dependence of the plasmonic photocurrent is the same as for a bare electrode in visible light suggests that the photocurrent may originate from the absorption of photons in the space charge layer of TiO_2, which could in turn indicate that the LSPRs decayed as UV photons rather than hot electrons. It could also indicate that the efficiency of the process is directly linked to the distribution of the electrical field and potential within the space charge region of TiO_2. To be more certain about the potential dependence, and the reactions to which the plasmonic photocurrent corresponds, a AuNP-TiO_2 system with a greater quantum yield would be required. The system could then be used with complimentary techniques such as TAS and GC.

- The overall effect of the AuNPs on the UV response of the TiO_2 was to severely decrease the overall incident photon-to-electron efficiency (IPCE) of the Au-TiO_2 to the solar spectrum.

The large decrease in the efficiency between 300 and 400 nm was not compensated by the additional plasmonic photocurrent above 400 nm. Without understanding the mechanism and how to optimize it, the addition of gold nanoparticles to TiO_2 should not be advocated.

5.3 Future Work

Fundamentally, as fabricated, the gold nanoparticle arrays did not allow the question of the mechanism of plasmonic water splitting on TiO_2 to be addressed in much detail. While an effect was seen, it was too small for practical purposes.

Any further investigation should concentrate on improving the performance of the Au-TiO_2 system in the following ways:

- A process to ensure that substrates are not reduced, or display unwanted variation from sample to sample. This could be an annealing process. The temperature and partial oxygen pressure would have to be determined precisely to ensure the conductivity of doped TiO_2 is preserved.

This would ensure that any increase or decrease in performance of nanoparticle arrays is due to the nanoparticles rather than slight differences in the substrate.

- A method for fabricating arrays of gold nanoparticles that can control the size, shape, and spacing of the nanoparticles in a reproducible manner on a large area. In particular, e-beam lithography and nano-imprint lithography are ideal candidates.

Generally, the choice of a nanoparticle technique that could cover a large area (one capable of up scaling) should not have been used, considering the lack of understanding of the mechanism. It did not allow sufficient control of the properties of the nanoparticles. The presence of a range of difference sizes, shapes, and inter-particle distances meant that it was not possible to attribute the measured plasmonic photocurrent to any particular feature of the AuNP array. The small magnitude of the measured current suggested that only a small proportion of the AuNPs were contributing. But it is impossible to identify which with the approach taken.

Given the lack of understanding of the mechanism of plasmonic photocurrent, it is not possible to predict the performance of a given nanoparticle array on a given substrate. Instead, a trial and error process is required to provide evidence for evaluating theory. It is recommended that nanoparticle arrays with well-defined, repeatable dimensions be used. This would result in arrays that could be studied by computational methods such as finite difference time domain (FDTD) electro-dynamics [5]. Changes in the experimental performance could be correlated to changes in the magnitude and distribution of the electric fields, and the resonant wavelength, as predicted by FDTD. The computational approach would also allow the precise nature of the surface plasmon (LSPR, SPP, or a combination) to be determined. Optical measurements such as UV-Vis cannot distinguish between plasmon modes.

An attempt was made to use NSL to provide a more controlled AuNP arrays. However, at the time of carrying out the experiments detailed in this thesis, the technique had not been developed to provide the level of uniformity required.

- Investigating the effect of anatase versus rutile and the crystal orientation.

This could be done using the TiO_2 thin films fabricated by pulsed laser deposition for which results are presented in Chaps. 7 and 8 of this thesis. Investigation of the effect of the substrate is considered of secondary importance to establishing the nature of the mechanism of plasmonic photocurrent and optimizing the fabrication of nanoparticle arrays.

5.4 Sample Production—General

The method of preparation of the single crystal rutile (110) substrates and the nanoparticle arrays is described in Chap. 4 (Experimental Methods) Sect. 4.2.

5.5 Nanoparticle Array Architecture

Based on the literature review of techniques to form nanoparticle arrays, and the existing expertise at Imperial College London (ICL), a number of feasible techniques were identified:

- Annealed metal thin film (AMTF)
- Micellar Nanolithography (MNL)
- Nano-sphere Lithography (NSL).

While E-beam Lithography (EBL) can give excellent control over size and spacing of nanoparticle arrays, it is typically used for areas up to 1 mm^2. Given the electrochemical characterization equipment available, a technique giving greater coverage of 25–100 mm^2 was considered preferable at the time. Additionally, at the beginning of the study the in-house capacity to do nanoparticle arrays by EBL and by Focussed Ion Beam Milling (FIB) was not well established.

There was also a project running concurrently on the same Grant using plasmonic nanoparticles as the antenna for an optical rectenna device. As the focus of that project was developing metal nanoparticle deposition processes, a decision was taken to start with the simple technique of AMTF and also MNL, a technique used regularly by the Dunlop group at Department of Materials, Imperial College London. The intention was to take advantage of the processing experience gained as the rectenna project progressed, to produce more complex structures.

The NSL technique was of particular interest as it could also be used to deposit AuNPs with a thin titanium inter-layer, potentially allowing the effect of this interlayer to be investigated. This is discussed further in in Sect. 5.5.4 below.

5.5.1 Annealed Gold Thin Films on Indium Tin Oxide (ITO)

To establish the conditions to form a homogeneous (mono-disperse) layer of gold nanoparticles by annealing, a study of initial film thickness and annealing temperature was done. The experimental method was described in Chap. 4 (Experimental Methods).

Thin films of Au of 2, 4, 8, and 16 nm thicknesses were sputtered onto ITO. The films were annealed at a range of temperatures: 300, 400, 500, and 600 °C giving 16 combinations in all. The aim was to obtain an indication of the uniformity of size and distribution of nanoparticles formed under different conditions before applying this technique to an expensive single crystal of TiO_2. Additionally, the high conductivity of the ITO film meant that an additional layer of gold or chromium would not be necessary to obtain high resolution images by scanning electron microscopy (SEM). Typically, a film of several nanometres of Au or Cr is sputtering onto the sample of interest. As the gold film thickness ranged from 2 to 16 nm, an additional layer was undesirable as, being of similar thickness, it could have obscured the results.

The results of this study are shown below. However, as will be shown later in this chapter, using ITO as the substrate gave a markedly different morphology to nanoparticles formed on TiO_2, so this initial experiment may have been misleading.

The effect of initial film thickness and annealing temperature on the morphology can be seen in Figs. 5.1, 5.2, 5.3 and 5.4. For the thinner gold films of 2 and 4 nm, the annealing resulted in the formation of discrete nanoparticles of about 50–100 nm over the whole range of annealing temperatures (Figs. 5.1 and 5.2).

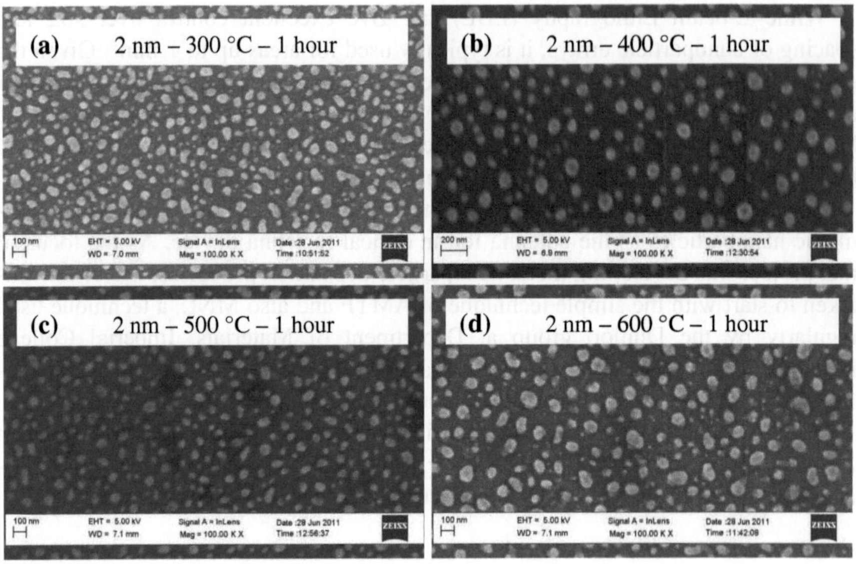

Fig. 5.1 SEM image of nanoparticle arrays formed by annealing a 2 nm thick gold film deposited by sputtering. Films annealed for 1 h at 300, 400, 500, and 600 °C. The magnification is 100 k. The scale bar is 100 nm in panels **a**, **c**, and **d** and 200 nm in panel **b**

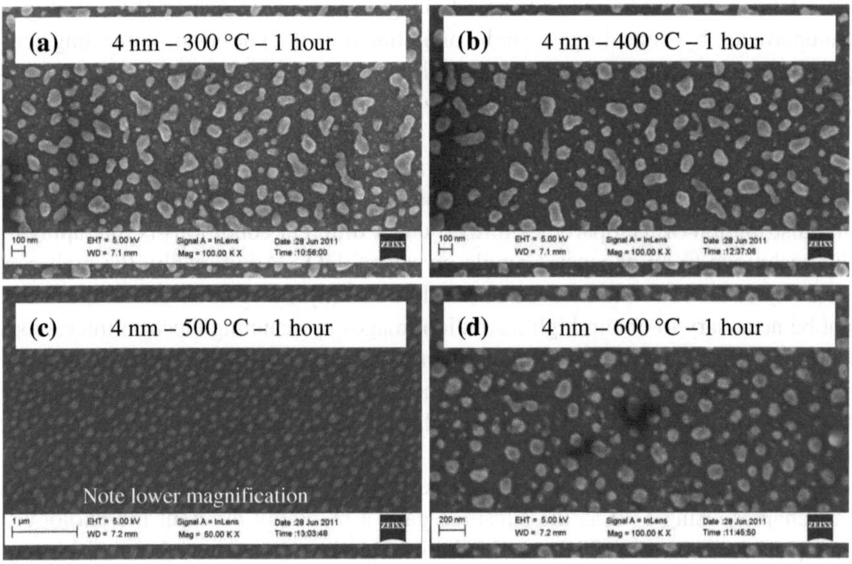

Fig. 5.2 SEM image of nanoparticle arrays formed by annealing a 4 nm thick gold film deposited by sputtering. Films annealed for 1 h at 300, 400, 500, and 600 °C. The magnification is 100 k except in panel **c** where it is ×50 k. The scale bar is 100 nm in panels **a** and **b**, 1000 nm in panel **c**, and 200 nm in panel **d**

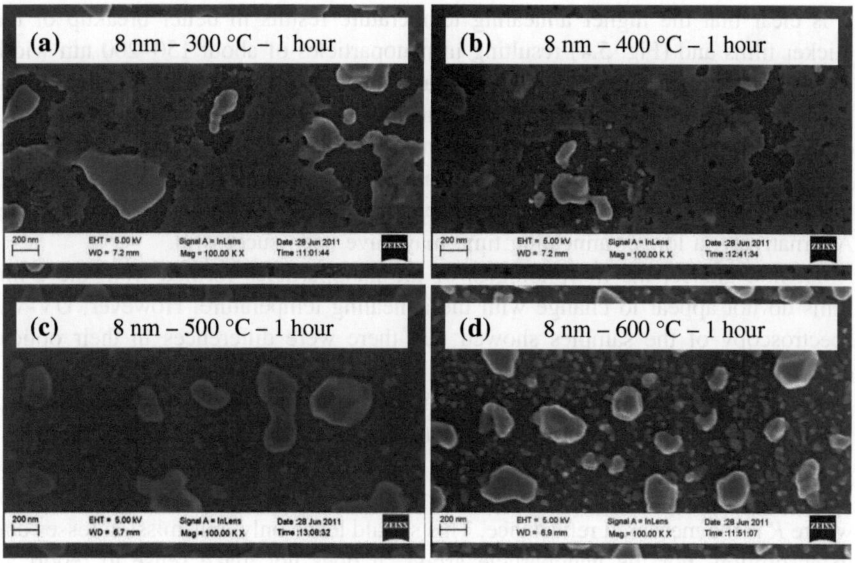

Fig. 5.3 SEM image of nanoparticle arrays formed by annealing an 8 nm thick gold film deposited by sputtering. Films annealed for 1 h at 300, 400, 500, and 600 °C. The magnification is 100 k. The scale bar is 200 nm in all panels

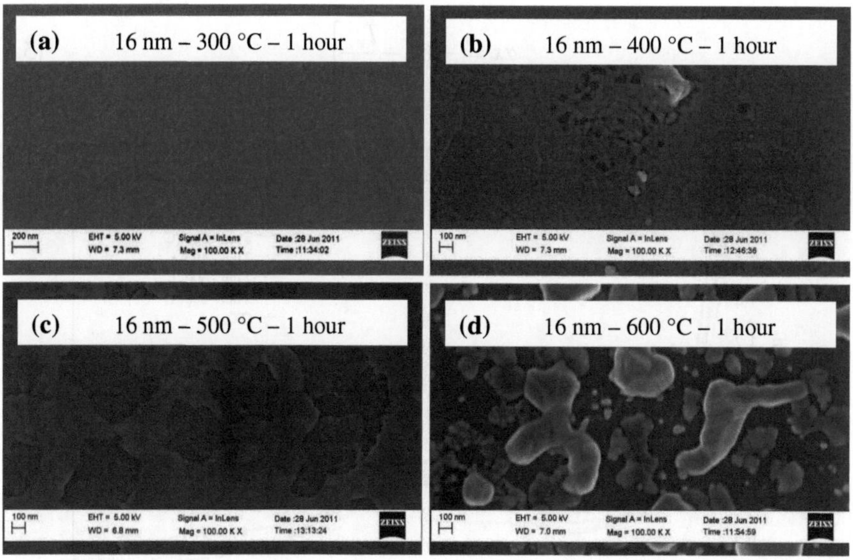

Fig. 5.4 SEM image of nanoparticle arrays formed by annealing a 16 nm thick gold film deposited by sputtering. Films annealed for 1 h at 300, 400, 500, and 600 °C. The magnification is 100 k. The scale bar is 200 nm in panel **a**, and 100 nm in the other panels

It is clear that the higher annealing temperature results in better breakup of the thicker films and (Fig. 5.4) resulting in nanoparticles of about 150–200 nm width for the 8 nm film (Fig. 5.3) and irregular shapes of up to 500 nm for the 16 nm film. This demonstrates that if a low annealing temperature is required (to avoid diffusion of gold into the substrate for instance), then a thinner film is preferable. For the 16 nm film, it appeared that the annealing temperature was too low to result in complete break-up of the film into discrete nanoparticles of equal size and shape. Alternatively, a longer annealing time may have been successful.

At first glance, the morphologies of the nanoparticles formed from the 2 nm films do not appear to change with the annealing temperature. However, UV-Vis spectroscopy of the samples showed that there were differences in their optical characteristics.

Figure 5.5 shows the UV-Vis absorption spectrum for the nanoparticles formed from 2 nm films. The absorption was calculated from measurements of the transmittance and reflectance using Lambert's law (Chap. 4, Sect. 4.5.6). The transmittance is first corrected for the reflectance by dividing by the factor $(1 - R)$, where R is the measured reflectance. This should leave only transmission losses due to absorption. For the nanoparticle arrays, it does not make sense to report an absorption coefficient α (a measure of the absorption per unit length into the material) because there is only a single layer of nanoparticles, so the reported absorption is equivalent to αx, where x is, usually, the thickness of the sample for a crystal or thin film.

$$\alpha x = -\ln\left[\frac{T}{1 - R}\right] \tag{5.1}$$

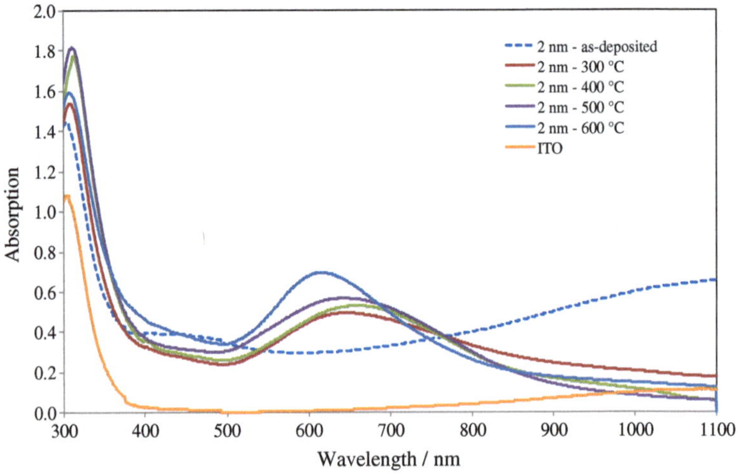

Fig. 5.5 Effect of annealing temperature on UV-Vis absorption of gold film and nanoparticles formed by annealing on ITO. Dashed line: as-deposited Au film

As evident from data in Fig. 5.5, the absorption of the ITO substrate above 400 nm was negligible. The as-deposited gold thin film resulted in increased absorption, with a small peak at 450 nm (2.75 eV), followed by increasing absorption into the near infrared, typical of a metal. The annealed films exhibited an absorption peak between 600 and 650 nm, typical of localized surface plasmon resonance (LSPR). The absorption peak was most pronounced for the film annealed at 600 °C, occurring at 615 nm, blue-shifted from 640 to 650 nm for the films annealed at lower temperatures. This is consistent with a decrease in the diameter of the nanoparticles [6]. The measured reflectance (Fig. 5.6) shows that there was a peak in the reflectance at approximately the same wavelengths as in the transmittance data. It is interesting to note that the ITO also seems to have a reflectance peak here. It is possible that this was due to thin film interference. Applying the above correction should remove the effect of reflection from the transmittance data, suggesting that there was both reflection and absorption due to the AuNPs (and by implication due to LSPR). However, the UV-Vis measurement was made only for specular reflections at twice the angle of incidence. It is possible that the absorption peak at 600–650 nm arose due to reflection by the AuNP at non-specular angles i.e. the AuNP does not behave like a planar interface between two materials of different refractive index. For example, an AuNP might absorb the photon and re-radiated it at an angle that does not coincide with a detector arranged to measure specular reflections. The use of a detector able to measure diffuse reflections in addition to specular reflections would confirm whether this was the case. At the time of the experiment, the available spectrometer was configured for specular reflectance only.

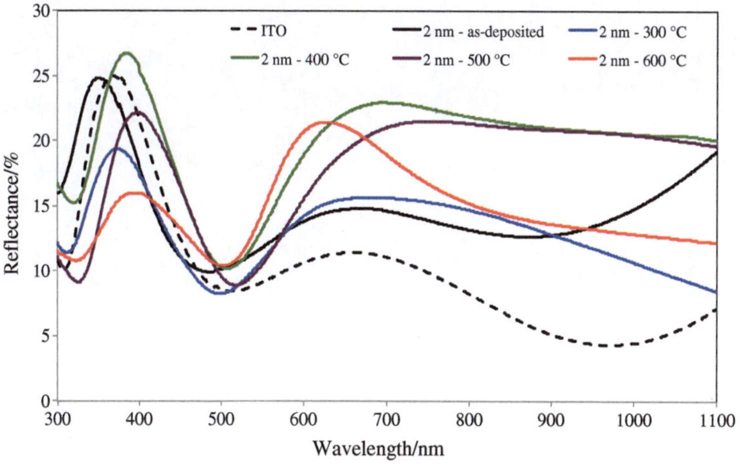

Fig. 5.6 Effect of annealing temperature on UV-Vis reflectance of gold film and nanoparticles formed by annealing on ITO substrate. Dashed line: absorption of ITO substrate

5.5.2 AuNPs on Rutile (110)—Annealing Gold Thin Films

On the basis that an array of nanoparticles of uniform size and shape should be preferable in determining the impact of size and shape, a 2 nm gold film was sputtered on both the un-doped and Nb-doped TiO_2 single crystals and subsequently annealed at 600 °C. The morphology of the resulting nanoparticle array is shown in Fig. 5.7 for the Nb-doped rutile (110) substrate. The nanoparticles had diameters of about 20 ± 10 nm with a surface coverage of about 15–20 %. The SEM images were analysed using the ImageJ software [7]. Extensive processing of contrast was required, which introduced error in the measurements, so the image processing process was repeated several times; the diameters reported are typical rather than exact. The nanoparticles on un-doped TiO_2 formed a similar size distribution of about 20 nm on average with larger particles approaching 50 nm in diameter. As well as ellipsoidal nanoparticles, irregular areas of gold film can also be seen.

The inset of Fig. 5.7 is a crop from Fig. 5.1; comparison shows that the AuNPs formed on the rutile (110) substrate were significantly smaller than those formed on ITO. Additionally, the AuNPs on rutile appeared to be somewhat rounder and have a larger spacing relative to the dimension of the nanoparticles. This suggested that

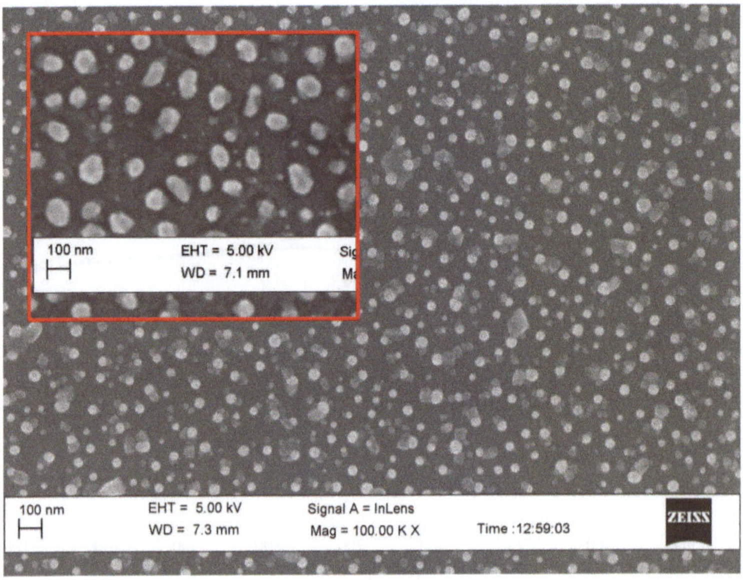

Fig. 5.7 SEM image of an array of gold nanoparticles on Nb-doped TiO_2 formed by annealing a 2 nm gold thin film at 600 °C for 1 h. The magnification is 100 k and the scale bar is 100 nm. The *inset* shows a film deposited with the same parameters on ITO, taken from Fig. 5.1. The scales are approximately the same, so comparison enables the effect of the substrate on the particle size to be inferred

ITO was a poor choice of substrate for the initial test. This could be explained by the fact that ITO is metallic in nature and TiO_2 is semiconducting, so that the contact between metal and substrate is of a different nature. The Au-TiO_2 interface should form a Schottky barrier, and as such interface is energetically unfavourable for wetting of the TiO_2 by Au. This would result in the nanoparticles minimizing their interfacial area, forming small nano-spheres (possibly faceted). On ITO, the contact appears favourable, resulting in wider, flatter nano-islands.

The optical characteristics of the AuNP coated TiO_2 are presented later, in Sect. 5.6 and the photo-electrochemical performance in Sect. 5.7.

5.5.3 *AuNPs on Rutile (110)—Micellar Nanolithography (MNL)*

To provide an alternative distribution of size and shape of nanoparticle, gold nanoparticle arrays were also deposited by MNL, carried out by Derfogail Delcassian of the Dunlop Group at Imperial College London. The methodology is outlined in Chap. 4 (Experimental Methods) and was based on a process optimized for glass rather than TiO_2 [8]. Arrays were deposited on both Nb-doped and un-doped TiO_2 rutile (110). The arrays are shown below in Fig. 5.8. The nanoparticles were about 10 nm in diameter, and formed a quasi-hexagonal pattern.

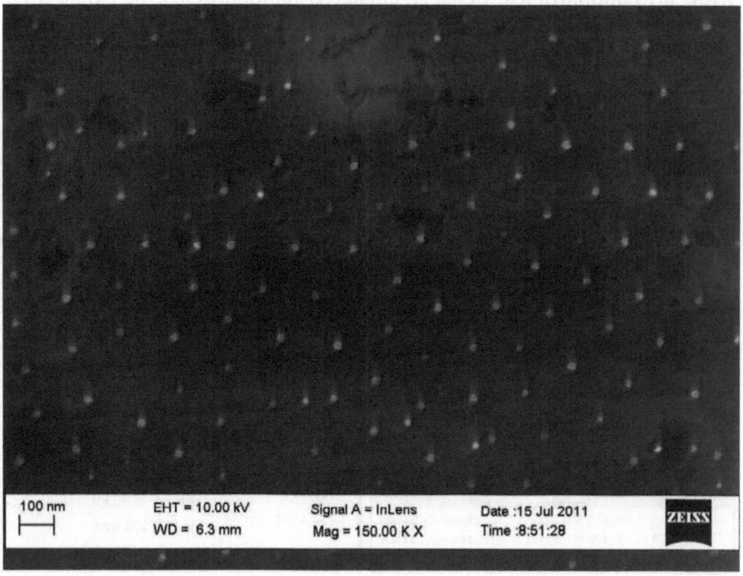

Fig. 5.8 SEM images of AuNP arrays formed by micellar nanolithography. The depositions on both un-doped and Nb-doped rutile are identical, with nanoparticle sizes ca. 10 nm and spacing 100–150 nm (image provided by D. Delcassian, Department of Materials, ICL)

The inter-particle spacing was on the order of 100 to 150 nm, so the surface coverage was very low. The depositions made on both the doped and un-doped substrates are essentially identical.

The optical characteristics are presented later, in Sect. 5.6 and the photo-electrochemical performance in Sect. 5.7.

5.5.4 Nano-sphere Lithography (NSL)

Nano-sphere lithography was also investigated briefly as a method to deposit gold nano-islands (AuNIs). This technique was being developed as part of a parallel work package under the same funding. The advantage is the ability to deposit a Ti interlayer between the TiO_2 and Au. This was of interest as Nishijima et al. [1] used an adhesive layer of Ti in their study. However, as Figs. 5.9 and 5.10 show, the process used was not well optimized. Between the AuNIs (Fig. 5.9b), there appeared to be either damage to the underlying substrate or un-milled Au or Ti metal film. As the substrate also contained Ti, it is difficult to distinguish between the two possibilities by standard analytical methods such as EDS/EDX. The resolution of the technique was also lower than required, i.e. a measurement will include both particles and the space between. However, if it is only a thin Ti layer that remains, it will oxidize on contact with oxygen/air. It also seemed that the initial deposition of polystyrene nano-spheres gave an uneven coverage, which resulted in areas with: no nanoparticles, densely clustered nanoparticles, or isolated particles,

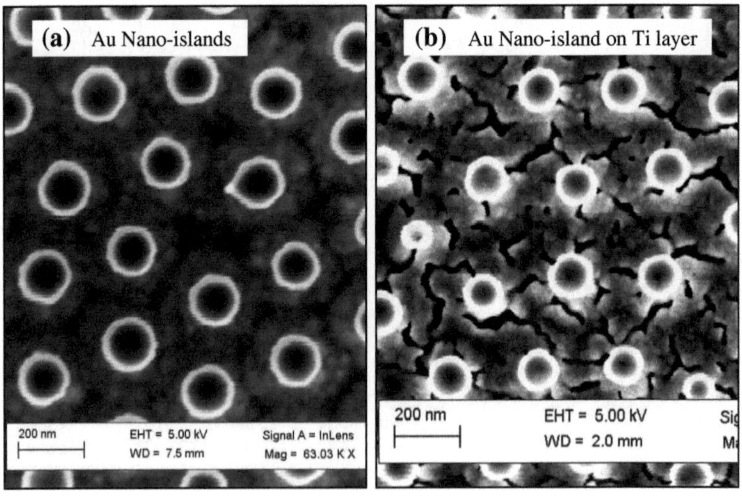

Fig. 5.9 Au Nano-islands formed by NSL. The *left hand panel* shows AuNIs formed directly on the TiO_2 substrate. The *right hand panel* shows AuNIs formed with a layer of Ti between the Au and TiO_2. This appears to have been removed only partially by the process

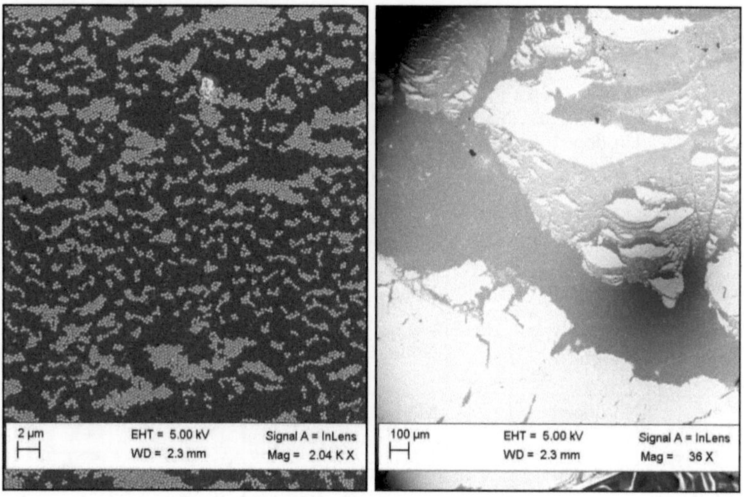

Fig. 5.10 SEM image showing the inhomogeneous coverage of AuNIs deposited using NSL. The *left panel* has a magnification of 2000 and a scale of 2 μm. The *right panel* has a magnification of 36 and a scale of 100 μm

as evident in Fig. 5.10. The process clearly required further optimization for a TiO_2 substrate. Due to the variation in coverage across the sample, UV-Vis spectra contained contributions from several types of array, but it was not possible to distinguish between those contributions. This technique was not developed further in this work.

The method used to produce the nano-islands is described in Chap. 4 (Experimental Methods).

The optical characteristics are presented later, in Sect. 5.6 and the photoelectrochemical performance in 5.7.

5.5.5 *Sample Identification—Batch Numbers*

Given the various combinations of doped, un-doped and reduced substrates, and the different types of nanoparticle deposition, it was useful to define short names that can be used to identify each sample for data handling. The following sample names given in Table 5.1 are used in the presentation of results and accompanying discussion. The table identifies the type of substrate and the nanoparticle deposition technique. For example, "$NbTiO_2$" represents Nb-doped TiO_2. The sample number is also given. Most sample numbers are "1" denoting that they are the only combination of substrate doping and nanoparticle deposition studied. Four samples were investigated, created by annealing gold thin films on Nb-doped TiO_2. The

Table 5.1 Sample identification table—gold nanoparticles on rutile (110) single crystals

Identification	Substrate Doping	Nanoparticle deposition technique	Sample # (of sample type)	Batch # (of crystal type)
unTiO_2	Un-doped	None	1	1
unTiO_2 AuNP	Un-doped	Annealing	1	1
unTiO_2 AuMNL	Un-doped	MNL	1	1
NbTiO_2 S1 B1	Nb-doped	None	1	1
NbTiO_2 S2 B2	Nb-doped	None	2	2
NbTiO_2 AuMNL	Nb-doped	MNL	1	1
NbTiO_2 AuNP S1 B1	Nb-doped	Annealing	1	1
NbTiO_2 AuNP S2 B1	Nb-doped	Annealing	2	1
NbTiO_2 AuNP S3 B2	Nb-doped	Annealing	3	2
NbTiO_2 AuNP S4 B2	Nb-doped	Annealing	4	2
redTiO_2	Un-doped/reduced	None	1	1
redTiO_2 AuNP	Un-doped/reduced	Annealing	1	1
NbTiO_2 Au NSL	Nb-doped	NSL	1	2
NbTiO_2 AuTi NSL	Nb-doped	NSL	1	2

batch number refers to the batch of crystals received from the supplier, rather than the batch of samples made. The sample identification is recorded here to allow interrogation of the body of data collected, if necessary (Table 5.2).

5.6 Optical Measurements

Measurements of the transmittance and reflectance of the nanoparticle films were done using the UV-Vis equipment outlined in Chap. 4 (Experimental methods). Transmittance measurements were made with a normal incidence beam, while reflectance measurements were made with a 10° angle of incidence, the minimum possible with the given configuration. The results are presented in Figs. 5.11, 5.12, 5.13 and 5.14.

Table 5.2 Summary of plasmonic photocurrent measurement on samples

Identification	Notes
unTiO_2	Not measurable
unTiO_2 AuNP	Not measurable
unTiO_2 MNL	Not measurable
NbTiO_2 S1 B1	No visible light response
NbTiO_2 S2 B2	No visible light response
NbTiO_2 AuMNL	No visible light response
NbTiO_2 AuNP S1 B1	Plasmonic photocurrent
NbTiO_2 AuNP S2 B1	Plasmonic photocurrent
NbTiO_2 AuNP S3 B2	No visible light response
NbTiO_2 AuNP S4 B2	No visible light response
redTiO_2	No visible light response
redTiO_2 AuNP	No visible light response
NbTiO_2 Au NSL	No visible light response
NbTiO_2 AuTi NSL	No visible light response

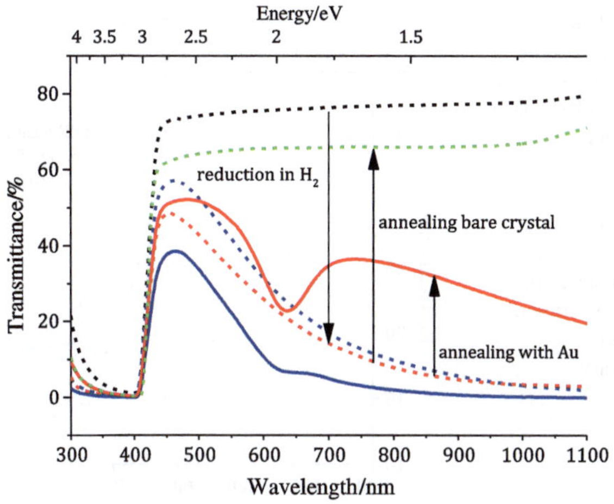

Fig. 5.11 Transmittance spectra of rutile (110) substrates with and without gold nanoparticle coatings. The *black-dashed line* shows the un-doped crystal. The *red-dashed line* shows a reduced un-doped crystal. The *green-dashed line* shows the effect of re-annealing a reduced crystal in air. The *blue-dashed line* shows an Nb-doped crystal. The *red* and *blue solid lines* show the reduced and doped crystals coated in AuNPs, respectively

5.6.1 Transmittance and Reflectance

Figure 5.11 shows that the level of reduction and doping used had negligible effect on the typical band gap of 3.05 eV (405 nm) for rutile TiO_2. The detail is discussed further in Chap. 6. The feature of interest here is the localized surface plasmon resonance (LSPR) peak between 500 and 800 nm.

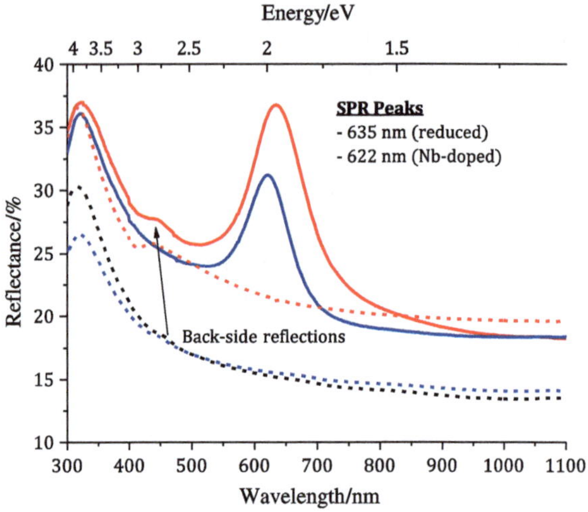

Fig. 5.12 Reflectance spectra of rutile (110) substrates with and without gold nanoparticle coatings. Line colour and style as for Fig. 5.11

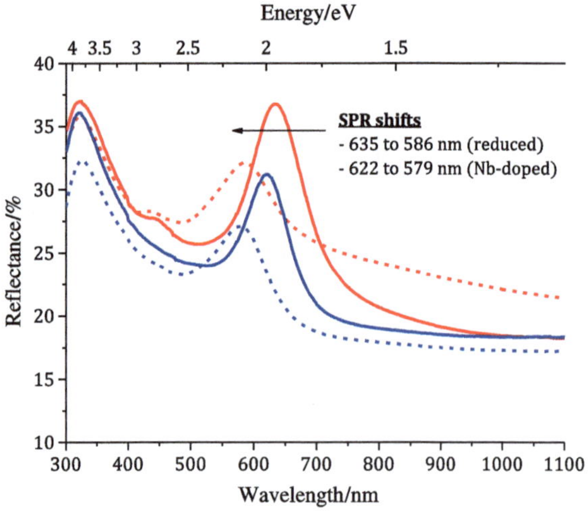

Fig. 5.13 Reflectance spectra of Nb-doped (*blue lines*) and reduced un-doped (*red lines*) rutile (110) coated in gold nanoparticles. The *solid lines* show spectra measured following nanoparticle deposition. The *dashed lines* show spectra measured after a series of electrochemical experiments (voltammetry and chronoamperometry)

The un-doped, un-reduced rutile exhibited high transmittance of 75–80 % throughout the visible (400–700 nm) and into the near infrared (NIR); correcting for reflectance gave a value of ca. 90 %. Reduction of the un-doped rutile resulted in decreased transmittance in the Vis-NIR range. The peak transmittance of 48.8 % at 455 nm then decreased with increasing wavelength, approaching zero towards 1000 nm. The Nb-doped rutile demonstrated the same behaviour, with a peak of 57.2 % at 462 nm. Both the reduced and Nb-doped crystals exhibited a blue colour when viewed by eye in both reflection and transmission. Un-doped material is typically colourless when viewed in transmission.

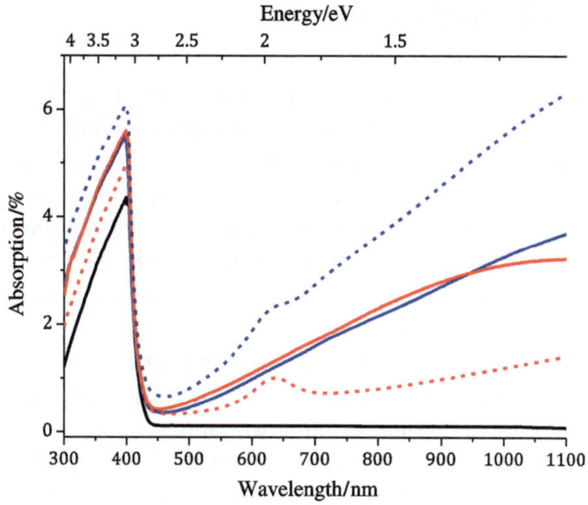

Fig. 5.14 Absorption spectra of the Nb-doped (*blue*) and reduced (*red*) rutile (110). The *dashed lines* show the nanoparticle coated samples, while the *solid lines* show the bare substrates. The absorption of the as-bought, un-doped, rutile (110) crystal is shown by a *solid black line*

The transmittance of the reduced rutile covered with gold nanoparticles exhibited a local minimum at 637 nm, attributed to reflection (mainly) and absorption by the localized surface plasmon resonance of the gold nanoparticles. The Nb-doped sample exhibited a similar feature between 600 and 650 nm, although the peak was not as well defined.

An interesting aspect is that transmittance of the reduced sample increased away from the plasmon resonance when compared to the uncoated reduced substrate. As the nanoparticles were formed by annealing in air at 600 °C for 1 h, it is probable that the rutile substrate partially re-oxidized. Annealing a freshly reduced rutile crystal, at the same conditions at which the nanoparticles were formed, recovered the general transmittance spectra (green dashed line of Fig. 5.11) of the un-doped un-reduced material (black dashed line of Fig. 5.11). It was noted that the sample measured still had a pale blue tinge, which would explain the lower transmittance of 65–70 % compared to the 75–80 % of the original un-doped sample. The transmittance beyond 400 nm can be taken as an indication of how reduced the sample was. The gold covered sample did not re-oxidize to the same extent as the bare crystal. It is probable that re-oxidation of the substrate was inhibited until the gold film became discontinuous, exposing the underlying rutile. It is unknown whether the rutile directly below the gold nanoparticles remained reduced. The Nb-doped rutile did not exhibit this effect on annealing. This was thought to be a result of not having been reduced, i.e. the Nb defects were responsible for the colouring and were not affected by the heat treatment in air. However, the literature on doping TiO_2 with Nb (see Appendix A) suggests that a reducing environment is necessary to form conductive TiO_2. So while the annealing of the Nb-doped crystals may not have significantly altered the transmittance, it may have caused the conductivity to decrease.

5.6.2 Reflectance

The reflectance spectra of the bare crystals (Fig. 5.12, dashed lines) showed that the un-doped rutile had a peak reflectance of 30.3 % at 317 nm, and the Nb-doped rutile 26.5 % at 320 nm, whereas the reduced rutile had a higher reflectance of 36.7 % at 315 nm. The higher reflectance may have been due to the presence of a higher Ti to O ratio at the reduced surface. During reduction, it is likely that the surface loses oxygen faster than the bulk, resulting in an abundance of Ti. This would result in a more metallic surface (closer to Ti_2O_3, TiO, or even Ti) that would be more reflective.

The spectra of the AuNP coated rutile crystals clearly demonstrated a strong reflection due to the LSPR of the AuNP. The peak reflectance of the reduced sample occurred at 635 nm, while it was at 622 nm on the Nb-doped sample. This value is red-shifted from 540 nm, the standard SPR peak of a spherical nanoparticle in vacuum as calculated using Mie Theory [6]. The shift will have been due to a combination of the higher refractive index of the TiO_2 substrate, and the non-ideal shape of the nanoparticles. It is notable that the LSPR peak was more pronounced for the AuNPs on TiO_2 than for those on ITO. This could be due to the smaller size or the more symmetrical nature of the AuNPs on TiO_2 as identified in Sect. 5.5.2.

Following photo-electrochemical measurements on the samples, the UV-Vis spectral measurement was repeated (Fig. 5.13), and the position of the reflectance peak maximum was found to have blue-shifted to 586 nm from 635 nm for the reduced sample, and to 579 nm from 622 nm for the Nb-doped sample. As well as a shift in the peak position, the magnitude of the peak reflectance had decreased. Fewer nanoparticles on the surface might explain a decreased reflectance alone, but the blue shift suggested a change in the nanoparticle size and/or shape. In particular, the reflectance increased at wavelengths shorter than that of the peak. One possibility is that local heating (during illumination) has caused further annealing of the nanoparticles. However, this is improbable in aqueous solution with a large heat capacity under a relatively low irradiance. Another possibility is that during the voltammetric scans, the electrode passed through a range of potential at which gold is predicted to form a soluble species in aqueous solution at high pH. This may have resulted in the gold dissolving into solution and being re-deposited, modifying the shape and size distribution of the nanoparticles.

The transmittance and reflectance data of the gold nanoparticles deposited by micellar nanolithography is not shown, as the spectra did not show any features other than the spectra of the underlying rutile TiO_2 substrate. This was probably due a combination of the small size and the low surface coverage, giving an insufficient signal. The small diameter (ca. <10 nm) resulted in poor coupling between the nanoparticle and incident light [9].

5.6.3 Absorption

To determine the absorption due to both the rutile crystal and the LSPR of the AuNPs, a nominal absorption was calculated from Lambert's law that includes the losses due to reflection. This is described in Chap. 4 (Experimental Methods) Sect. 4.5.6 and earlier in this chapter, Sect. 5.5.1. The absorption shown in Fig. 5.14 is given by $-\ln\left(\frac{T}{(1-R)}\right)$.

The reason for presenting the absorption in this unconventional manner is due to the combination of absorption from a relatively thick single crystal with a single layer of nanoparticles. If the absorption coefficient α were calculated for a 0.5 mm crystal, then the absorption due to the single layer of nanoparticles would be misrepresented. As the focus here is on the plasmonic behaviour, the absorption is left as a total value. The absorption coefficient is presented elsewhere for single crystal rutile (Chap. 6) and thin films of anatase and rutile (Chaps. 7 and 8). Additionally, the thickness term would simply scale the entire spectrum without altering the shape.

It is noted that most of the absorption around the SPR wavelength was due to the rutile substrate, and that the SPR accounted only for a slight increase in absorption. Overall, the AuNP arrays resulted in more reflective samples by 5–10 % over the bare substrate, but only a moderate increase in absorption, so it might be expected that they would not perform efficiently for solar harvesting.

It should be noted that the transmittance and reflectance were measured using a specular arrangement, so would not have detected any diffuse reflectance or scattering at other angles than incident. Hence, the reflectance may have been underestimated, and the absorption overestimated, which makes the performance of the system even poorer. On this basis, it would be interesting to repeat the measurements with larger nanoparticles that appear to have a weaker LSPR, such as those grown from thicker gold films.

It is also worth highlighting the differences in the nanoparticle size and distribution between those formed on TiO_2 and those on ITO. For the nanoparticles formed at 600 °C, from a 2 nm film, the absorption on an ITO substrate (detailed earlier in Sect. 5.5.1) appeared to be greater than for those on TiO_2. However, this was only because the absorption of ITO was small compared to TiO_2: the absorption peak appeared large relative to the substrate. In both cases, the absorption peak of the LSPR was on the order of 0.2 to 0.4 above the substrate absorption.

5.7 Photo-Electrochemical Measurements

This section describes a series of electrochemical measurements done in the dark and under illumination to establish whether the coating of the TiO_2 electrodes in various types of gold nanoparticles can result in sensitization of the photo-anode to visible light and the generation of photocurrent due to water splitting.

In brief, an effect was indeed observed for two samples of Nb-doped TiO_2 coated in AuNPs that were formed by annealing gold thin films. However, the effect was negligible. Much of the work detailed below was done to confirm that the effect seen was due to the LSPR of the AuNPs, and also to confirm that the effect was indeed completely absent from samples that did not appear to exhibit it (some Nb-doped samples and the reduced rutile). In essence, this section documents the efforts made to measure an effect that was either vanishingly small or absent, rather than presenting results of relative performance. The possible reasons for the poor performance, lack of effect in some cases, and the proposed next steps are discussed in Sect. 5.8. Overall, the performance of the system as fabricated is insufficient to really probe the mechanism of plasmonic enhancement of water splitting.

The difficulties of measuring the performance using chronoamperometry are discussed below in Sect. 5.7.1. This expands on the basic technique outlined in Chap. 4 (Experimental Methods).

The first experimental results presented below (Sect. 5.7.2) are chronoamperometric measurements, made at a single potential, comparing the performance of the gold nanoparticles deposited onto Nb-TiO_2 by both annealing thin films (AuNPs) and by MNL (AuMNLs). Initial voltammetric measurements on the un-doped TiO_2 (with and without Au) indicated that the samples were insufficiently conductive, so only the Nb-doped samples were measured with both types of nanoparticle array: AuNPs and AuMNLs.

Chronoamperometric measurements were also done for Nb-TiO_2 with AuNPs at a range of potentials (Sect. 5.7.3). Following this, the un-doped rutile (110) substrates were reduced and further chronoamperometry was done for both the Nb-TiO_2 and reduced-TiO_2 crystals, with and without AuNPs (Sect. 5.7.4).

Next, a lock-in amplifier was used to remove the dark current signal from the measured current under illumination in order to obtain the potential dependent behaviour of the plasmonic photocurrent of the samples (Sect. 5.7.5). The dependence of the photocurrent on potential is discussed in terms of possible mechanisms for the plasmonic photocurrent, and several new possibilities are highlighted.

Finally, the voltammetry of the bare Nb-doped and reduced substrates is introduced briefly and leads into the next chapter on the detailed electrochemistry of rutile (110).

All measurements were made in 1 M NaOH at approximately pH 13.6.

5.7.1 Measuring the Photocurrent in Chronoamperometry

The total current is the sum of the dark current and the photocurrent. While the photocurrent appears to reach a stable value over time, the dark current exhibited a continuous decay that complicated the calculation of the photocurrent. A typical set of raw current-time data is shown in Fig. 5.15. Thus, taking a time-averaged value for any of the dark, photo, or total current at each wavelength would be inaccurate.

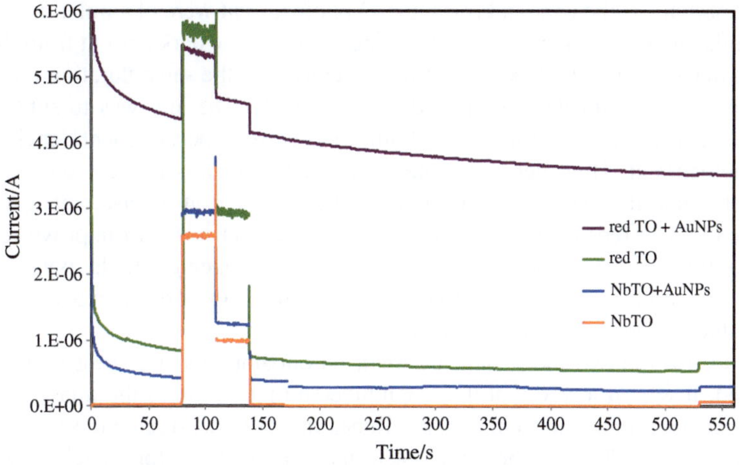

Fig. 5.15 Time dependence of raw current data at +1.0 V (HgO|Hg) showing the decay of the underlying dark current with time. Data is shown for Nb-TiO_2 and reduced TiO_2 with and without AuNPs. The discontinuities in the current are due to changing the wavelength of illumination

The reason for the changing dark current was the establishment of concentration gradients in the electrolyte due to changes in the working electrode potential. For example, in the measurements presented below, a potential step was applied in the measurements made. The open circuit potential (OCP) of most of the samples was found to be in the potential range −0.1 to +0.1 V (HgO|Hg), so making the measurement at +1.0 V (HgO|Hg) resulted in a step. In hindsight, these measurements may have been better made at or near the OCP, as it is possible (but unlikely) that the conduction band of highly doped (degenerate) TiO_2 could oxidize water at potentials of ca. +0.35 V (HgO|Hg).

To calculate the incident photon to current efficiency (IPCE), it was necessary to extract just the value of the photocurrent. Several problems were identified. For certain wavelengths of illumination, i.e. close to the band gap, the photocurrent is no longer orders of magnitude larger than the dark current, so it was not possible to simply neglect the dark current; it must be subtracted from the total in a rigorous manner. This was also important for measuring plasmonic photocurrents, for which the photo-response was of a similar order of magnitude as the dark current. This was particularly difficult considering the time dependent dark currents detected with the measured samples. The methods considered in correcting the data will be discussed below.

Despite allowing the initial transients to settle at the beginning of each measurement, there was still a considerable transient decay in the current measurements, superimposed on top of the steps in the measured current, which was the sum of the dark and photo- currents. This was not a serious issue for the bare TiO_2 electrodes illuminated by UV light, as the photocurrent was orders of magnitude greater than the dark current. However, for the gold nanoparticle coated TiO_2, it was important, as the dark and (plasmonic) photocurrent were of similar magnitude. The initial approach taken to correct for and eliminate the dark current from the data was to make a separate dark current measurement for the same duration and at the same potential as for the illuminated measurements. The aim was to subtract the dark current measurement from the illuminated current measurement, leaving only the photocurrent. However, calculations derived from this process proved to be unreliable due to variation in the transient behaviour from measurement to measurement. Even repeatedly re-starting the measurement to try to improve repeatability of transients was not successful. Bubbling of nitrogen gas to de-oxygenate the electrolyte solution during experiments also seemed to make the experiments less repeatable and was therefore stopped.

To try to improve the match, the dark chronoamperometry measurement was repeated. It was observed that the magnitude and the decay of the transient varied from one measurement to the next. Repeating the measurement several times decreased the difference between each measurement. Unfortunately, the time dependence (magnitude and slope) of the independently measured dark and illuminated current measurements were not the same, so the subtraction of the transient current measured in the dark from the transient current measured under illumination could not adequately correct the data.

It is probable that with each measurement, the concentration profiles of the various possible species (including H_2O, O_2, H_2, OH^-, and HO_2^-) at both the solid|liquid interface and in aqueous solution were changing slightly, resulting in different starting conditions. It is also possible that the condition of the electrode surface was changing, although the measurements were made in a region of predicted stability of TiO_2.

Following the initial attempts described above, an alternative method was used to estimate the photocurrent from the chronoamperometric measurements. The steps/discontinuities in the current at each change of wavelength were assumed to be equal to the change in the photocurrent, and it was assumed that the dark current decayed continuously (and was unaffected by the changing illumination). The step was calculated from the difference between average currents before and after the change in illumination. At the starting wavelength of 350 nm, there was no photo-response of the electrode, due to the particular UV filter employed (Fig. 5.16) blocking all photons with wavelengths from ca. 340 to 360 nm (Fig. 5.17). Taking the initial photocurrent at 350 nm to be zero, the photocurrent at each wavelength was then calculated by cumulatively adding the steps in the measured current, which were both positive and negative. As shown later in this chapter, in Fig. 5.28, these did result in a negative photocurrent at longer wavelengths. It is unlikely that a negative photocurrent was really present and that this is due to errors in the process of subtracting the transient dark current background. If the negative photocurrent at longer wavelengths were real, this would suggest that the nature of the plasmonic mechanism changes with wavelength. For example, the mechanism might have switched from donating an electron from the AuNP to accepting an electron, with the redox reaction switching from an oxidation to a reduction. This demonstrates the limited accuracy of this approach, although it was the best

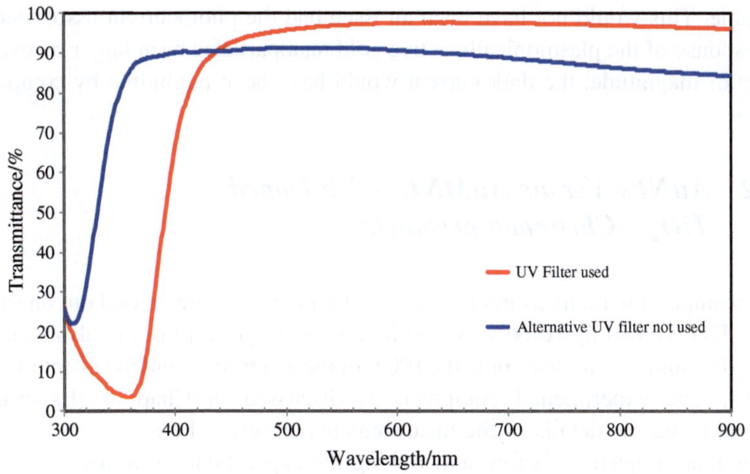

Fig. 5.16 Transmittance spectra of UV filters used for photocurrent and IPCE measurements. The *red line* shows the transmittance spectrum of the UV filter used in all experiments. The *blue line* shows the transmittance of another UV filter considered for use: it does not block UV photons as well. When the poorer UV filter was used, the TiO_2 could still be excited by harmonics produced in the monochromator. The poorer performing filter also had little impact on photocurrent under 375 nm illumination

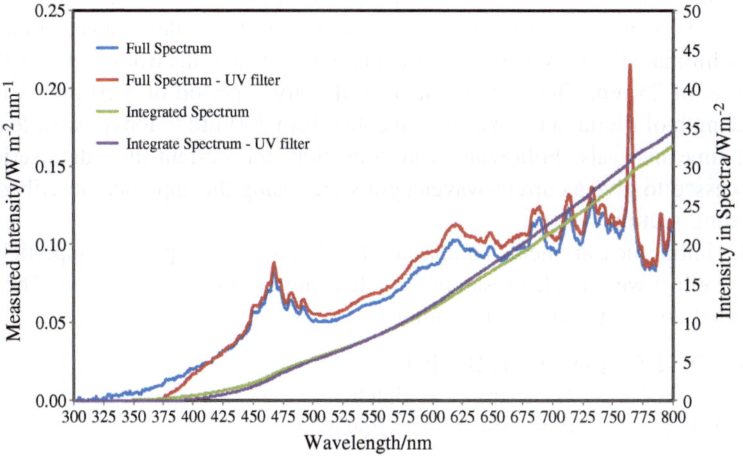

Fig. 5.17 Spectrum of lamp used for photocurrent measurements. The spectrum of the lamp is shown in *blue*. The effect of the UV filter on the spectrum is shown in *red*. The UV filter cuts out all photons below 375 nm. The cumulative intensity of light is shown. The intensity of the white light falling on the sample is ca. 35 W m^{-2} and is found by integrating the spectra

available. This would not have been an issue had the photocurrent associated with the response of the plasmonically active gold nanoparticles been larger by even two orders of magnitude; the dark current would have been negligible by comparison.

5.7.2 AuNPs Versus AuMNLs—Nb-Doped TiO_2—Chronoamperometry

Chronoamperometric measurements under illumination were carried out on samples of Nb-TiO_2 coated by AuNPs formed by annealing gold films and also formed by MNL. The aim was to determine the IPCE of the electrodes, and the effect of adding AuNPs. The experimental equipment is discussed in Chap. 4 (Experimental Methods), and the details of the measurements are given here.

The main objectives before undertaking the experiments were to:

- Identify any photocurrent attributable to the surface plasmon resonance observed in the optical measurements.
- Compare the effect of the nanoparticle fabrication techniques on the photocurrent generated on Nb-doped rutile (110).
- Compare the results with doped and un-doped substrates with AuNPs formed by annealing.

The initial measurements were made with an applied working electrode potential of +1.0 V versus an HgO|Hg reference electrode (unless stated otherwise). Monochromated light was used over a range of wavelengths from 350 to 900 nm at intervals of 25 nm. The current was recorded for a period of 5.10 min, while the wavelength of illumination was incremented from 350 nm to longer wavelengths at equal time intervals. Following data collection, the current-time data series was reprocessed to give a current-wavelength series using the approach described in the preceding section.

The initial measurements comparing the effects of the type of nanoparticle array and the effect were made on samples made using crystals from Batch 1. The results presented are for the following samples

- NbTO S1 B1 (Sample 1, Batch 1).
- NbTO AuNP S1 B1 (Sample 1, Batch 1).
- NbTO AuMNL S1 B1 (Sample 1, Batch 1).

Figure 5.18 and show the total current measured for the first samples (S1) from the first batch (B1) of Nb-doped TiO_2 electrodes (0.5–1.0 cm^2 active area), with and without the addition of a UV-filter. Results include the bare substrate, and those covered in gold nanoparticles from annealing films (AuNPs) and micellar nanolithography (AuMNLs).

Figure 5.19 shows an enlarged view of the data of Fig. 5.18 in the range 400–900 nm. This demonstrates the importance of using an appropriate UV filter (or cut-off filter) of known transmittance. Initially, the measurements were made

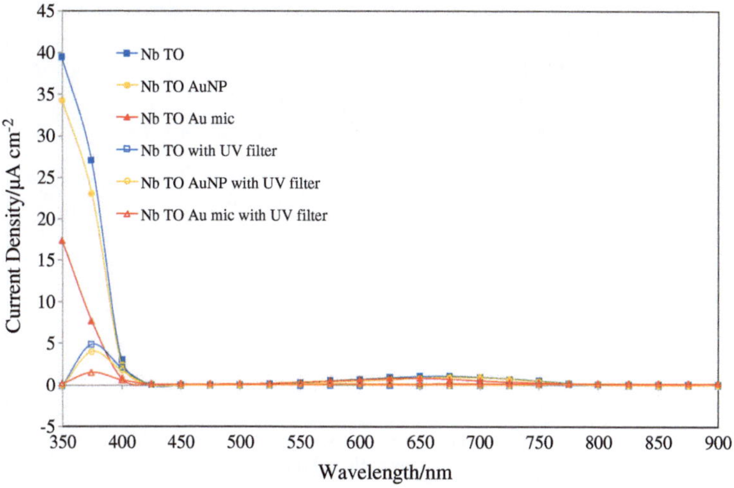

Fig. 5.18 Measured total current for samples; NbTO S1 B1 (*blue line, square icon*); NbTO AuNP S1 B1 (*yellow line, circle icon*); NbTO AuMNL S1 B1 (*red line, triangle icon*). *Open* and *solid symbols* correspond to measurements with and without the presence of a UV filter after the monochromator

without a UV filter, resulting in the observation of a current between 500 and 800 nm that could potentially be misidentified as deriving from a plasmonic excitation. This current was present for the bare Nb-doped rutile (110) crystal as well as the nanoparticle-covered samples. Figure 5.19 demonstrates that the addition of a UV-filter removed a large component of the current between 500 and 800 nm, so this current was attributed to second harmonic generation of UV photons in the monochromator that directly excite electron-hole pairs in the TiO_2.

A measurement of the intensity of light output by the monochromator is shown in Fig. 5.20. The measurement was made using a Stellarnet integrating cube to measure the intensity of the monochromated light. It is clear that even though the light is "monochromated", that second order harmonics are generated. For example, the light monochromated to 700 nm shows a small intensity at 350 nm. The effect of placing the UV filter between the monochromator and the detector is to remove harmonics at 375 nm and below, as shown in Fig. 5.21.

The IPCE (Fig. 5.22) was calculated from the output intensity of the lamp-monochromator as measured with a calibrated silicon photodiode. Initially, the IPCE calculation was based on the total current, assuming the dark current to be negligible. This was reasonable for determining the response of the photoanodes to above band gap photons (300–400 nm). However, as discussed in the preceding sections, this assumption was not strictly appropriate for the calculations based on the plasmonic photocurrent.

The bare Nb-doped rutile (sample $NbTiO_2$ S1 B1) had an IPCE of 33 and 16 % at 350 and 375 nm, respectively, decreasing to 1.5 % at 400 nm. At 425 nm and above, the dark current was of the same order as the photocurrent, i.e. there is no

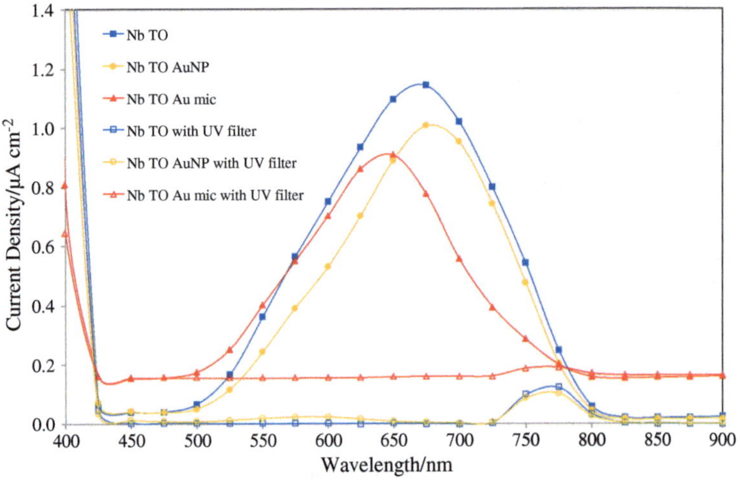

Fig. 5.19 Effect of UV filter on measured current between 400 and 900 nm. The current is shown for AuNPs and AuMNLs on Nb-TiO_2 at a potential of +1.0 V (HgO|Hg). Indicates that majority of current in visible region is due to UV harmonics generated in the monochromator. This is only fully removed by the UV filter for wavelengths of 725 nm or shorter. The *solid symbols* are for measurements made without the UV filter

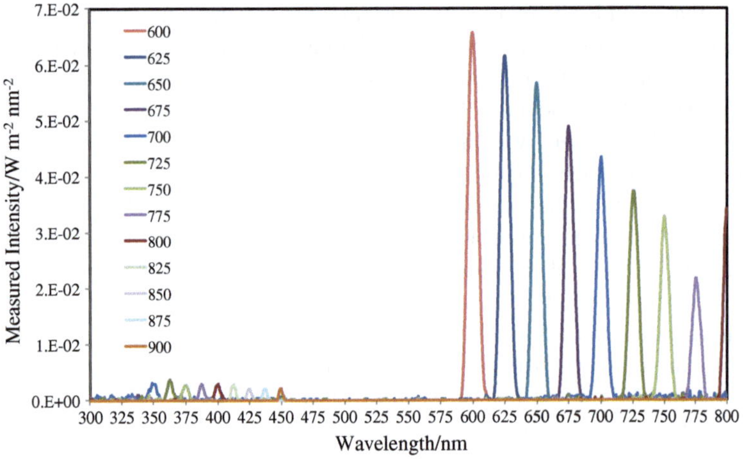

Fig. 5.20 Spectral intensity of monochromated beam incident on TiO_2 samples. The monochromator generates second harmonics as demonstrated by the presence of photons at half the wavelength of the main beam

photo-response. This placed the band gap of rutile (110) between 400 and 425 nm (3.1–2.9 eV), as expected.

The UV-filter reduces the performance of all samples equally as confirmed by multiplying the un-filtered photocurrent curve by the transmission curve of the filter

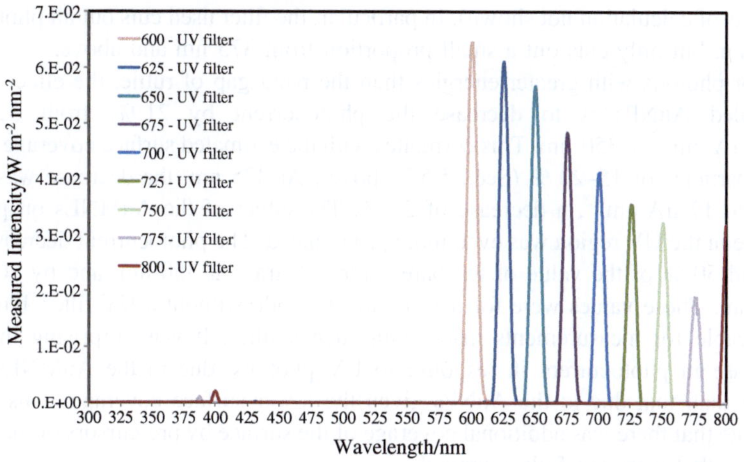

Fig. 5.21 Effect of UV filter on the spectral intensity of the monochromated beam incident on TiO_2 samples. The UV filter removes all UV harmonics of beams of 725 nm and below. Any excitation of TiO_2 by light of wavelength 425–725 nm cannot be attributed to harmonics

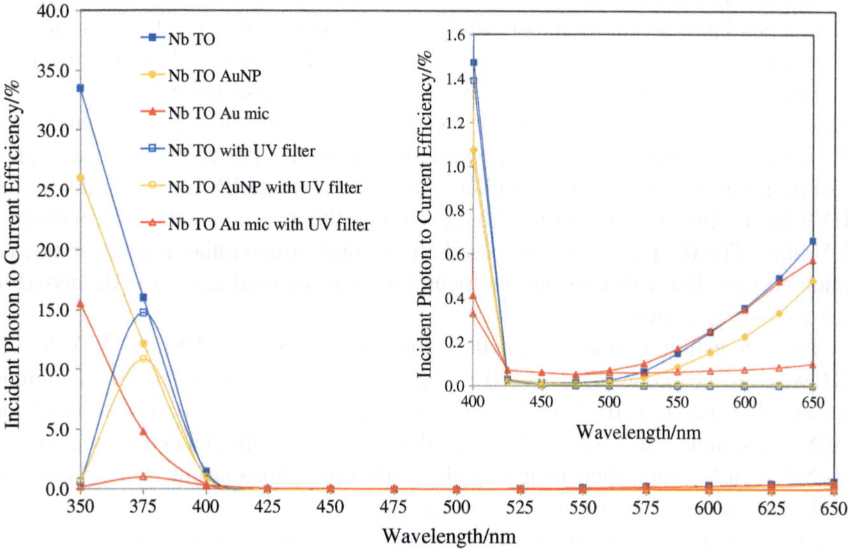

Fig. 5.22 Calculated IPCE for samples; NbTO S1 B1 (*blue line, square icon*); NbTO AuNP S1 B1 (*yellow line, circle icon*); NbTO AuMNL S1 B1 (*red line, triangle icon*). *Open* and *solid symbols* correspond to measurements with and without the presence of a UV filter after the monochromator. The *inset* shows the IPCE between 400 and 650 nm on a more relevant scale

(results of calculation not shown). In particular, the filter used cuts out all photons at 350 nm, but only cuts out a small proportion from 375 nm and above.

For photons with greater energies than the band gap of rutile, the effect of the annealed AuNPs is to decrease the photocurrent by 21 %, from 32.3 to 25.5 μA cm^{-2} at 350 nm. This correlates with the estimated surface coverage of the nanoparticles of 15–20 % (Sect. 5.5.2 above). At 375 nm, the decrease was from 22.5 to 17 μA cm^{-2}, a decrease of 25 %. The effect of the AuMNLs on performance in the UV region was even more pronounced. The photocurrent decreased by around 50 % of the value of the bare rutile substrate at 350 nm, and by 70 % at 375 nm. These values were for measurements made without a UV-filter, but were applicable for measurements made with such a filter. It was surprising that the decrease in photocurrent in response to UV photons, due to the AuMNLs, was larger than that due to the AuNPs given the lower surface coverage. This might indicate that there was additional coverage of the surface by pre-cursors of the MNL process that were not fully removed.

The photocurrent response between 500 and 800 nm (Fig. 5.19) of the Nb-doped TiO_2 electrode measured without a UV filter was also decreased by the presence of the AuNPs. This further demonstrated that this current was due to harmonic generation of UV rather than a plasmonic response.

Interestingly, in the range 500–700 nm of Fig. 5.19, the response of the AuNP coated Nb-TiO_2 sample (number $NbTiO_2$ AuNP S1 B1) was greater than the bare substrate for measurements made with a UV filter. This is believed to be photocurrent generated by the LSPR of the AuNPs and is presented more clearly in the following section. The magnitude of the photocurrent peak was approximately 20–30 nA cm^{-2} at 575–600 nm. The AuMNL sample did not demonstrate any photocurrent under visible light. The photocurrent density and IPCE measured with the UV filter is shown again in Fig. 5.23, excluding the measurements made without a UV filter. The IPCE is calculated based on the total current rather than the extracted photocurrent. Extraction of the photocurrent from the total current is discussed in the following section.

It is interesting to note that the dark current of the sample $NbTiO_2$ AuMNL S1 B1 was on the order of 1.4×10^{-7} A cm^{-2}, which is two orders of magnitude greater than the 1×10^{-9} A cm^{-2} for the bare substrate. The dark current of the AuMNL sample was also much larger than that of the illuminated current of the AuNP sample even when including the plasmonic photocurrent. The increase in dark current and decrease in photocurrent of the AuMNL sample may have been due to the plasma treatment used to reduce the $HAuCl_4$ salt to metallic Au. This may have reduced the surface of the TiO_2, enhancing conductivity and the dark current.

In short, the observed plasmonic performance of the AuNP coated electrode was vanishingly small (Fig. 5.24).

To demonstrate more clearly that this was indeed a photocurrent and can be attributed to the LSPR of the gold nanoparticle, it is instructive to review the raw

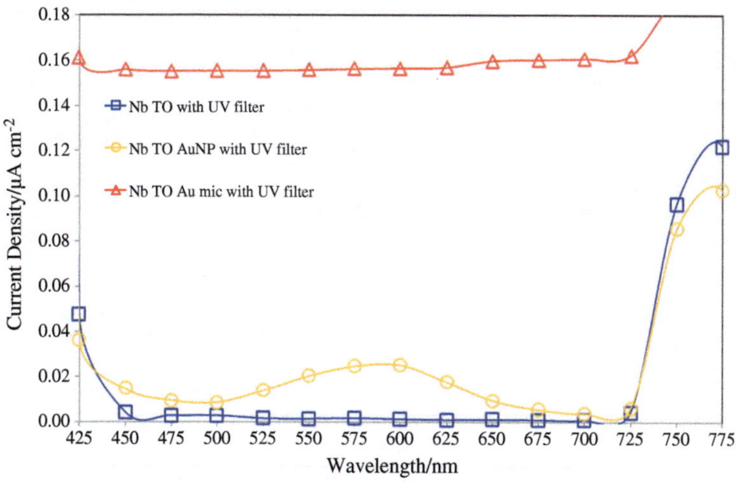

Fig. 5.23 Total current density of samples measured with a UV filter in place. Samples $NbTiO_2$ S1 B1 (*Blue squares*), $NbTiO_2$ AuNP S1 B1 (*yellow circles*), and $NbTiO_2$ AuMNL S1 B1 (*red triangles*). It shows the photocurrent attributed to the plasmonic resonance of the AuNPs (*yellow circles, inset*) compared to the bare TiO_2

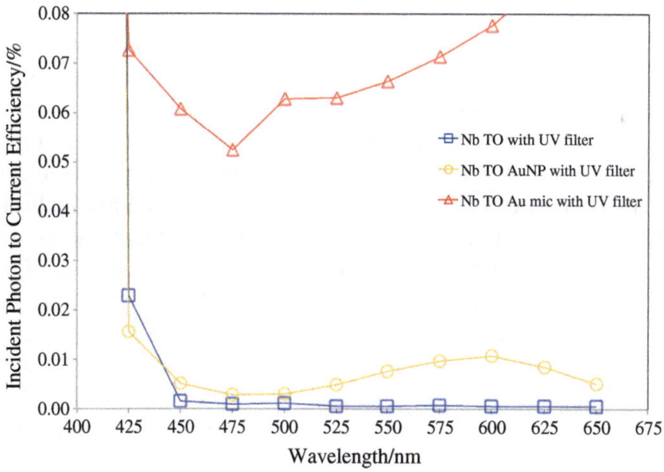

Fig. 5.24 IPCE of samples measured with a UV filter in place. Samples $NbTiO_2$ S1 B1 (*Blue squares*), $NbTiO_2$ AuNP S1 B1 (*yellow circles*), and $NbTiO_2$ AuMNL S1 B1 (*red triangles*). It shows the photocurrent attributed to the plasmonic resonance of the AuNPs (*black triangles, inset*) compared to the bare TiO_2. This is not strictly IPCE as it is calculated from the total current not just the photocurrent

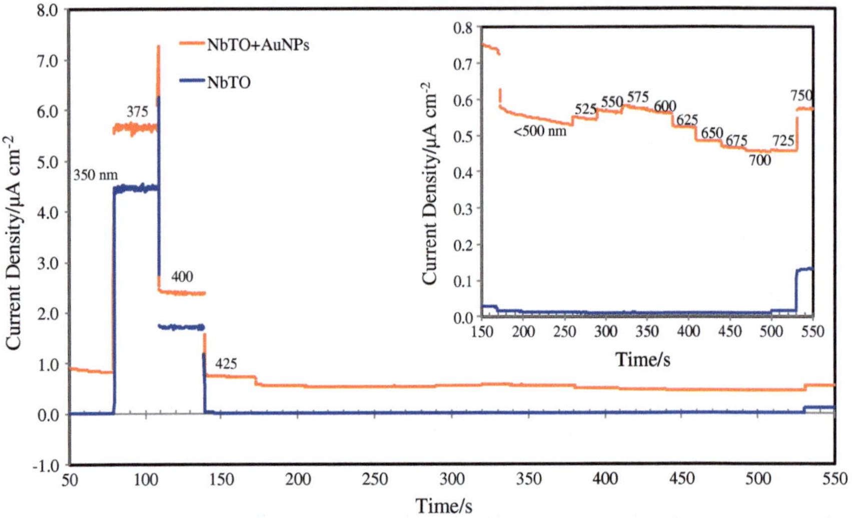

Fig. 5.25 Raw current-time data for samples NbTO AuNP S2 B1 and NbTO S1 B1. Measurements made with UV filter in place. The wavelength of illuminating light is indicated from 425 to 750 nm. There is a clear photo-response of the sample with AuNPs from 525 to 675 nm

current-time data (Fig. 5.25), for sample NbTO AuNP S2 B1. This second sample showing plasmonic photocurrent was required, as the first sample (results presented in preceding paragraphs) had become inactive. This may have been due to the loss of AuNPs (or at least the few providing a photocurrent) from the surface. Under illumination by light of 425 nm, there was a small photocurrent due to excitation of TiO_2. The current between 450 and 500 nm decreased smoothly, indicating that it was purely dark current and was unaffected by the illumination wavelength. However, at 525 nm, there was a positive step in the current of about 10–15 nA cm^{-2}. With each increase of wavelength up to 575 nm, the current increased. It then decreased again, approaching zero around 700 nm. This photocurrent was superimposed on the on-going transient decay of the dark current. Correcting the dark current by adding and subtracting the wavelength dependent steps in the current gave an approximately continuous decay profile. This wavelength-dependent photocurrent from 525 to 675 nm was attributed to the action of the surface plasmon resonance of the gold nanoparticles, based on the close matching of the current-wavelength curve with the UV-Vis transmittance and reflectance spectra of the sample with wavelength. The position of the peak photocurrent around 575 nm matched closely with the reflectance peak at 579 nm measured following the electrochemical experiments (Fig. 5.13).

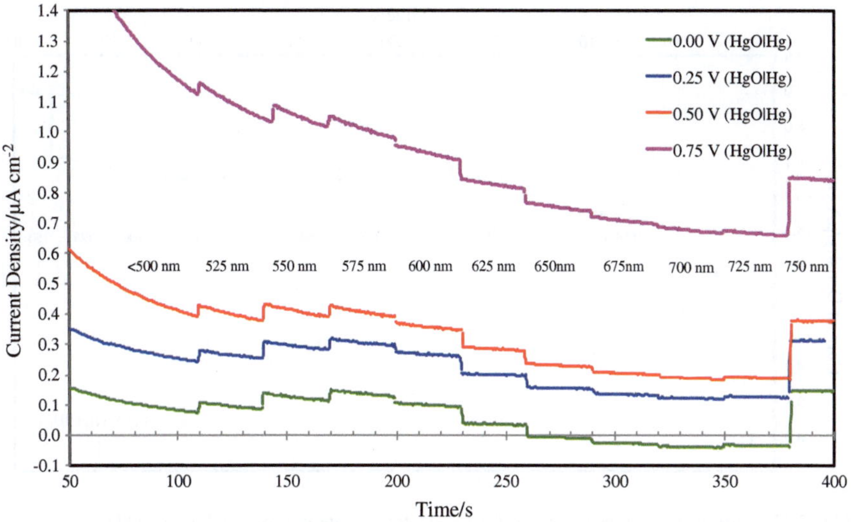

Fig. 5.26 Current-time measurements for a range of electrode potentials of sample NbTO AuNP S2 B1. The potentials used are from 0.0 to +0.75 V (HgO|Hg). The wavelength of illumination is shown from 500 to 750 nm

5.7.3 *Effect of Electrode Potential on Plasmonic Photocurrent on Nb-TiO_2 AuNPs*

The first sets of measurements in which the plasmonic photocurrent was observed were all made at an electrode potential of +1.0 V (HgO|Hg), corresponding to ca. +1.10 V (SHE). Given that the measurements were made in pH 13.6 solution in which the equilibrium electrode potential of O_2/H_2O is +0.43 V (SHE), the potential of the electrode was, in theory, sufficiently positive to oxidize water in the dark. Hence, it was possible that the observed current peak in visible light was due to the action of the LSPR of the gold nanoparticles enhancing the dark current by Au_2O_3|Au catalysing the oxygen evolution kinetics, rather than a true photocurrent. To confirm that this was not the case, chronoamperometric measurements were made over a range of electrode potentials from −0.2 to +0.75 V. The raw current-time data is shown in Figs. 5.26 and 5.27, while the photocurrent density and IPCE are shown in Figs. 5.28 and 5.29, respectively.

For the measurements investigating the effect of working electrode potential, the original sample NbTO AuNP S1 B1 was no longer operational,[1] so the following sample NbTO AuNP S2 B1 (Sample 2, Batch 1) was used.

[1]This also suggests that gold nanoparticles deposited directly on single crystal rutile are not suitable for a device that is stable in the long-term. This could have been due to subjecting the electrode to unfavourable potentials. However, as metals are immiscible with metal oxides, with

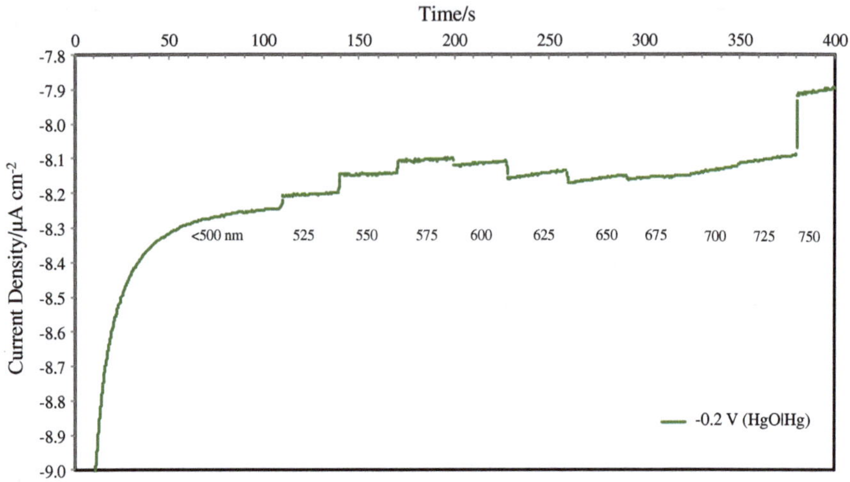

Fig. 5.27 Current-time measurements for an electrode potential of −0.2 V (HgO|Hg) on sample NbTO AuNP S2 B1. The wavelength of illumination is shown from 500 to 750 nm

Fig. 5.28 Plasmonic photocurrent spectra for a range of electrode potentials. Sample NbTO AuNP S2 B1. Measurements shown for potentials of: −0.20, 0.0, +0.25, +0.50, +0.75, and +1.0 V (HgO|Hg)

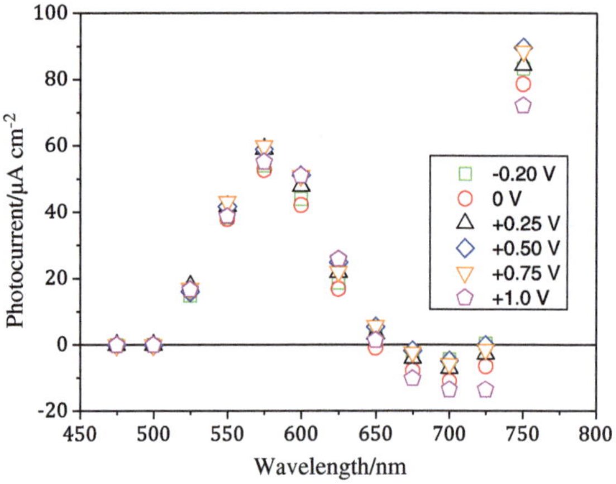

It is clear from Figs. 5.26 and 5.27 that the sample NbTO AuNP S2 B1 demonstrated the visible light photocurrent attributed to the plasmon resonance of the AuNPs, as the total currents varied time for a range of working electrode potentials. The wavelength of the illuminating light was changed every 30s, as indicated on the plot. The magnitude of the underlying dark current changed significantly with potential and in all cases demonstrated a transient decay. The

(Footnote 1 continued)

adherence caused by van der Waals forces only, it is possible that the AuNPs were detached physically by, for example, agitation of the electrolyte.

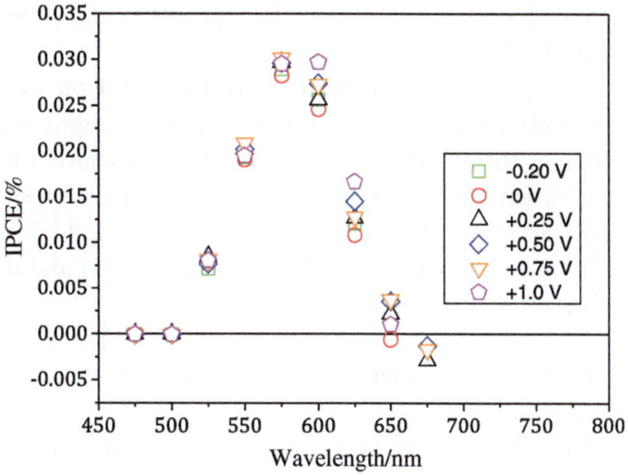

Fig. 5.29 IPCE of plasmonic photocurrent versus wavelength for a range of electrode potentials. Sample NbTO AuNP S2 B1. Measurements shown for potentials of: −0.20, 0.0, +0.25, +0.50, +0.75, and +1.0 V (HgO|Hg)

photo-response between 525 and 675 nm was similar in all cases, suggesting that the photocurrent was potential independent (within the accuracy of the measurement). In Fig. 5.26, the dark current was decreasing from positive values and the photocurrent was a small increase. As noted earlier, the OCP of the samples is approximately 0.0 V (HgO|Hg), and the dark current of the measurement made at 0.0 V (HgO|Hg) decreased from positive values to be slightly negative. At 0 V (HgO|Hg), the reaction on the electrode may be reduction of oxygen in the dark. In Fig. 5.27, the electrode potential was −0.2 V (HgO|Hg) and the current was relatively large and negative compared to the current at potentials positive of the OCP. The transient decay in the dark current was towards less negative values over time, but still assuredly negative. Super-imposed on top of this decay was the positive photocurrent peak attributed to the LSPR of the AuNPs. This was the best indication that this effect was not due to Au_2O_3|Au catalysing the oxygen evolution kinetics in the dark. While the dark current was negative (possibly due to oxygen reduction), the photocurrent was positive, resulting in a positive step in the measured current, assumed to be oxygen evolution.

Based on the methods for extracting the photocurrent outlined previously in Sect. 5.7.1, the wavelength dependent photocurrent density and IPCE were calculated at each electrode potential and are shown in Figs. 5.28 and 5.29, respectively. The peak photocurrent occurred at a wavelength of 575–600 nm, which, as seen in the UV-Vis spectra, corresponds to the plasmon resonance of the AuNPs. Interestingly, the photocurrent appeared to be independent of the potential: there was no overall trend. This confirmed that the measured current (if due to oxygen evolution) was not due to the potential of the electrode being positive enough to oxidize water or hydroxide ions. A potential dependence would indicate a dark process limited by the kinetics, whereas a potential-independent dark current might imply mass transport control over the potential range investigated, though were that current density due to hydroxide ion oxidation to oxygen, its magnitude would be: $j_{L,\mathrm{OH}^-} = Fk_{m,\mathrm{OH}^-}[\mathrm{OH}^-] \cong 0.1\,\mathrm{A\,cm}^{-2}$, i.e. orders of magnitude greater than detected current densities. The potential dependent behaviour is discussed in Sect. 5.7.5 below.

The peak photocurrent density measured on sample NbTO AuNP S2 B1 was about 60 nA cm^{-2} at an illuminating wavelength of 575 nm. This was of the same order as the current (20–30 nA cm^{-2}) measured on sample NbTO AuNP S1 B1. The differences were likely to be explained by slight differences in the size of nanoparticles, due to the thickness of the gold films deposited (it is difficult to control the thickness of a 2 nm film precisely) and the exact annealing temperature achieved in the furnace.

The peak IPCE calculated for this photocurrent is of the order of 0.03 % (i.e. 0.0003), so while there appeared to be a measurable effect, it was well below a level of practical use.

In particular, the photocurrent was so small that it was impossible to carry out any meaningful measurement of the reaction products by a technique such as gas chromatography. Hence, it was impossible to assign the photocurrent to oxygen production by water splitting with any certainty, rather than e.g. oxidation of impurity species in the electrolyte solution.

Furthermore, the low IPCE means that the as-fabricated Au-TiO_2 system is unsuitable for investigation by other techniques such as transient absorption spectroscopy (TAS), which requires a yield of at least 1 % (ideally more) to achieve an adequate signal to noise ratio for tracking the dynamics of the excitation of the electrons.

On this basis, one cannot recommend this as-fabricated Au-TiO_2 system to investigate the mechanisms of water splitting by plasmonic metal nanoparticles. Of course, this statement holds true only for this combination of gold nanoparticle fabrication and substrate. As shown in the literature review (Chap. 2), other approaches to the nanoparticle formation have yielded impressive values for the IPCE.

5.7.4 AuNPs—Reduced Versus Nb-Doped TiO_2—Chronoamperometry

The same measurements as made on the Nb-doped rutile (110) were made on reduced rutile (110) substrates with and without AuNPs formed by annealing a gold thin film. There was no wavelength dependent visible light photocurrent observed in either the chronoamperometry or the voltammetry. The time dependent chronoamperometric data was examined in detail. If there was a photocurrent response to visible light, then it was smaller than the 1 nA resolution of the measurements. The results are not presented in graphical form, as they do not add anything to the discussion. The dark current of the reduced rutile (with and without AuNPs) was also greater than that of the Nb-doped rutile as seen in Fig. 5.15. The higher dark current may also have made it more difficult to identify a small photocurrent.

Neither the second batch of Nb-doped rutile nor the reduced rutile demonstrated the 'plasmonic' photocurrents. As will be discussed in the following chapter, both sets of substrates (reduced and second batch of Nb) demonstrated qualitatively similar behaviour in the voltammetry. It seems likely that the second batch of Nb-doped substrates were also reduced to some extent. This indicated that whatever the nature of the plasmonic photocurrents, the condition of the substrate may be important. Based on the results presented here, it could be said that the reduction of the rutile (110) substrates prevents the plasmonic water splitting effect from occurring. However, in the cases that the effect did occur, it was vanishingly small, suggesting that the overall system had poor efficiencies. It may be that the reduction of the substrate is less important for an Au-TiO_2 system that shows a larger effect (such as that by Nishijima et al. [1]). In that case, using a reduced TiO_2 substrate may result in a smaller photocurrent density, rather than none at all. This could be determined by using a nanoparticle architecture that is known to work, on a reduced substrate. The nanoparticle architecture used here was not well suited to plasmonic water splitting. This is addressed further in the discussion at the end of this chapter.

5.7.5 Chopped Light Voltammetry of Nb-TiO_2 AuNPs

The 'plasmonic' photocurrents measured on the Nb-doped samples by chronoamperometry were extremely small, in some cases an order of magnitude smaller than the dark current. As such, it was impossible to determine the potential dependence of the photocurrent by conventional voltammetry. In the usual case of a TiO_2 photoanode illuminated by UV light, the photocurrent is orders of magnitude larger than the dark current and it is straightforward to detect the difference in current on illumination.

To determine the potential dependence of the photocurrent attributed to the LSPR or the AuNPS, the voltammetry was carried out using mechanically chopped light and a lock-in amplifier, referencing the frequency of the mechanical chopper. This technique is described in Chap. 4, Experimental Methods. The parameters used were; 227 Hz chopping frequency; 10 ms time constant; 90° phase; negative edge, based on the experience of other users. A UV-filter was used to prevent UV harmonics exciting the TiO_2 directly. The effect of varying the chopping parameters was discussed in Chap. 4.

The potential dependent photocurrent is presented in Fig. 5.30, which shows there was a photocurrent response of the AuNP-coated TiO_2 (sample NbTO AuNP S2 B1) to illumination by visible light of both 575 and 600 nm. The response to illumination by 575 nm (shown by the green line) was slightly greater than that for 600 nm (shown by the blue line), which was also evident in the chronoamperometry. The magnitude of the photocurrent in the potential range −0.2 to +1.0 V (HgO|Hg) was about 40 nA cm^{-2}, approximately the photocurrent density

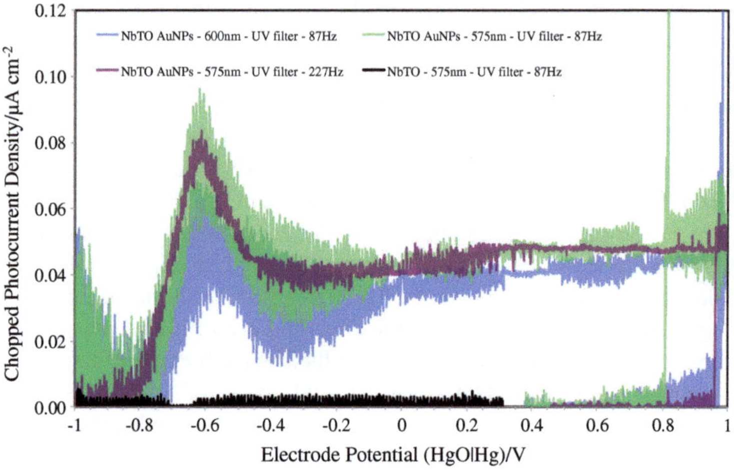

Fig. 5.30 Chopped light voltammetry of the samples NbTO AuNP S2 B1 (*blue*, *green* and *purple lines*). The potential was scanned at 10 mV s^{-1} from −1.0 to +1.0 V (HgO|Hg) and back again. The wavelengths of illumination were 575 nm (*green*) and 600 nm (*blue*) chopped at 87 Hz. The *purple line* is for illumination at 575 nm chopped at 227 Hz, and shows the decreased noise. For comparison the photocurrent of the bare substrate NbTO S2 B1 is also shown for light at 575 nm (*black lines*). The illumination was blocked on the reverse scan to demonstrate that the current is indeed zero in the absence of visible light

measured by chronoamperometry. This may have been due to the nature of measurements employing a lock-in amplifier (as discussed in Experimental Methods). An identical measurement on the bare Nb-doped rutile (sample NbTO S1 B1) did not show any photo-response to visible light (data for 575 nm shown in black).

The chopped light voltammetry exhibited additional features that were not captured in the chronoamperometric measurements, which suggested that the plasmonic photocurrent was independent of potential. The voltammetry showed that this was only approximately true for potentials positive of −0.1 V (HgO|Hg). Between −1.0 and −0.1 V (HgO|Hg) there were a number of features in the chopped light voltammetry. The photocurrent onset occurred at about −0.8 V, there is a current peak at −0.6 V, and a local minimum at −0.3 V, followed finally by a rise to an approximately constant current at potentials positive of −0.1 V (HgO|Hg).

The equilibrium potentials for water reduction (H^+/H_2) and oxidation (H_2O/O_2) are shown at −0.8 V and +0.43 V (HgO|Hg). The photocurrent onset was positive of the water reduction potential, but negative of that for water/hydroxide ion oxidation, reinforcing the view that this was not a dark process (if it was indeed water splitting).

Furthermore, as shown in Fig. 5.31, the onset of the plasmonic photocurrent on the AuNP covered TiO$_2$ coincided with the photocurrent onset on bare TiO$_2$ in UV light (375 nm). The shoulder of the photocurrent on the bare TiO$_2$ occurred at the same potential as the peak on the AuNP-TiO$_2$ sample. As described in Chap. 3

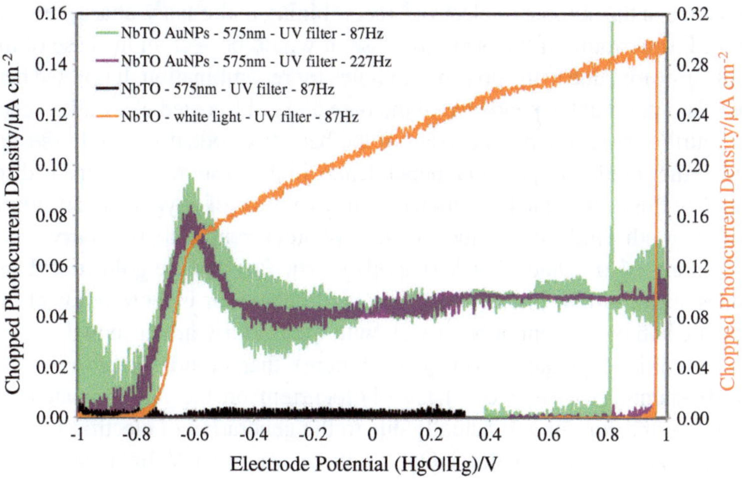

Fig. 5.31 Chopped light voltammetry of Nb-doped TiO_2 sample NbTO S2 B1 (*orange line*) superimposed on results of plasmonic photocurrent from preceding figure. The wavelength of illumination of the $NbTiO_2$ is 375 nm. The scale of the *purple line* is twice that of the other data. All measurements made with a UV filter

(Principles of Photo-Electrochemical Systems), the photocurrent-potential relationship of a semiconductor electrode is dependent on the width of the space charge region (dependent on the band bending), and the absorption coefficient. That the current-potential relationship of the photocurrent of the bare TiO_2 in UV and of the AuNP-TiO_2 electrode in visible light were similar, suggested that the plasmonic photocurrent was determined somehow by the band bending and a characteristic absorption length.

The dominant theory of plasmonic photocurrents at present (as discussed in the Literature Review) is that the gold nanoparticle absorbs a photon as a localized surface plasmon, which then creates a hot electron that is injected into the conduction band of TiO_2. In general terms, it would be reasonable to suppose that band bending results in an electric field that drives the hot electron away from the solid | solution interface, and as such, greater band bending might result in an increase in the plasmonic photocurrent. This might have been due to driving a greater proportion of hot electrons away from the interface more quickly, the electric field strength increasing with the band bending, and decreasing recombination rates of the electrons with the nanoparticle.

However, the photocurrent-potential relationship of the photocurrent of a normal TiO_2 photo-electrode is determined by the exponential decrease in concentration of un-absorbed photons as they pass through the semiconductor. It is surprising that the hot electron model would follow this behaviour. All electrons injected into the TiO_2 conduction band should contribute to the current. In the simple Gärtner model of electron-hole pair generation, all electrons generated in the space charge region are assumed to contribute. It is difficult to see why there would be an exponential decrease with distance in the number of electrons recombining in the space charge region of the Au-TiO_2. Furthermore, if this were the case, it would be seen in the case of the bare TiO_2 substrate too, due to the presence of holes for recombination. It is noted here, that the onset and potential dependence of the bare and gold-coated electrodes seem to be similar. Furthermore, the photocurrent of the bare electrode does not fit the standard Butler-Gärtner model for potential dependence, in the onset region. This is discussed in Chap. 6 of this thesis, and is attributed to the evolution of oxygen (and its immediate reduction) modifying the shape of the photocurrent-potential curve between approximately −1.0 V and −0.4 V (HgO|Hg). The fact that the gold coated and bare electrodes show similar photocurrent-potential behaviour is more relevant than the fact that the behaviour cannot be fit well with the Butler-Gärtner model.

One possible explanation (introduced here) that could be proposed for the observed potential dependence of the photocurrent on the AuNP coated samples might be that the current was actually due to the generation of electron hole pairs in the space charge region of the TiO_2. As there was no UV light incident on the sample (see the lack of photocurrent on the bare TiO_2 sample), this would require the transfer of energy equivalent to the band gap (ca. 3 eV) from the plasmon to the TiO_2.

This could be from the generation of UV light by the decay of the plasmon, i.e. generation of higher order harmonics such as 300 nm photons from 600 nm photons. Given that gold nanoparticles are highly non-linear [10], the generation of harmonics by the LSPR seems possible. However, the efficiency of this process is likely to be very low. It is a possible explanation for the results presented here (with a low IPCE), but is not consistent with the higher efficiency systems reported by others with IPCEs up to about 5 % [1, 11]. Additionally, the response of rutile to light of 300 nm is low (as seen in the IPCE of the bare electrodes, Sect. 5.7.2) due to the lower absorption coefficient at 300 nm (Fig. 5.14). Conversely, Nishijima et al. [1] reported resonances in the range of 700–1400 nm that could have harmonics in a stronger absorbing region of TiO_2, although the harmonics would need to be of third or fourth order for the LSPRs in the infrared. This merits further investigation.

Another possibility, proposed here, is the coupled decay of nearby LSPRs with the emission of a virtual photon with the combined energy of several LSPRs (energy pooling), and the subsequent absorption of a UV photon by the TiO_2. A similar effect might be seen if a single nanoparticle was to absorb multiple photons and re-radiate a single photon of higher energy. In both cases, the absorption of the emitted UV photons would be seen to obey the exponential potential dependence.

Finally, it is proposed that the oscillating electric field of the LSPRs could be involved in directly generating electron-hole pairs in the TiO_2 close to the nanoparticles. This effect might be driven by the extremely high electric fields predicted to occur at the interface between a metal nanoparticle [12] and the dielectric substrate.

An interesting feature of the photocurrent is the peak soon after onset, followed by a local minimum and then a plateauing of the current. This might be explained if two separate photo-driven processes occurred over different ranges of potential. The first would run from say −0.8 V to about −0.3 V (HgO|Hg) with a bell curve shape, followed by a second process running from about −0.5 V (HgO|Hg) with a similar potential dependence to the photocurrent of the bare TiO_2 electrode. In fact, there was a dark reduction process observed for AuNPs on rutile (100), peaking at around −0.7 V (HgO|Hg); this is discussed in a following chapter and also appeared to be enhanced by illumination with UV light. However, that process resulted in a cathodic current, while the plasmonic photocurrent observed here appears to be anodic. This could be an issue with the phasing of the lock-in amplifier and the chopper and would bear investigating further if time were available.

One possible (but untested) reaction is that the loss of an electron from the AuNP to the TiO_2 results in the oxidation of Au as shown in Eq. (5.2). The first process may be the oxidation of the surface of the AuNP, starting at −0.8 V and complete by about −0.3 V (HgO|Hg). The Au_2O_3 might then be reduced by the injected electron, and therefore act as a recombination pathway. While this might also explain the low IPCEs, it would probably occur for all Au-TiO_2 systems, and may not be the primary reason for the low IPCE for the AuNP-TiO_2 combination investigated in this thesis.

$$Au_2O_3 + 6H^+ \leftarrow Au + 2H_2O + 6h^+ \tag{5.2}$$

Finally, it should be noted that while there is definitely a visible light photocurrent that is not seen (nor would be expected) on bare TiO_2, whether it is attributable to water oxidation could not be verified due to the very low IPCE. The wavelength response matches the reflectance of the LSPR of the gold nanoparticle well, so it is termed a plasmonic photocurrent. The effect of the gold nanoparticles on the electrochemistry of TiO_2 in general will be discussed in a separate chapter, the focus here being on the formation of the gold nanoparticles and any plasmonic effects.

5.7.6 Effect of Reduction of Rutile (110) on Plasmonic Photocurrent

As noted in Sects. 5.7.2–5.7.4, the plasmonic photocurrent was only observed on Nb-doped rutile (110) crystals from the first batch received from the supplier. Subsequent measurements on a second batch of Nb-doped rutile (110) did not demonstrate the plasmonic photocurrent. Likewise, the reduced un-doped rutile (110) did not demonstrate the plasmonic photocurrent.

Based on the analysis of the bare rutile (110) crystals, there was a clear difference in the behaviour of the crystals from Batch 1 and Batch 2, as evident in Fig. 5.32. The second batch of crystals exhibited redox processes between −0.2 and +1.5 V

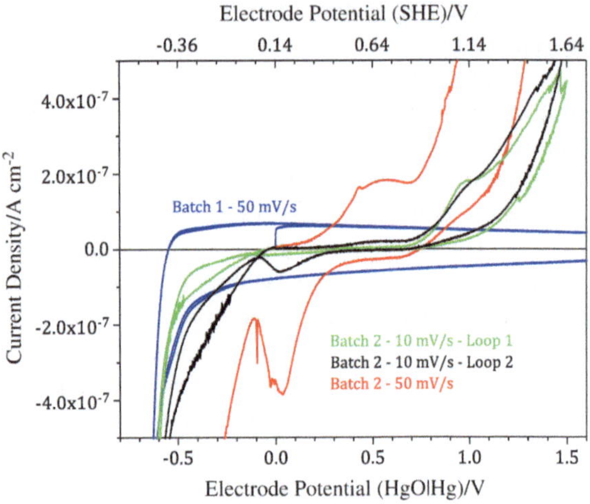

Fig. 5.32 Current-potential measurements in the dark showing the different behaviour between the batches of the Nb-doped rutile. A scan rate of 50 mV/s for Batch 1 was used (*blue line*) and Batch 2 (*red line*). The Batch 2 scan at 50 mV/s is partial. Scans at 10 mV/s are also shown for Batch 2 (*green* and *black lines*) to demonstrate the range of behaviour

(HgO|Hg) that were not evident in the first batch of crystals. The detailed analysis of the bare Nb-doped and reduced substrates (discussion follows in Chap. 6) showed that these redox processes were typical of reduced rutile (110) and the conclusion was made that the second batch of Nb-doped crystals were, in all likelihood, partially reduced. The details of the voltammetric scans are elaborated in the following chapter.

This suggests that some characteristic of the reduced rutile prevented the plasmonic photocurrent or inhibited it, such that the small current measured on the first batch of Nb-doped rutile was decreased to below detectable levels. From the results of the following chapter and the literature review of the defect chemistry of TiO_2 (see Chap. 2), it appeared that the reduction of rutile can result in the formation of conductive filaments or planes that cut the rutile (110) surface. These may act as recombination pathways that transport any hot electron injected into the TiO_2 back to the gold nanoparticle.

5.8 Discussion of Results

The results of the experiments to fabricate gold nanoparticles on rutile (110) single crystals and to investigate incident visible light generating photocurrent attributed to surface plasmon resonance, is discussed in the context of the literature review. The focus of this chapter was on the formation of the gold nanoparticles and determination of any plasmonic effect in visible light. The detailed electrochemistry of rutile (110) is discussed in Chap. 6.

In summary, a number of techniques were investigated for forming gold nanoparticle arrays on rutile (110) single crystals, with the aim of observing visible light-induced water splitting due to the LSPR. The effect was observed for several samples, but was vanishingly small. The main questions that arose during the literature review and experimental work are:

- What is the mechanism?
- Why is an effect evident on only some rutile substrates?
- Why is the effect so small in comparison with reports in the literature?

The mechanism of plasmonic water splitting with Au-TiO_2 was discussed in detail in the Literature Review, and trying to understand it was one of the principal objectives of the experiments. However, as discussed in the results of this chapter, the performance (IPCE) of the fabricated Au-TiO_2 system was insufficient for any meaningful further investigation of the mechanism, beyond what is detected in the chopped light voltammetry (for which results were reported in Sect. 5.7.5). As such, a better performing combination of TiO_2 and gold nanoparticle arrays is required to study the mechanism, for example by techniques such as transient absorption spectroscopy (TAS) that can probe the charge transfer dynamics. As discussed further below, although the electrode configuration investigated did not perform well, it also served to provide information on what may be happening, and suggests design rules for future studies. At the least, the investigation demonstrates a number of approaches to gold nanoparticle formation that do not result in sufficiently high performance, and should be avoided by other researchers.

The effect of the substrate was discussed above and it appears that the reduction of the rutile (110) surface inhibited the plasmonic photocurrent. It is suggested that this was due to the reduction process resulting in TiO_{2-x} phases running through the rutile (110) crystal that can act as recombination pathways. While it is not completely clear that this did indeed occur, it is suggested that reduced substrates are avoided for any further investigations. This may require the development of a repeatable annealing process to ensure as-received single crystals are not partially reduced.

In the case where an effect was detected, it could be assumed that the substrate was in the appropriate state, i.e. doped with Nb but not reduced. If so, then it seems likely that the reason for the low IPCE and small photocurrents may have been due to the particular formation of the nanoparticle array. The nanoparticle array formed by annealing consisted of a random distribution of particle sizes of about 20 ± 10 nm in diameter. By comparison, the array used by Nishijima et al. [1] consisted of an ordered array of nano-rods of ca. 250 nm by 100 nm by 40 nm on a 5 nm layer of Ti; the behaviour of 175 nm diameter nano-dots was also investigated. The other study showing reasonably good performance (by Tian and Tatsuma [11]) used nanoparticles of similar size to those in this study (ca. 20 nm), but used them in a relatively thick porous TiO_2 film, so that there would have been a larger total area of Au|TiO_2 interface than for a single layer of nanoparticles on a single crystal. Additionally, the Tian and Tatsuma study used a $Fe^{3+}|Fe^{2+}$ couple, with more energetically favourable kinetics than those of the O_2/OH^- system.

These studies suggest that either a large number of nanoparticles are required or a single layer of high-performance nanoparticles.

There could be a number of reasons for the high performance which are now discussed:

- The Ti interlayer
- The larger size of the nanoparticles
- The ordered structure of the array.

The effect of the Ti interlayer between the Au and the TiO_2 may have a number of effects. The first is to improve the adherence of the Au to the TiO_2. Improving the contact of the Au to the TiO_2 will increase the contact area and, one might suppose, the rate of hot electron injection from the gold nanoparticle to the TiO_2. However, it is hard to see how this could be responsible for a change in IPCE of approximately four orders of magnitude (comparing these results to those of Nishijima et al.). An alternative that might be proposed is that the Ti layer could change the electronic structure of the interface in such a way that electron injection is more energetically favourable. However, the work function of Ti is typically less than that of TiO_2 (4.1 eV vs. 4.3 eV) so any hot electron injected into the TiO_2 would be expected to be driven back into the Ti, i.e. back towards the gold nanoparticle. It is for this reason that Ti is used as an ohmic contact for TiO_2 crystals [13]. Therefore, one might expect the Ti layer to reduce the overall efficiency. Finally, the effect of the Ti layer might be to increase the ratio of surface plasmons that decay by hot electron generation compared to re-radiation (reflection) of the incident photon. It is not clear at present how this might be determined in practice.

To try to investigate the impact of the Ti layer on performance, samples were made with nanoparticle arrays of only Au, and others with a 5 nm inter-layer of Ti. The technique used was nano-sphere lithography (NSL), as described in Experimental Methods. The resulting nanoparticle arrays were presented in Sect. 5.5.4 of this chapter. Unfortunately, the process was not well optimized at the time of use and resulted in arrays with inhomogeneous coverage of the surface. The UV-Vis spectroscopy (not shown) did not present any features that could be attributed to surface plasmons. Furthermore, the samples did not survive long enough for electrochemical analysis to be undertaken. During the first voltammetric scans, the gold was found to come away from the surface and appeared to re-deposit as a single flake of gold film. It was not clear whether this was an agglomeration of gold nanoparticles held together by electrostatic forces, or whether the nanoparticles had dissolved and been re-deposited as a solid film. Therefore, it was not possible to determine if the samples were able to generate plasmonic photocurrent. Given further time, it would be worth repeating this experiment with an optimized NSL technique. Likewise, a lower pH would have decreased the solubility of Au^{III} species and may preclude such dissolution/re-deposition behaviour. It is also noted that many of the other studies discussed in the literature made the measurements at neutral pH.

The size of the nanoparticles will affect a number of aspects of the surface plasmon resonance, including the wavelength at which the LSPR occurs and the ratio of scattering to absorption. It is likely that larger nanoparticles were more absorbing [6, 9], and so may have resulted in better performance, i.e. photons to hot electrons. However, the nano-rods used by Nishijima et al. seemed to perform well over a range of sizes and the corresponding resonant wavelengths, which suggested that nanoparticle size may not be that important once a certain size is exceeded. Ideally, a method of controlling the nanoparticle size and distribution would be developed, allowing the systematic study of the effect. The nanoparticle array fabrication techniques examined in this chapter do not (currently) give sufficient control. The technique of e-beam lithography (EBL) gives the required level of control; however, it is slow, expensive and appropriate only for small areas. For future work, it is recommended that a technique like nano-imprint lithography (NIL) be used with a range of templates for size and distribution.

Furthermore, it is not even clear that the plasmons excited on the nano-rods discussed above are LSPRs, rather than say surface plasmon polaritons (SPPs), travelling at the interface of the nanoparticle and substrate. If the high performance water splitting were due to SPPs, then of course the system of AuNPs investigated in this study would not be expected to perform well, as the plasmon resonance is consistent with the localized surface plasmon resonance at the surface of a spheroidal gold nanoparticle. Based on the preliminary results (not yet published) of some computational electrodynamics modelling of gold nanoparticles undertaken by Anthony Centeno (Universiti Teknologi Malaysia, formerly Department of Materials, Imperial College London) in relation to this work, it appears that the resonances may indeed be an SPP rather than an LSPR. This might explain why the peak of the photocurrents (of the Nishijima et al. work) seems to follow the length of the nano-rods, and has multiple peaks corresponding to each of the length and width dimensions.

One of the main differences between the work of this thesis and the study by Nishijima et al. is the control of the nanoparticle size and spacing. In this work, the nanoparticles are randomly distributed and have a random size with an average of about 20 nm. This means that there was no long-range order in the nanoparticle array. In a talk at Department of Physics, Imperial College London ("Designing plasmonic nanoparticles for light trapping applications in solar cells", 18/06/2012), Beck [14] highlighted that the optical characteristics (reflection and absorption peaks) of a random array of metal nanoparticles are equivalent to the characteristics of a single nanoparticle of the average size. Therefore an array as fabricated by annealing metal thin films provides an optical response that is in principle straightforward to model by methods such as finite difference time domain (FDTD).

However, while the optical response is straightforward to model, the plasmonic photocurrent response is not. It is likely that only a proportion of a random array of nanoparticles has the correct properties to result in plasmonic water splitting, so while the overall current response is the average of the whole array, that response will be coming from a small subset of nanoparticles. In fact, further work by Nishijima [2] shows that this randomness has an effect of increasing the enhancement of the electric field for some particles, but reduces the overall coverage of the surface by nanoparticles with enhanced electric fields. Therefore, understanding the average optical properties of the array may not give any useful information on what the required properties for water splitting are. By comparison with the literature and the difference in performance between the AuNPs and AuMNLs here, it is most likely that the plasmonic photocurrent is coming from only the larger individual nanoparticles, which may even sustain SPPs rather than LSPRs. With the experimental analysis carried out here, it is impossible to identify precisely which nanoparticles contribute. In particular, each of the nanoparticles of slightly different size and shape will exhibit different patterns of the electromagnetic field around the surface, and have different resonances. It would be difficult (if not impossible) to identify which plasmon behaviour was responsible, given the lack of understanding of the mechanism. Hence, the random arrays were not particularly helpful in understanding the mechanism, due to the range of performance of nanoparticles exhibited simultaneously on illumination.

As mentioned previously, it is recommended that in any future work, that a well-defined array of nanoparticles of a single size and shape are employed. In this way, it will be possible to tune a computational model (say using FDTD) to replicate the optical characteristics of the array and identify the plasmon modes occurring over a range of photon wavelengths. This was the original intention in this work; however, the focus of the project was to be TiO_2 rather than developing nanoparticle fabrication techniques, which was to be done by others. By measuring the performance of each type of array, it should be possible to correlate the performance with the type and magnitude of plasmon modes determined by simulation.

To date, none of the experiments reported here or in the literature have been able to prove conclusively what the mechanism of plasmonic photocurrent is, or isn't. There is no established way to distinguish between the various proposed plasmon decay mechanisms: only the resulting current can be detected and an attempt made to try to observe the dynamics for systems with good performance. Despite these uncertainties, the injection of hot electrons from a metal nanoparticle into a semiconductor has been shown to occur [15]. The question is whether this effect can be responsible for the water splitting photocurrents observed. It is still not clear how the injection of hot electrons can generate holes at potentials sufficiently positive to oxidize water. This would require new physics, allowing long-lived holes being generated by LSPR or SPPs on metal nanoparticles.

Finally, while injection of hot electrons is currently the dominant mechanism proposed in the literature, there is a particular phenomena that has been reported, suggesting that something else may be happening, or at least that the electron injection theory needs to be modified: the observation of photocurrent in the work of Nishijima et al., at energies less than the equilibrium water splitting energy of 1.23 eV. This strongly suggests that there is something like energy pooling [16] occurring, such as multiple photon absorption by the nanoparticle or simultaneous decay of plasmons on multiple adjacent nanoparticles, as proposed earlier in this chapter. This is also more likely to occur in an ordered array of nanoparticles than in a randomly distributed array where there is no symmetry and a low probability of finding two or more nanoparticles with the same resonance close enough to couple via virtual photons.

References

1. Nishijima, Y., Ueno, K., Yokota, Y., Murakoshi, K., Misawa, H.: Plasmon-assisted photocurrent generation from visible to near-infrared wavelength using a au-nanorods/TiO_2 electrode. J. Phys. Chem. Lett. **1**, 2031–2036 (2010)
2. Nishijima, Y., Nigorinuma, H., Rosa, L., Juodkazis, S.: Selective enhancement of infrared absorption with metal hole arrays. Opt. Mater. Express **2**, 1367–1377 (2012)
3. Tian, Y., Tatsuma, T.: Plasmon-induced photoelectrochemistry at metal nanoparticles supported on nanoporous TiO_2. Chem. Commun. **16**, 1810 (2004). doi:10.1039/b405061d
4. Liu, Z., Hou, W., Pavaskar, P., Aykol, M., Cronin, S.B.: Plasmon resonant enhancement of photocatalytic water splitting under visible illumination. Nano Lett. **11**, 1111–1116 (2010)
5. Centeno, A., Breeze, J., Ahmed, B., Reehal, H., Alford, N.: Scattering of light into silicon by spherical and hemispherical silver nanoparticles. Opt. Lett. **35**, 76–78 (2009)
6. García, M.A.: Surface plasmons in metallic nanoparticles: fundamentals and applications. J. Phys. D Appl. Phys. **44**, 283001 (2011)
7. Abràmoff, M.D., Magalhães, P.J., Ram, S.J.: Image processing with ImageJ. Biophotonics Int. **11**, 36–43 (2004)
8. Delcassian, D., et al.: Nanoscale ligand spacing influences receptor triggering in T cells and NK cells. Nano Lett. **13**, 5608–5614 (2013)
9. Ghosh, S.K., Pal, T.: Interparticle coupling effect on the surface plasmon resonance of gold nanoparticles: from theory to applications. Chem. Rev. **107**, 4797–4862 (2007)
10. Danckwerts, M., Novotny, L.: Optical frequency mixing at coupled gold nanoparticles. Phys. Rev. Lett. **98**, 026104 (2007)
11. Tian, Y., Tatsuma, T.: Mechanisms and applications of plasmon-induced charge separation at TiO_2 films loaded with gold nanoparticles. J. Am. Chem. Soc. **127**, 7632–7637 (2005)
12. Beck, F.J., Verhagen, E., Mokkapati, S., Polman, A., Catchpole, K.R.: Resonant SPP modes supported by discrete metal nanoparticles on high-index substrates. Opt. Express. **19**, A146–A156 (2011)
13. Fuke, N., et al.: Influence of TiO_2/electrode interface on electron transport properties in back contact dye-sensitized solar cells. Sol. Energ. Mat. Sol. C. **93**, 720–724 (2009)
14. Mokkapati, S., Beck, F.J., Polman, A., Catchpole, K.R.: Designing periodic arrays of metal nanoparticles for light-trapping applications in solar cells. Appl. Phys. Lett. **95**, 053115 (2009)
15. Giugni, A., et al.: Hot-electron nanoscopy using adiabatic compression of surface plasmons. Nature Nanotech **8**, 845–852 (2013)
16. Jenkins, R.D., Andrews, D.L.: Three-center systems for energy pooling: quantum electrodynamical theory. J. Phys. Chem. A **102**, 10834–10842 (1998)

Chapter 6
Electrochemistry of TiO_2—Rutile (110)

This chapter details the results of a series of experiments on Nb-doped and reduced (undoped) rutile (110) TiO_2 single crystals. The structure of the chapter is outlined below, followed by summaries of both the experimental aims and results.

First, the voltammetry of two batches of Nb-doped rutile (110) single crystals are presented. The differences in the current-potential behaviour of the batches are suggested to be indicative of physical differences between the batches. As presented in the preceding chapter, the first batch demonstrated a small plasmonic photocurrent when coated in gold nanoparticles. This plasmonic photocurrent was not seen with samples made from the second batch of crystals. The batches of Nb-doped TiO_2 and reduced rutile (110) are investigated in this chapter to try to understand what the differences between batches were and how this affected the observation (or not) of plasmonic photocurrent as presented in the preceding chapter of results.

The structure of the space charge region is determined and discussed for the various samples, making use of the theoretical models reported in Chap. 3 (Principles of Photoelectrochemical Cells). This includes analysis of both capacitance and photocurrent data to determine characteristic parameters including the flat band potential, the charge carrier concentration, the space charge width constant, and the interfacial capacitance. An attempt is made to fit the photocurrent with the Gärtner model.

Following this, an attempt is outlined to better characterize the impedance of the electrodes over a range of both potential and frequency. This includes the presentation and discussion of a range of equivalent circuits required to fit the data at each potential.

The voltammetry of both batches of Nb-doped rutile is discussed and compared to the voltammetry of a reduced TiO_2 single crystal. The potential-pH phase diagram of TiO_2 is also introduced in Appendix B as a tool for the analysis of the voltammetry.

© Springer International Publishing Switzerland 2016

J.C. Alexander, *Surface Modifications and Growth of Titanium Dioxide for Photo-Electrochemical Water Splitting*, Springer Theses,
DOI 10.1007/978-3-319-34229-0_6

6.1 Experimental Aims

While there are numerous publications relating to the electrochemistry of TiO_2, it is argued here that the voltammetry and impedance spectroscopy (for example) has not been related to the micro-structural characteristics of the material in much detail. This issue was encountered during this project while trying to establish why individual single crystals (that should have been the same) were displaying qualitatively different behaviour. The existing literature proved to be of minimal help in understanding the detail.

It should be noted that the difference in behaviour in the dark is typically of minimal interest compared to the photo-response, and of a much smaller magnitude. However, the small magnitude of the plasmonic photocurrent[1] detected (and discussed in the previous chapter) meant that the dark current was of approximately equal (or greater) magnitude, and hence could not be dismissed.

In particular, this line of investigation grew out of the observed effect of the substrate condition on the performance of plasmonic AuNPs. As discussed in the following chapter, a small visible light photocurrent, attributed to the surface plasmon resonance (SPR) of the AuNPs, was observed for the first batch of Nb-doped crystals. However, samples from a different batch of single crystals did not exhibit this effect. Voltammetry on crystals from both batches showed different behaviour over a range of potential. This gave an indication of the condition of the surface. The reduced-undoped TiO_2 also failed to demonstrate the plasmonic photocurrent. The voltammetry of the reduced TiO_2 and the ineffective Nb-doped TiO_2 appeared to be more similar to one another than to the original batch of doped TiO_2.

The focus of the work outlined in this chapter was in trying to understand in what ways the two batches differed, and why this had an impact on the ability to generate plasmonic current with AuNPs.

Furthermore, a theoretical work package—running in parallel with the experimental work reported here—was focussed on the rutile (110) surface in a water splitting context [1]. As such, a more detailed treatment of rutile (110) in general is appropriate.

6.2 Summary of Experimental Results and Conclusions

The structures of the space charge regions of rutile (110) single crystals were determined using a combination of voltammetric and impedance measurements under illumination and in the dark. This was repeated for 0.05 wt% Nb-doped crystals from two separate batches and a reduced, but undoped, single crystal.

[1]As outlined in the previous chapter, plasmonic photocurrent is taken to mean a photocurrent that may be attributed to the action of the plasmonic metal (gold) nanoparticles, but not to the substrate (TiO_2) alone, i.e. photocurrent from photons of sub-band-gap energies.

The qualitatively different behaviour was then considered in the light of a review of the defect chemistry of TiO_2 as presented in Chap. 2 (Literature Review).

The research identified a number of novel features and conclusions.

The voltammetry of the second batch of Nb-doped crystals demonstrated qualitatively similar behaviour to the reduced crystals. There are additional reduction/oxidation processes occurring at anodic potentials compared to the first batch of Nb-doped crystals.

This suggested that the second batch of Nb-doped crystals were in fact reduced as well as doped, and that the reduction resulted in qualitatively different behaviour, i.e. the reduction determines the overall performance. This is important because it suggested that the growth or preparation of the photo-anode had a critical impact on the behaviour of the electrode. Given the discussion of defect chemistry (in the literature review), it seemed that for dopants to be incorporated as electronically compensated defects, that a reducing environment is required. However, if the reduction is such that all the Nb^{V} dopant is electronically compensated and the material is further reduced, then the electrode may behave in an un-desirable way.

The presence of the additional reduction/oxidation processes between −0.2 and +1.0 V (HgO|Hg) indicated that the structure of the rutile (110) electrode and surface might be fundamentally different to the un-reduced rutile (110) electrode.

The additional redox processes on the reduced electrodes suggested several possibilities. The potential was consistent with the presence of a second phase at the surface of the electrode. Based on a review of the literature of the defect chemistry of TiO_2, reduction was expected to result in the presence of interstitial Ti organized into planes or filaments. These should cut the rutile (110) surface, resulting in planes or filaments of Ti_2O_3-like material, which is expected to be metallic in nature [2]. It is suggested here that these metallic filaments may act like a metallic electrode, and that the redox processes seen on the reduced material (but not the un-reduced material) were occurring here. Essentially, the reduced electrodes act like a semiconducting TiO_2 element in parallel with a metallic Ti_2O_3 like element. The metallic element dominates the behaviour in the dark, while the semiconducting element provides photocurrent under illumination.

The difference in behaviour may also be credited to higher charge carrier concentration resulting in a narrower space charge region, and tunnelling of electrons that was precluded in the first batch of Nb^{V}-doped rutile.

Impedance spectroscopy for a range of frequencies at each potential, for a wide range of reducing and oxidizing potentials, demonstrated that the TiO_2 electrode cannot be modelled by a simple equivalent circuit composed of only resistance and capacitance elements.

The need to include constant phase elements (CPEs) in the equivalent circuits for the electrodes is an indication that there was a spatially distributed element present. This was consistent with the proposal that there were metallic planes or filaments cutting the TiO_2 crystal. This may also indicate surface roughness, as the exponent of the CPE approaches that of an ideal capacitor with increased band bending, i.e. with a depletion depth in the bulk.

The doped and reduced material have a relatively deep-lying defect state about 0.5 eV below the conduction band minimum.

The presence of two slopes in the Mott-Schottky plot of $1/C^2$ versus potential indicated that as the potential became positive by about 0.2 V of the flat band potential of −0.8 V (HgO|Hg), the charge carrier density of the semiconductor N_D increased. This suggested that there was either a bulk or surface state present. As the Fermi level of the electrode was programmed to pass through this energy level, the electrons trapped in this state were ionized and contributed to the conductivity.

This implies that the width of the space charge region (in this case a depletion layer) should contract. This would be expected to result in a decrease in the photocurrent. This was detected in the case of the plasmonic photocurrent (Chap. 5), but not clearly for the bare electrodes discussed in this chapter. However, the inability to fit the photocurrent data in the region just positive of the flat band potential was consistent with a more complex structure than assumed by the Gärtner model. Alternatively, the changing slope of the Mott-Schottky plot may indicate that different processes were occurring and that the equivalent circuit used was inappropriate.

6.3 Further Work

The next stages of the research suggested are as follows.

Given that the as-received single crystals can demonstrate qualitatively, as well as quantitatively, different behaviour, it is suggested that a pre-treatment process be developed to ensure that all as-received crystals can be conditioned into the same initial state, i.e. a single phase of rutile TiO_2. This would likely be a process of annealing in air or oxygen to re-oxidize any reduced phases present. Given the effect of the oxygen pressure on the conductivity of TiO_2, this is not as trivial an exercise at it may seem. If the oxygen pressure is too high, it is likely that the electronic conductivity due to a dopant, such as Nb^V, would be lost. If the pressure is too low, then the material will not be oxidized. Therefore, the next step would be a series of experiments investigating the effect of pre-annealing at a range of partial oxygen pressures on the qualitative behaviour of the voltammetry.

Following the development of a process to ensure identical crystals, a more controlled range of experiments on the effect of reduction conditions on the voltammetry would be recommended. Given that electrochemical reduction of the rutile (110) single crystals was detected during experiments (but not reported here), this may provide an alternative approach to annealing in a reducing atmosphere at temperature. In particular, the precise electrode potential at which the electrode becomes unstable needs to be identified. From the experiments done to date, it was between −1.0 and −1.6 V (HgO|Hg) in 1 M NaOH. Following this, chronoamperometry at fixed potential might be used to precisely control the degree of reduction and relate this to changes in the voltammetry and impedance spectroscopy.

Finally, the effect of the reduction on the structure of the crystal should be determined using a technique such as transmission electron microscopy (TEM) to prove conclusively whether or not the reduced phases are present through the crystal and at the surface.

It could be argued that the steps outlined above are excessively detailed and would not have any benefit to the application of water splitting. Indeed, these steps would do little to change the fundamental issue with TiO_2 for solar applications: the large band gap energy.

However, there are other uses for TiO_2 that would benefit from a better understanding of the relationship between structure and functional properties, in particular, the recent application of TiO_2 to memristor applications [3]. The use of TiO_2 for photoelectrochemical oxidation of pollutants under UV light would also benefit, as in these circumstances, unwanted dark redox processes may limit the performance of an otherwise economically viable device.

6.4 Comparing Nb-Doped to Reduced Rutile (110)

Figure 6.1 shows the qualitatively different behaviour detected for two separate batches of Nb-doped rutile (110) single crystals that inspired this line of enquiry. The figure is also shown as Fig. 5.32 in the chapter on plasmonic photocurrent, but

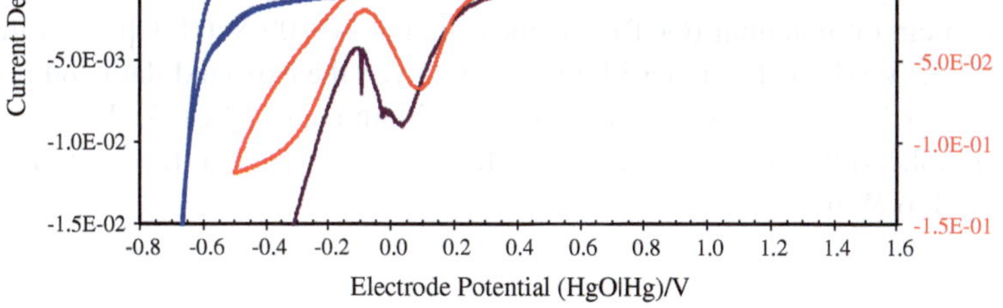

Fig. 6.1 Comparison of voltammograms of batches 1 (*blue line*) and 2 (*purple line*) of the Nb-doped single crystals in 1 M NaOH; potential scan rate used for the original batch was 50 mV s^{-1}. A partial scan is shown for the second batch (*purple line*). The second batch exhibited redox processes at potentials positive of −0.2 V that do not occur for the first batch. A voltammogram for the reduced TiO_2 is also shown (*red line*) on a larger scale for current density

is repeated here for ease of reference. The figure shows the similar behaviour of the second batch of Nb-doped rutile to the reduced but undoped rutile.

The figure shows that despite both single crystals of rutile (110) having nominally the same manufacturing specification with Nb-doping of 0.05 Wt%, and being received from the same manufacture, there was clearly a fundamental qualitative difference in behaviour between batches. It could be argued that under illumination by UV light, that the photocurrent density in both cases was a similar order of magnitude and much greater than the dark current density, and as such there is little interest in the qualitative differences shown. From a solar water splitting application, this is essentially correct. However, it is still of interest in terms of beginning to be able to relate the surface chemistry of the electrode to the structural properties of the surface: or at least demonstrating that there can be subtle differences.

The as-received 0.05 Wt% Nb-doped single crystals were investigated using cyclic voltammetry before any other experiments. The aim was to benchmark the performance for comparison to undoped, reduced, and AuNP covered samples. The area of the single crystals, as-received, was 1 cm^2. The crystals were sealed using silicone sealant (RS Components), resulting in slightly different electrode areas. The electrode assembly and general details of the voltammetry are described in Chap. 4 (Experimental Methods). All electrode potentials presented are referenced to HgO|Hg unless stated otherwise.

6.5 Phase Diagrams

The phase diagrams of the Ti-H-O and O-H systems are presented and discussed in Appendix B.

6.6 Nb-Doped Rutile (110)—Batch 1—Voltammetry

The single crystal measured from batch 1 had an active electrode area of 0.58 cm^2. The open circuit potential (OCP) was measured at +0.100 V (HgO|Hg) in the dark and −0.247 V (HgO|Hg) under illumination. Given the expected flat band potential around −0.8 V, this shows that the white light illumination of ca. 35 Wm^{-2} was too weak to unbend the bands. In fact, the intensity of UV light in the white light was less than 0.6 Wm^{-2}.

6.6.1 'Dark' Voltammetry

The behaviour of the as-received electrode was established by cyclic voltammetry over increasing potential ranges, as shown in Fig. 6.2. The starting potential was

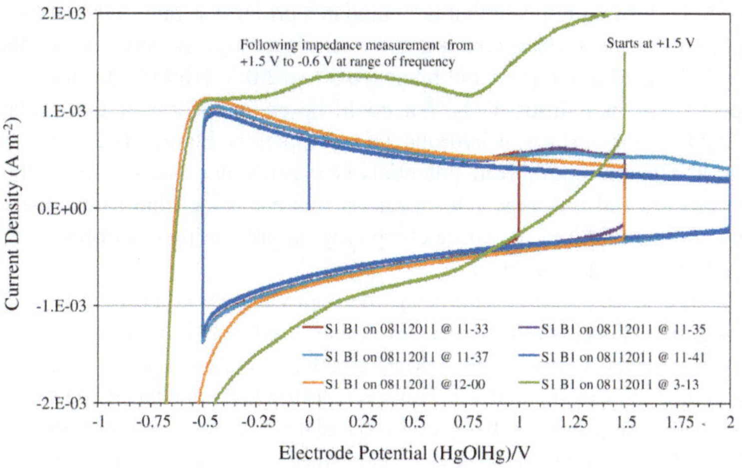

Fig. 6.2 Cyclic voltammograms of Nb-doped rutile (110) TiO_2 crystals (Sample 1, Batch 1). Scans started from 0 V at 50 mV s^{-1}, with increasingly positive limit variable upper limit with a lower limit of −0.5 V (*red*, *blue*, *purple*). The scans following illuminated voltammetry (*orange*) and impedance spectroscopy (*green*) were between −1.0 and +1.5 V

0.0 V, with a steadily increasingly positive potential limit of +1.0 V (red line), +1.5 V (purple line), and +2.0 V (turquoise and blue lines). The lower potential limit was −0.5 V and the initial scan rate was 50 mV s^{-1}, but it was later decided to use 10 mV s^{-1} to closer approximate to steady-state conditions. The dark voltammetry was also repeated following several measurements under white light (orange line), and after impedance spectroscopy (green line). The main feature of the first scans was that on positive-going potential scans, current densities decreased steadily. When the positive potential limit was reached and the scan was reversed, the current changed sign instantaneously (shown by the vertical line in the data). An approximately equal current flowed at each potential on both the positive going and negative going scans, indicating charge balance expected if surface transformations only and no solution species were involved in the reactions.

The behaviour in Fig. 6.2 is indicative of the charging of the space charge region of a semiconductor electrode (see Chap. 2 of this thesis covering the theory of photo-electrochemical systems). As the potential became more positive (increased band bending), the width of the space charge region (in this case a depletion region) increased. As the capacitance decreases with increased band bending, the current decreased. When the band bending was decreased, the stored charge was reduced, resulting in a change in the sign of the current density. Interestingly, there was also a component of Faradaic current in the voltammograms with positive potential limits of +1.5 V (purple line) and +2.0 V (turquoise line). There is an anodic current wave between +0.75 and +1.5 V, and another between +1.5 and +2.0 V. A subsequent scan between −0.5 and +2.0 V (blue line) was run for five cycles. The Faradaic processes disappeared, leaving only the capacitive charging of the electrode. This might indicate that the Faradaic current was due to the adsorption or desorption of some surface contaminants. Alternatively, the structure of the electrode had changed in the potential range −0.5 to +2.0 V (HgO|Hg). This would not be expected for anhydrous TiO_2, but could be an indication of the oxidation of some Ti_4O_7 at the surface of hydrous TiO_2 (Appendix Figures B.2 and B.3)

It is noted that the equilibrium potential of oxygen/water is ca. +0.35 V (HgO|Hg), but the process did not appear to occur on this electrode, due to extensive band bending and the width of the space charge region preventing electron transfer from solution to the conduction band.

Following some photo-voltammetry, the scan was repeated (orange line), with a more negative lower limit of −1.0 V (also used for illuminated scans). The photo-voltammetry is discussed below. The behaviour of the electrode between −0.5 and +1.5 V was unchanged. However, following impedance spectroscopy in the dark at several potentials between −0.6 and +1.5 V, the qualitative behaviour of the electrode appeared to change (green line), as several oxidation processes become apparent, peaking at +0.25 and +1.25 V. Superimposed on the capacitive behaviour are a number of unidentified redox waves. The capacitive behaviour was still evident from the sudden decrease in current at the upper potential limit. On the negative-going potential scan, the space charge layer was discharging its positive charge (negative current), while an anodic current flowed due to oxidation or adsorption processes (Fig. 6.2).

It is clear that reduction processes in the dark occurred from −0.5 V and below, due to the onset of asymmetry in the dark current. Positive of −0.5 V (HgO|Hg) the currents of the positive and negative going scans are equal and opposite, due to

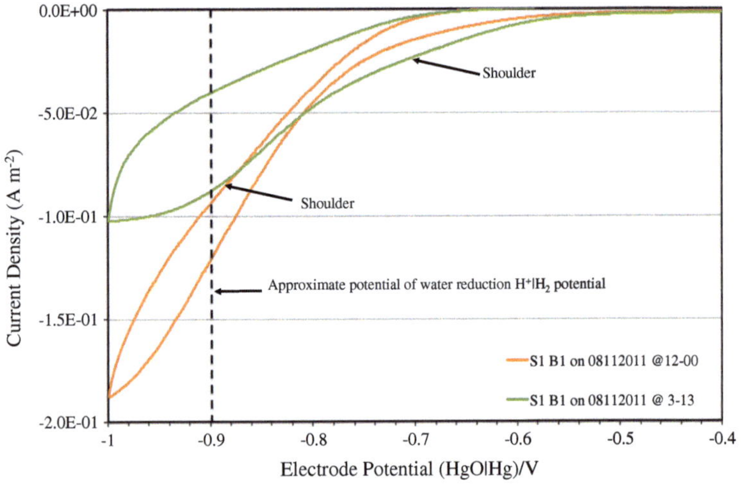

Fig. 6.3 Cathodic range of scans shown in Fig. 6.2. Both scans have shoulders around −0.7 and −0.9 V (HgO|Hg) which are positive of the equilibrium water/hydrogen potential (indicated by a *dashed line*)

charging of the space charge layer. Negative of −0.5 V, reductions occur and dominate the charging currents, leading to negative currents on both negative and positive going scans. At the operation pH of 13.6, the hydrogen/water equilibrium potential $H^+|H_2$ is −0.8024 V (SHE) or about −0.9 V (HgO|Hg), while the equilibrium oxygen/water potential is +0.4276 V (SHE) or about +0.33 V (HgO|Hg). Hence, the reduction evident in the data between −0.5 and −0.9 V (HgO|Hg) could not have been due to water reduction. Therefore, the reduction must have been of either the electrode itself, or another reducible species in solution, such as dissolved oxygen.

The voltammograms were repeated several times (following various other experiments) to check repeatability; results are shown in Fig. 6.4. All the scans except one were done at 50 mV s^{-1} and show gradually changing behaviour. Each scan performed at 50 mV s^{-1} showed more asymmetry (around the potential axis) than earlier scans. The anodic current curve of the positive-going potential scan started to increase, suggesting an oxidation process occurring on the electrode in addition to the capacitive charging and discharging. It appeared that the surface of the electrode had been modified, and that the magnitude of the Faradaic oxidation process began to dominate. This can be seen more clearly at a potential scan rate of 10 mV s^{-1} (red line, Fig. 6.4) in which the magnitude of the non-Faradaic current decreased to less than 10^{-4} A m^{-2}. It is then possible to detect an oxidation process beginning around +0.8 V, and plateauing at +1.0 V with a maximum current density of 3×10^{-4} A m^{-2}. A second oxidation process then began. It is unlikely that the first oxidation was water oxidation due to the presence of the plateau. This may be an adsorption process or an oxidation of the surface of the electrode. A reduction process was also evident on the negative-going sweep, from around

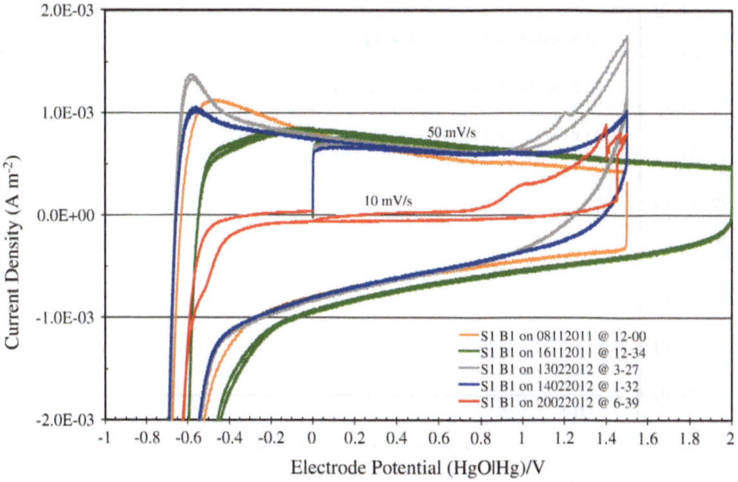

Fig. 6.4 Repeatability of 'dark' voltammetry on Nb-doped rutile (110) crystal (Sample 1, Batch 1) in 1 M NaOH; 50 mV s^{-1} except for one scan at 10 mV s^{-1} (*red line*)

−0.2 V, with currents beginning to plateau at ca. −0.5 V. This reduction was then dominated by another reduction process at potentials negative of ca. −0.6 V. This is most clearly seen by comparing the subsequent negative-going potential sweep, which did not display the smaller reduction peak. This may suggest that the first species being reduced was no longer present at the surface on the subsequent negative-going scan.

The positive-going sweep at a rate of 10 mV s^{-1} (red line Fig. 6.4) did not show the current peak at −0.6 V. The peak at −0.6 V may be due to the charging of the space charge layer. Close to the flat band potential, the capacitance is largest and the charging current will drop quickly as capacitance decreases. At lower scan rates, the capacitive charging may be less apparent. It is also possible that this could also be due to desorption of a reduced from the surface before it can be oxidized. The peak is also absent from the scan shown by the green line in Fig. 6.4. That scan was not preceded by a multiple-loop scan and so it is likely that any oxygen initially present at the surface was previously reduced. That the peak is not seen at low scan rates suggests that all the adsorbed oxygen may be reduced and have time to desorb, while at higher scan rates, the oxygen species remain at the surface. Alternately, the reduced species may be unstable, and decompose.

6.6.2 Photocurrent

Figure 6.5 shows the effect on voltammograms of illuminating the electrode with white light (blue-grey line) and with a UV filter between the lamp and the electrode

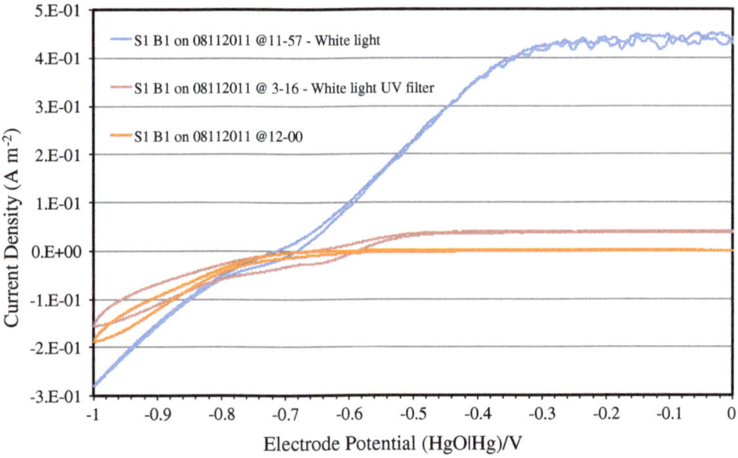

Fig. 6.5 Photo-voltammograms of Nb-doped rutile (110) (Sample 1, Batch 1) in 1 M NaOH; scan rate 10 mV s^{-1}, illumination with white light (*light purple*), white light with a UV filter in-line (*pink*), and dark current (*orange*, same data as Fig. 6.2)

(pink line); the dark current is shown for reference (orange). The white light spectrum contained a UV component of energies above the band gap of rutile (3.05 eV), producing the measured photocurrent of ca. 0.45 A m^{-2} (45 μA cm^{-2}), which decreased to 0.04 A m^{-2} when a UV filter was inserted between lamp and cell. While low in absolute terms, the photocurrent, attributed to the oxidation of water to generate gaseous oxygen, was equivalent to an incident photon to current efficiency in the range of 30–40 % (see preceding chapter on plasmonic photocurrent for IPCE curves).

In addition to decreasing the maximum photocurrent at positive potentials, the UV filter appeared to affect the dark current on the negative-going sweep, as shown in Fig. 6.6. At potentials less negative than −0.5 V, the currents on the positive-going and negative-going potential sweeps were approximately equal. Negative of −0.5 V, the positive- and negative-going sweep current curves separated. At −0.59 V, the current became negative on the negative-going scan (Fig. 6.6). The current of the positive-going sweep was positive, as was the current when the electrode was under more intense UV illumination (ca. 0.6 Wm^{-2} versus ca. 0.06 Wm^{-2}).

This suggests that there was a reduction process occurring on the negative-going sweep that was greater than the photocurrent. The dark current was also negative at these potentials; however, the current is less negative than for the illuminated electrode. Initially, it is surprising that, despite the contribution of the positive photocurrent, the total current was more negative than the dark current. This may be explained if the effect of the positive-going sweep was to generate oxygen at the surface of the electrode that remains adsorbed (or close to the surface) and was reduced on the subsequent negative-going potential scan. In the case of the dark scan, the concentration of adsorbed oxygen would have been lower, resulting in a

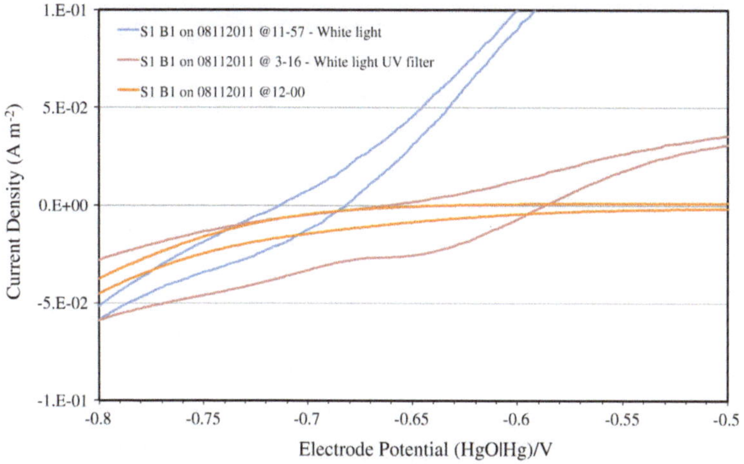

Fig. 6.6 Zoomed view of Fig. 6.5 showing the enhanced reductive current under weak illumination leading to net negative current on negative scan (*pink line*)

less negative current on the negative-going sweep. The O-H phase diagram suggests that this is possible as the equilibrium reduction potential of oxygen is +0.025 V (SHE) or about −0.075 V (HgO|Hg) in 1 M NaOH.

6.6.3 Photocurrent—Reproducibility

The repeatability of the photocurrent measurements were investigated by repeating the voltammetry several times, and with UV filters of different transmittance to change the intensity (Fig. 6.7). As noted above, the addition of UV filters modified both the oxidative photocurrent and the reductive 'dark' current occurring together at the same potentials. Additionally, the start potential of the scan appeared to have a significant effect on the current. Most scans were begun at 0 V and the potential swept to the positive limit. Two scans (green and light purple lines, Fig. 6.7) were begun at +1.5 V and the potential was swept to the negative limit. The maximum current was the same in both cases between −0.2 and +1.5 V. The current for the scans begun at +1.5 V decayed at a potential of −0.2 V, while the current in the scans begun at 0 V decayed at a potential of −0.3 V.

This behaviour was somewhat strange, as at first glance it might appear that the flat band potential of the sample had changed, and hence the onset of photocurrent, by 0.1 V. However, later scans (also starting at close to 0 V) and those with UV filters in line seemed to agree more closely with the first set of data; i.e. the current was scaled rather than shifted along the potential axis.

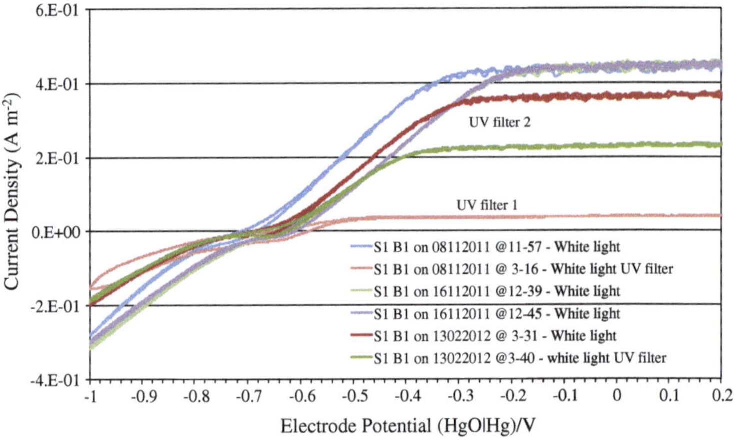

Fig. 6.7 Repeatability of photocurrent on Nb-doped rutile (110) crystal (Sample 1, Batch 1) in 1 M NaOH; scan rate 50 mV s^{-1}. The *pink* and *green lines* are measurements made with two different UV filters. The difference in the maximum photocurrent between the *red line* and the other measurements made under white light without filters was attributed to deterioration of the lamp. Voltammograms started at 0 V with positive-going scans, except for those on 16112011

This suggests that the voltammetry as shown was highly dependent on the potential history, and that even at −0.2 V there may have been significant dark processes occurring in addition to the photo-electrochemical oxidation of water. This may have been due to the accumulation or depletion of reactive species at the surface. The potential scan that began at 0 V was swept to +1.5 V, before passing back through zero to negative potentials. Therefore, the electrode was passing the maximum current for longer than in the case in which the scan was started at +1.5 V. Hence, it may be assumed that the concentration of oxygen at the interface was greater. This would imply that the electrode with the higher concentration of oxygen would reach a larger reduction current at a less negative potential, and result in a quicker drop in the total current. However, this was the opposite of what was detected, as the scan starting at 0 V was expected to have more oxygen present due to the longer time at oxygen evolving potentials. As such, it suggests that the reduction current that opposes the oxidation photocurrent was not due to oxygen, but a reduced species. In the case of the scan starting at +1.5 V less of the species was present.

This merits further investigation. This could be done by starting scans at a range of potentials between 0 and +1.5 V and detecting the shift in potential at which a given current occurs.

The voltammetry of the second batch of as-received Nb-doped crystals is discussed in the following section.

6.7 Nb-Doped Rutile (110)—Batch 2—Voltammetry

The Nb-doped single crystal measured from batch 2 had an active electrode area of 0.30 cm^2.

The open circuit potential (OCP) was measured as +0.013 V (HgO|Hg) in the dark.

6.7.1 Initial 'Dark' Voltammetry

As noted at the beginning of this chapter, a second batch of Nb-doped rutile (110) crystals displayed qualitatively different behaviour to the first batch, which displayed close to ideal capacitive behaviour over a wide range of potential.

Figure 6.8 shows the initial voltammetric measurements made with an electrode fabricated from the second batch of Nb-doped rutile (110), which exhibited additional redox processes between −0.4 and +0.8 V (HgO|Hg) that were not detected on the electrode from the first batch of doped rutile. In addition, current densities were much greater (10^{-3} vs. 10^{-5} A m^{-2}) despite the slower sweep rate of 10 mV s^{-1} versus the 50 mV s^{-1} used for the initial voltammograms of the first batch.

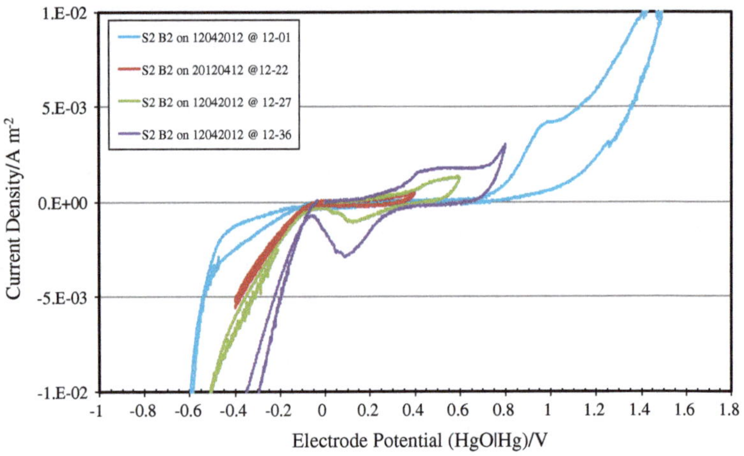

Fig. 6.8 Initial 'dark' voltammograms of Nb-doped rutile (110) crystal (Sample 2, Batch 2) in 1 M NaOH at 10 mV s^{-1}, with the positive potential limit increased between cycles. The voltammogram is truncated to show the detail between −0.2 and +1.0 V

These additional processes were also evident on the reduced undoped rutile (110) electrode, as presented in Sect. 6.8 below and are discussed therein. This similarity leads to the suggestion that the second batch of Nb-doped rutile (110) were received partially reduced from the manufacturer and that the processes exhibited by the electrode were characteristic of a reduced rutile (110) surface. An alternative possibility is that a higher doping level in the second batch resulted in a narrower space charge region, and tunnelling of electrons to and from the solution. The structure of the space charge layer of the Nb-doped and reduced electrodes is analysed in Sect. 6.9. The qualitative difference in behaviour was primarily of interest due to the different results obtained with electrodes from each batch coated in gold nanoparticles, as presented in the following chapter.

6.7.2 *Further 'Dark' Voltammetry*

Following a range of impedance spectroscopic measurements the voltammetry was repeated. The OCP was measured at about +0.14 V (HgO|Hg).

The qualitative behaviour of the electrode had changed and become more like the electrode from the first batch of Nb-doped rutile, demonstrating only capacitive behaviour at potentials positive of −0.4 V. A series of scans at higher sweep rates showed more clearly the redox process peaking at −0.6 V that was discussed for the first batch electrode, and which is believed to be the reduction of oxygen. At high scan rates, oxidation current was evident on the subsequent positive-going sweep, following reduction on the negative-going sweep.

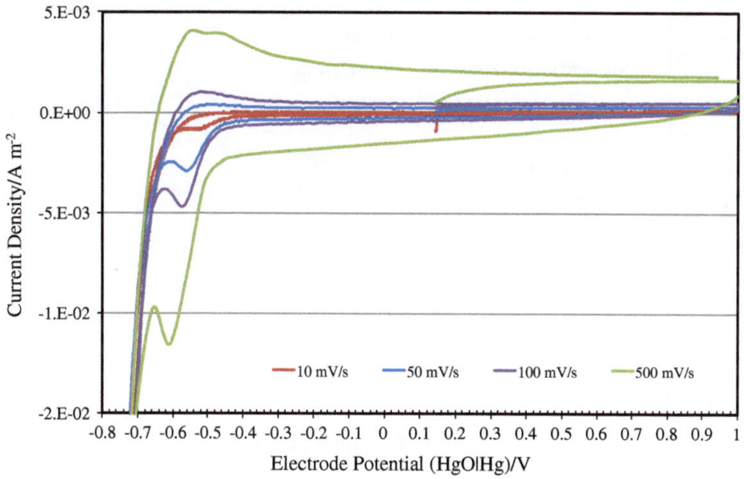

Fig. 6.9 Effect of potential sweep rate on 'dark' voltammograms of Nb-doped rutile (110) crystal (Sample 2, Batch 2) over a potential range of −1.0 to +1.0 V (HgO|Hg)

There was a further, presently unidentified process with an onset at −0.6 V that was insufficiently negative for water reduction at standard conditions. Based on the calculations reported in Appendix B.2, it is possible that in the open cell used to make measurements, that the water reduction potential was closer to −0.6 V due to the low activity of hydrogen in the atmosphere. However, this would indicate an unexpectedly rapid onset for the process, and there are additional processes with higher currents at more negative potentials that may be more easily attributed to water reduction (Fig. 6.9).

6.8 Voltammetry of Reduced (Undoped) Rutile (110)

The single crystal measured had an active electrode area of 0.56 cm^2. The open circuit potential (OCP) was measured as 0.103 V (HgO|Hg) in the dark.

6.8.1 'Dark' Voltammetry

Figure 6.10 shows the initial 'dark' and photo-voltammetry of the reduced rutile (110), which exhibited qualitatively different 'dark' voltammetric behaviour to that of the first batch of Nb-doped rutile. In particular, there was a reduction peak at 0.0 V (HgO|Hg) and oxidation process between +0.3 and +0.8 V, followed by a steep rise in the current at +1.0 V, none of which were present in the voltammograms of the first batch of Nb-doped rutile. The illuminated electrode did not demonstrate

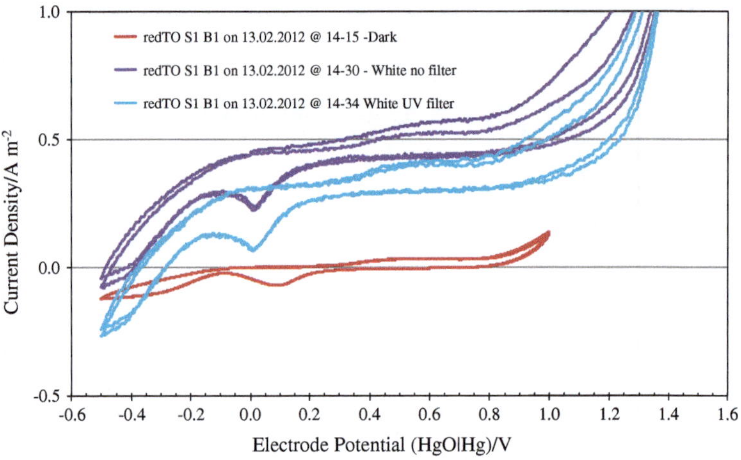

Fig. 6.10 'Dark' and photo-voltammograms of reduced rutile (110) crystal (Sample 1, Batch 1) in 1 M NaOH; potential scan rate 50 mV s^{-1}. The *purple line* is an illuminated measurement and the *blue line* is illuminated with a UV filter (not a good one)

the classic photocurrent behaviour of the Nb-doped crystal (Sample 1, Batch 1). Instead, in addition to the oxidative photocurrent, the magnitude of the reduction processes at potentials negative of +0.2 V appeared to be enhanced by the illumination, presumably due to the previous positive going scan generating oxygen that was subsequently reduced. On the positive-going scan from −0.5 to +1.5 V, there was no obvious oxidation at 0.0 V, and a small peak between +0.3 and +0.8 V. On the subsequent negative-going scan, a reduction process occurred with a current peak at ca. 0.0 V, decreasing the still positive photocurrent significantly. As the electrode should be forming a depletion region positive of −0.8 V, this suggests that electrons are able to be transferred to solution species despite the presence of an electric field driving them away from the interface with the electrolyte. This is an indication that either the space charge region is extremely narrow, or that there are charge transport pathways through the space charge layer, to the electrolyte. It is suggested that these pathways may be the Ti_2O_3 like Magnéli phases that have been detected in reduced rutile [2, 4]. These could trap electrons and then act as a conduit for metallic conduction of the electrons to the surface whereupon they can reduce a species in solution, bypassing the space charge region.

This demonstrates that reducing rutile in hydrogen reduces the charge-separation efficiency of the electrode, and its photo-electrochemical performance. If the electrons are lost to the reduction of species at the electrode, then they are unavailable for the hydrogen evolution reaction, regardless of whether the photo-generated holes oxidize water to evolve oxygen. The onset of an oxidation current at +1.0 V also indicated that the width of the space charge region was significantly thinner than for the first batch of Nb-doped crystals, which displayed only capacitive behaviour.

6.8.2 Redox Process on Reduced Rutile

To further investigate the redox processes occurring between −0.2 and +0.8 V (HgO|Hg), a set of voltammograms was run with progressively incremented positive potential limits, as shown in Fig. 6.11. The scans demonstrate that the oxidation between +0.3 and +0.8 V was linked to the reduction process that occurred between +0.2 and −0.2 V. The greater the current passed in the oxidation, the greater the reduction current. There is an additional oxidation process visible from +0.8 to +1.2 V, which was then overwhelmed by a further oxidation that was presumably due to oxygen evolution. The additional charge passed by the oxidation process from +0.8 to +1.2 V was greater than the additional charge passed on the subsequent reduction, compared to the reduction of a previous scan with a lower positive potential limit (for example the blue versus the red line). This suggested that the additional reduction current was due only to the first oxidation process from +0.3 to +0.8 V. The identity of this process is unknown, but based on the potential-pH diagrams of Sect. 6.5, it is suggested that this may be the oxidation and reduction of reduced phases at the surface of the electrode. The oxidation between +0.8 and +1.2 V may have been the oxidation of water by defect states or planes at the surface. Once a sufficiently positive potential of +1.2 V was reached, the oxidation process proceeded readily via the conduction band of rutile.

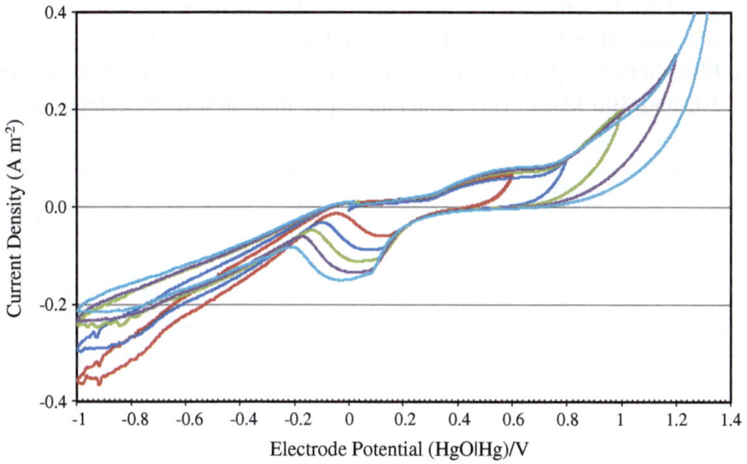

Fig. 6.11 'Dark' voltammetry of reduced rutile (110) crystal (Sample 1, Batch 1) in 1 M NaOH with positive potential limits incremented by 0.2 V between cycles; scan rate 10 mV s^{-1}. Structure of the Space Charge Region of Rutile (110)

6.9 Structure of the Space Charge Layer of Rutile (110)

This section presents the results of analysis that applied the Mott-Schottky equation to the capacitance of the space charge region. This was used to determine the charge carrier concentration of the electrodes and to estimate the flat band potential.

The expected charge carrier concentration of the as-bought electrodes was calculated to provide an indication of the expected doping concentration. The manufacturer specified the concentration of Nb doping as 0.05 % by total weight of the crystal; i.e. in $Nb_xTi_{1-x}O_2$, $x \approx 0.0005$.

The volume of the rutile TiO_2 unit cell is 62.43×10^{-30} m^3 (XRD reference pattern 00-021-1276). This means there are 1.55×10^{29} unit cells per m^3 ($\equiv 1.55 \times 10^{23}$ cm^{-3}). There are two Ti atoms per unit cell or a concentration of 3.1×10^{29} m^{-3}. Applying the doping ratio of 0.0005 Nb to Ti atoms results in a doping concentration of 1.6×10^{26} m^{-3} (1.6×10^{20} cm^{-3}). This would be expected to give a good indication of the doping concentration and conductivity.

For comparison, typical doping concentrations in Si or GaAs solar cells are in the range 10^{22}–10^{25} m^{-3} (10^{16}–10^{19} cm^{-3}) [5].

6.9.1 Mott-Schottky Analysis—Nb-Doped TiO_2 (Batch 1)

The flat band potential and actual charge carrier concentration were estimated by applying the Mott-Schottky equation to a measurement of the interfacial capacitance, as described in Chaps. 3 (Theory) and 4 (Experimental Methods).

The Mott-Schottky plot is shown in Fig. 6.12 (for 10 and 100 Hz) and Fig. 6.13 (for 1, 10, and 100 kHz). The capacitance per unit area of the electrode C_p was

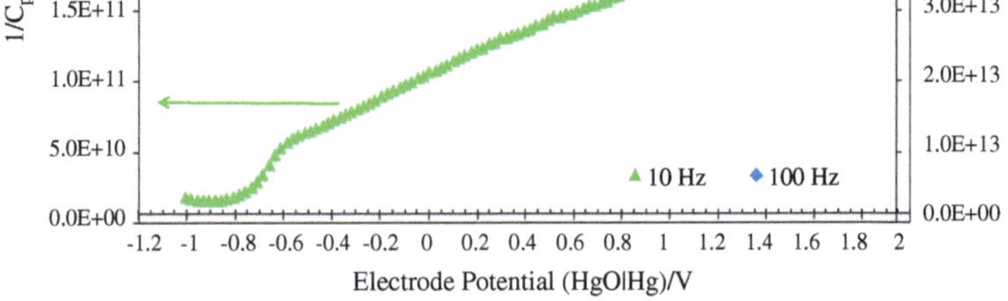

Fig. 6.12 Mott-Schottky plot of Nb^{V}-doped rutile (110) crystal (Sample 1, Batch 1) in 1 M NaOH. The frequency was fixed at 10 and 100 Hz and the potential varied from −1.0 to 2.0 V in 0.025 V increments. The capacitance, C_p, was calculated by assuming a model for the impedance

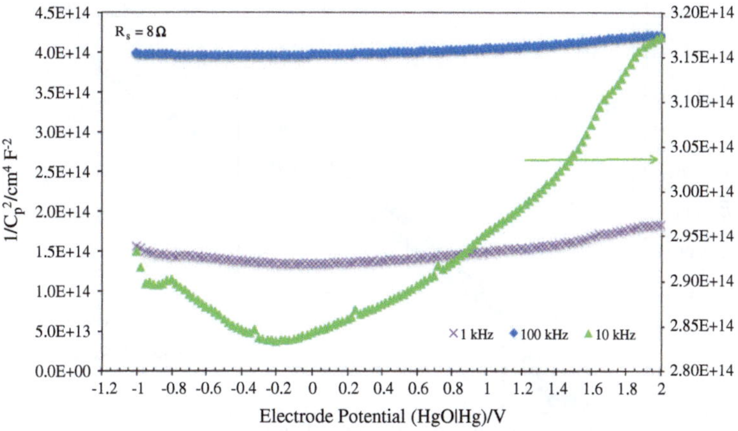

Fig. 6.13 Mott-Schottky plot for Nb^V-doped rutile (110) crystal (Sample 1, Batch 1) in 1 M NaOH. The frequency was fixed at 1, 10 or 100 kHz and the potential varied from −1.0 to 2.0 V in 0.025 V increments. The capacitance, C_p, was calculated by assuming an electronic analogue model for the impedance

originally calculated automatically by the NOVA software from a fit of the impedance data with an R_s-R_pC_p equivalent circuit (see Experimental Methods, Fig. 4.18). The measurement was repeated on several occasions and gave similar results. A typical Mott-Schottky plot was not possible for the data other than at 10 Hz, suggesting that the equivalent circuit assumed may have been unsuitable. A further point to note is the presence of two linear regions in the Mott-Schottky plot at 10 Hz. Extrapolating both linear regions to the potential axis gave two different estimates for the flat band potential. The flat band potential should be estimated from the initial, steeper region. This is because the doping concentration varies with potential, resulting in a change in slope [6]. If the impedance measurement does not reach the point at which $1/C_p^2$ tends to zero, then it is possible to miss the initial steep region, and hence any extrapolation for the flat band potential will result in an overly negative estimate. This is evident in the relevant literature in general and in particular on measurements made on hematite electrodes. This extrapolation error may result in a wide spread of reported flat band potentials [7].

The data gathered for the electrode did not display behaviour consistent with the Mott-Schottky equation, making analysis difficult. Only the measurements made at 10 Hz provided a reasonable set of data for analysis (Fig. 6.12). In particular, the measurements made between 1 and 100 kHz (Fig. 6.13) indicate an approximately constant capacitance across a large potential range. This is not expected for a semiconductor/dielectric electrode. It also precludes the estimation of the flat band potential. Extrapolating the 'linear' data leads to a flat band in the region of −40 V (HgO|Hg), which is unrealistic.

The data taken at 10 Hz does show a more typical Mott-Schottky type behaviour. Assuming for the moment that the data for 10 Hz is adequate, estimates for

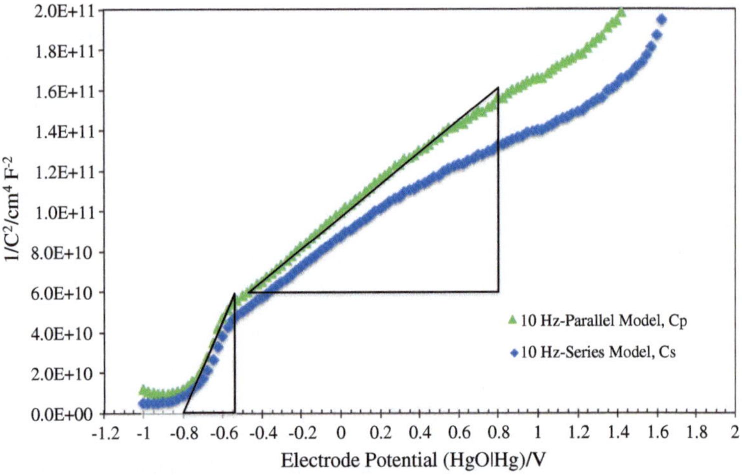

Fig. 6.14 Mott-Schottky plot of Nb^{V}-doped rutile (110) crystal (Sample 1, Batch 1) in 1 M NaOH. The frequency was fixed at 10 Hz and the potential varied from −1.0 to 2.0 V in 0.025 V increments. The capacitance was calculated by assuming a model for the impedance of both a series and parallel configuration. The assumed series resistance was R_s = 10 kΩ

both the flat band potential and the charge carrier concentration can be made. The capacitance was estimated using both a series model and a parallel model for the resistance and capacitance. The comparison is shown in Fig. 6.14.

The data showed a steep rise in $1/C_p^2$ from −1 to −0.6 V, including the characteristic linear region. There was a second linear region from −0.6 V to more positive potentials, with a shallower slope. Extrapolating the second linear region to the potential axis (Fig. 6.14) gives the flat band to be at about −1.2 V (HgO|Hg). This is unlikely as it is ca. 0.3 V negative of the hydrogen evolution potential (spontaneous water splitting on rutile (110) has not been observed). If the extrapolation is made from the first linear section of the plot, the flat band potential is estimated to between −0.8 and −0.9 V. This is more realistic, but accepted with caution due to the inability to fit other frequencies with the equivalent circuit assumed.

The slope of a Mott-Schottky plot should be given by $(2/\varepsilon q_e N_D)$, where ε is the permittivity, N_D is the charge carrier concentration, and q_e is the electronic charge. The slope of the linear region of the plot (Fig. 6.14) from −0.75 to −0.5 V is about 2.3×10^{11} F^{-2} cm^4 V^{-1}. Assuming a relative permittivity of ε_r = 170, the charge carrier density was calculated as 3.6×10^{18} cm^{-3} (3.46×10^{24} cm^{-3}). The slope of the plot positive of −0.6 V corresponds to a charge carrier concentration of 1.04×10^{19} m^{-3}. It is suggested that as the potential passes through −0.6 V, previously un-ionized impurities are ionized, and the charge carrier concentration increases by a factor of three. Comparing this to the theoretical estimate of 1.6×10^{20} cm^{-3} it seems that the actual charge carrier density is about two orders of magnitude lower than expected from the specified doping concentration.

Attempted measurements of the conductivity (by four-point probe and multi-meter) showed the crystal to be highly resistive.

As mentioned above, the simple equivalent circuit used was not suited to the analysis of the data. A more rigorous approach, whereby the Nyquist plot is fitted by a unique equivalent circuit at each potential was attempted and is discussed in Sect. 6.11.

6.9.2 *Mott-Schottky Analysis—Nb-Doped TiO_2 (Batch 2)*

The flat band potential and actual charge carrier concentration were also estimated for the second batch of Nb^V-doped rutile (110) crystals. As with the first batch of rutile, the impedance data was analysed by assuming an equivalent circuit of a resistor in series with a resistor and capacitor in parallel. It was necessary to assume a series resistance R_s of about 4 Ω to calculate the parallel components, where R_s represents the solution resistance.

The result of the calculated value $1/C_p^2$ against electrode potential is shown in Fig. 6.15. The measurements show characteristic Mott-Schottky behaviour over the whole potential range. As with the first batch, the data had two linear regions, which corresponded to potential ranges with different doping concentrations. Interestingly, the slopes at each frequency differed slightly, indicating frequency dependence. There was also frequency dispersion in the apparent flat band potential, ranging from −0.99 V at 100 kHz to −0.72 V at 10 Hz. The frequency dependence of the flat band potential and charge carrier concentration are shown in Figs. 6.16 and 6.17.

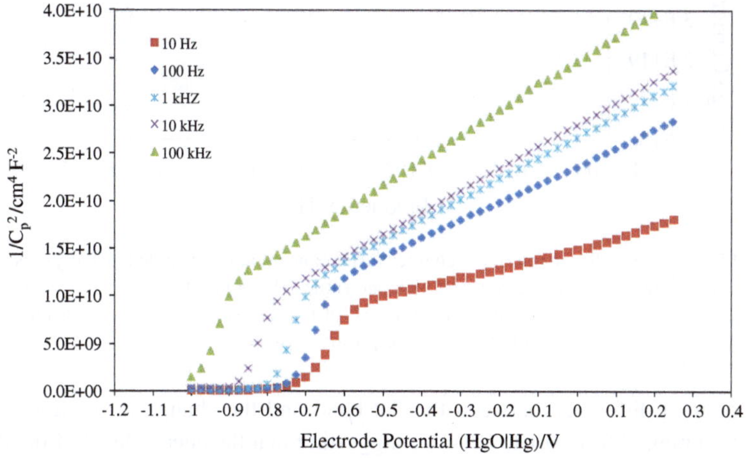

Fig. 6.15 Mott-Schottky plot of Nb-doped rutile (110) crystal (Sample 2, Batch 2) in 1 M NaOH. The frequency was fixed at 10, 100 Hz, 1, 10, or 100 kHz and the potential varied from −1.0 to 0.25 V in 0.025 V increments. The capacitance was calculated by assuming a model for the impedance of a parallel configuration, with R_s assumed to be 4 Ω

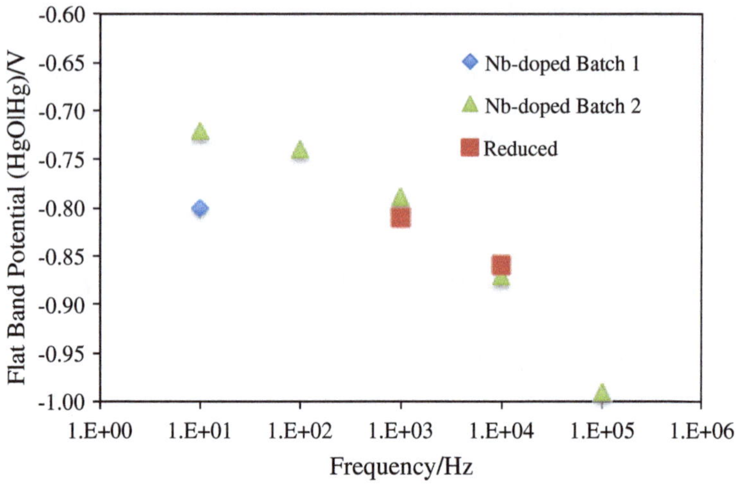

Fig. 6.16 Frequency dispersion of the flat band potential of rutile (110) crystals for both batches of Nb^{V}-doped rutile and reduced undoped rutile

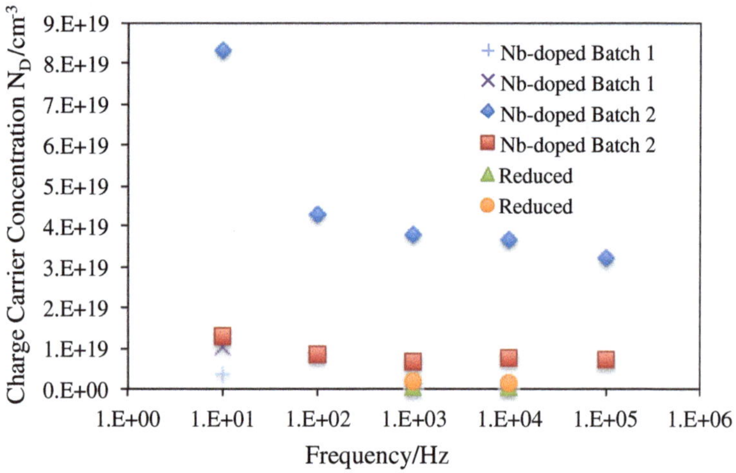

Fig. 6.17 Frequency dispersion of the charge carrier concentration of rutile (110) crystals for both batches of Nb^{V}-doped rutile and reduced undoped rutile. The concentration is shown for both the partially ionized dopant concentration just positive of the flat band potential (lower number) and the fully ionized dopant concentration at more positive potentials

The potential difference between the flat band and the change in the doping concentration ranged from 0.1 to 0.14 V, suggesting that the energy level of the dopant is 0.1–0.14 eV below the conduction band minimum. The low doping concentration means that the Fermi level will lie close to the conduction band.

Comparing the doping of the first batch of crystals at 10 Hz to the data for the second batch, it is clear that the concentration (and by implication the conductivity)

in the second batch was about ten times that of the first batch. However, in both cases the measured charge carrier concentration was an order of magnitude below the specified doping concentration of the material.

While the charge carrier concentration may vary with potential, due to ionization of impurities, the frequency dependence cannot be explained in the same way. A possible explanation is that it is due to frequency dependence of the relative permittivity. It is not possible to calculate both the permittivity and the charge carrier concentration exactly from the data without a separate and independent measurement of one or the other.

A likely possibility is that the frequency dispersion is introduced by using a non-physically representative equivalent circuit. The use of realistic equivalent circuits is discussed in Sect. 6.11 based on the work of others [8]. This would help to explain the dispersion in the flat band potential, as well as the slope of the Mott-Schottky plot.

6.9.3 Mott-Schottky Analysis—Reduced TiO_2

The flat band potential and actual charge carrier concentration was also estimated for an undoped rutile (110) crystal that had been reduced in a hydrogen atmosphere (see Experimental Methods for method of reduction). As with the Nb^{V}-doped material, the Mott-Schottky plot demonstrates two linear regions that should correspond to different charge carrier concentrations due to the ionization of an impurity.

The flat band potential was estimated at −0.78 V at 1 kHz and −0.86 V at 10 kHz. These values are within 0.01 V of the Nb^{V}-doped samples at the same frequency. The charge carrier concentration varies from 6.5×10^{17} m^{-3} to 1.6×10^{18} m^{-3} for the 1 kHz measurement and from 4.75×10^{17} m^{-3} to 1.3×10^{18} m^{-3} for the 10 kHz measurement. These values are an order of magnitude lower than the Nb^{V}-doped electrode (Sample 2, Batch 2), which would suggest lower conductivity, and lower current at a given potential. The potential at which the slope of the Mott-Schottky plot changed was also about 0.14–0.18 V from the flat band potential, as with the Nb^{V}-doped samples (Fig. 6.18).

Interestingly, the Mott-Schottky plot of the reduced sample was both qualitatively and quantitatively similar to the Nb^{V}-doped samples, but the crystal was not doped. The as-received undoped material was transparent and colourless, while the doped material was a deep semi-transparent blue. After reduction, the undoped material turned blue as expected [9].

Given the lack of doping in the reduced sample, and the similarity of the Mott-Schottky plots, this suggested that the energy level the donor impurity ionized to produce mobile charge carriers was the same in both cases. In the case of Nb^{V}-doped TiO_2, it is usually assumed that the Nb^{5+} impurity provides the electron to the conduction band. In reduced TiO_2, it is assumed that the electron comes from Ti^{3+}. Therefore, it is possible that the donor level of Nb^{5+} and Ti^{3+} are both at about 0.1–0.18 V below the conduction band of rutile. Another likely alternative is that

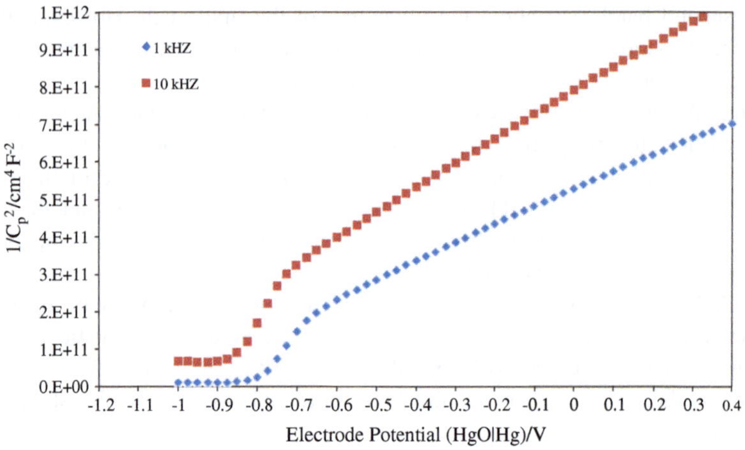

Fig. 6.18 Mott-Schottky plot of reduced undoped rutile (110) crystal (Sample 1, Batch 1) in 1 M NaOH. The frequency was fixed at 1 or 10 kHz and the potential varied from −1.0 to 1.0 V in 0.025 V increments. The capacitance was calculated by assuming a model for the impedance of a parallel configuration, with R_s assumed to be 7.5 Ω at 10 kHz and 20 Ω at 1 kHz

the additional electron associated with an Nb impurity does not reside on the Nb ion, but with a Ti ion. For example, it might be energetically more favourable for the electron to reduce a Ti^{4+} ion to Ti^{3+} rather than remain bound to the Nb^{5+} site. In this case, the donor level in both the reduced and doped samples would correspond to Ti^{3+}: an electron loosely bound to a Ti^{4+} lattice site.

In simple terms, this can be demonstrated to be favourable. The Nb atom has 41 protons and electrons, while the Ti atom has 22 of each. The ground state electron configuration of the Ti atom is [Ar] $3d^2$ 4 s^2, so Ti^{4+} is [Ar]. The ground state configuration of Nb is [Kr] $4d^4$ 5 s^1, so Nb^{5+} is [Kr]. Only four of the electrons in the outer shell of Nb can contribute to the ionic bonding as O^{2-} lattice sites in rutile, leaving a spare electron. It should be energetically more favourable for this electron to fill the 4 s shell of the Ti cation rather than the 4d shell of the Nb cation. Therefore, it seems probable that the donor impurity that is ionized with increasingly positive potentials is in fact the Ti^{3+} cation, losing an electron to the conduction band and resulting in Ti^{4+}.

The change in slope in the Mott-Schottky plot at around −0.7 V (HgO|Hg) also corresponds to the reversible potential of the reduction of TiO_2 (hydrated) to Ti_2O_3 (hydrated) as shown in Appendix Figure B.3. It could be suggested that the change in slope might correspond to the changes in the electrode surface. This possibility can be dismissed because the charge carrier concentration increases (slope of plot decreases) with increasingly positive potential. If Ti_2O_3 were oxidized to TiO_2, the charge carrier concentration would be expected to decrease, which contradicts the capacitance measurement.

6.9.4 Discussion of Mott-Schottky Analysis

Based on the Mott-Schottky analysis of the capacitance of the Nb^{V}-doped and reduced electrodes, it appears that the charge carrier concentration of all samples varies between 10^{18} and 10^{20} cm^{-3} depending on the ionization of impurities. This is between two orders of magnitude below and up to the expected doping concentration on the order of 10^{20} cm^{-3}. Given that it was not possible to image any of the material using scanning tunnelling microscopy (STM), it suggests that this might be an insufficient doping level to give a conductive sample.

The charge carrier density was used to calculate the width constant of the space charge layer, w_0 as shown in Fig. 6.19. For the second batch of Nb^{V}-doped rutile, this was on the order of 20–50 nm.

Given the similarity of the voltammetry for the reduced and second batch of Nb^{V}-doped rutile, it was surprising that the estimated charge carrier concentration of the reduced material was lower by an order of magnitude, and closer to the value for the first batch of Nb^{V}-doped rutile. This suggested that the distribution of charge carriers or the nanostructure of the reduced material may be slightly different to the doped material. It is probable that the charge carrier concentration is determined by the oxygen partial pressure during annealing of the crystals, and that the pressure used resulted in ionic compensation of some of the Nb^{5+} impurity. In the case of the first batch of Nb^{V}-doped rutile, it seems that the oxygen pressure may have been higher, resulting in only 10 % as many charge carriers.

It is also surprising that an order of magnitude difference in the charge carrier concentration had such a significant impact on the behaviour of the electrode, as measured by cyclic voltammetry. It is possible that the slightly narrower space charge layer (which varies with the square root of the charge carrier concentration) allows tunnelling of electrons from the solution species to the conduction band of the rutile (through the potential barrier of space charge region), while the slightly wider space charge layer does not. This suggests that the critical charge carrier concentration would be between 10^{19} and 10^{20} m^{-3}. Below the threshold, the electrode displays only non-Faradaic current, while above the threshold the oxidation of species in solution is possible.

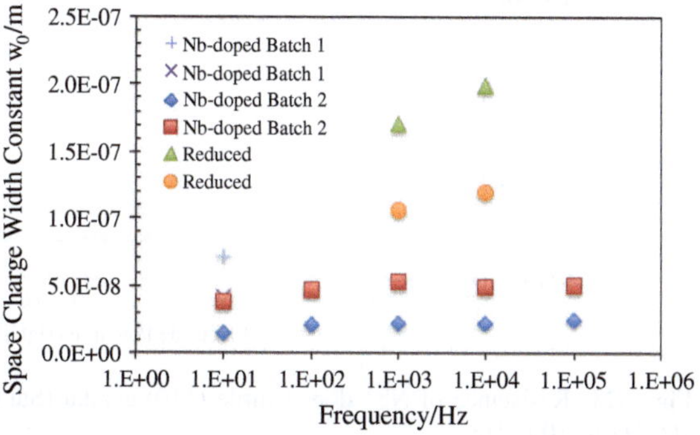

Fig. 6.19 Frequency dispersion of the space charge width constant of a Nb-doped rutile (110) crystal (Sample 2, Batch 2). Data is shown for 10 Hz to 100 kHz. The width constant is shown for both regions of the Mott-Schottky plot for partially and fully ionized dopants

6.9.5 *Resistance Measurements*

To further interrogate the initial impedance measurements, the frequency and potential dependence of the resistance of the electrodes was also calculated. The resistance per unit area of the electrode is shown in Figs. 6.20, 6.21, 6.22 and 6.23.

Comparing data at 1 kHz, the resistance per unit area of electrode of the first and second batch of Nb^V-doped rutile was $1 \times 10^{-4}\ \Omega\ cm^{-2}$ whereas for the reduced material, it is $3 \times 10^{-3}\ \Omega\ cm^{-2}$. For data at 10 kHz, the resistance of the first and

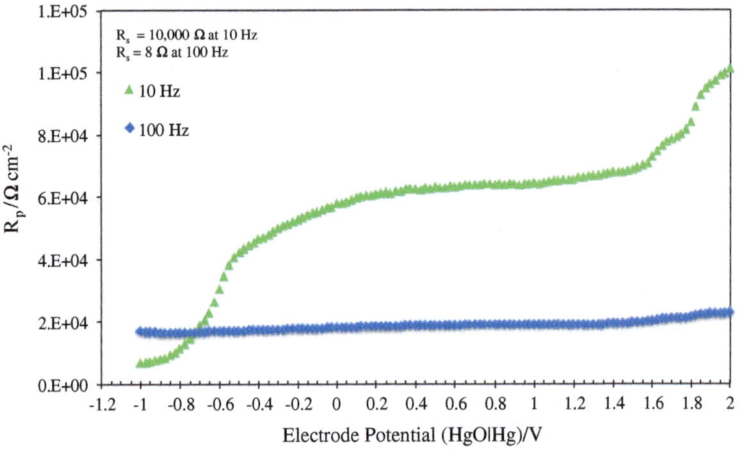

Fig. 6.20 Resistance of Nb^V-doped rutile (110) crystal (Sample 1, Batch 1) for 10 and 100 Hz

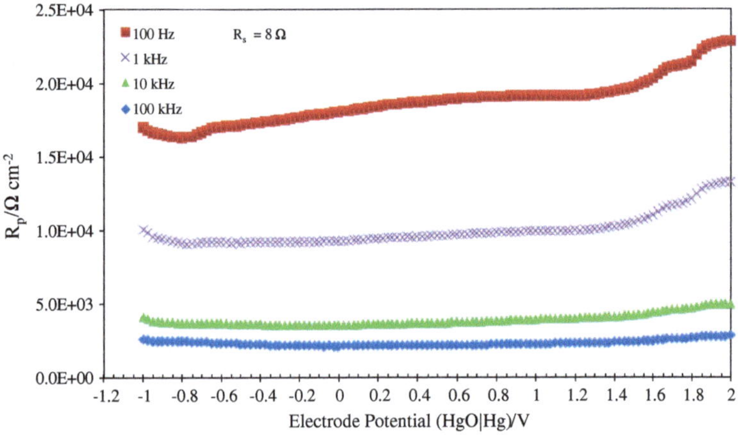

Fig. 6.21 Resistance of Nb^V-doped rutile (110) crystal (Sample 1, Batch 1). Data is shown for 100 Hz to 100 kHz

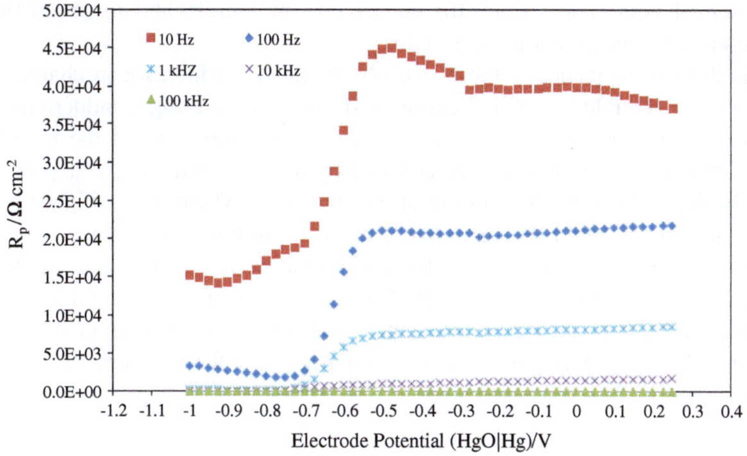

Fig. 6.22 Resistance of Nb^{V}-doped rutile (110) crystal (Sample 2, Batch 2). Data is shown for 10 Hz to 100 kHz

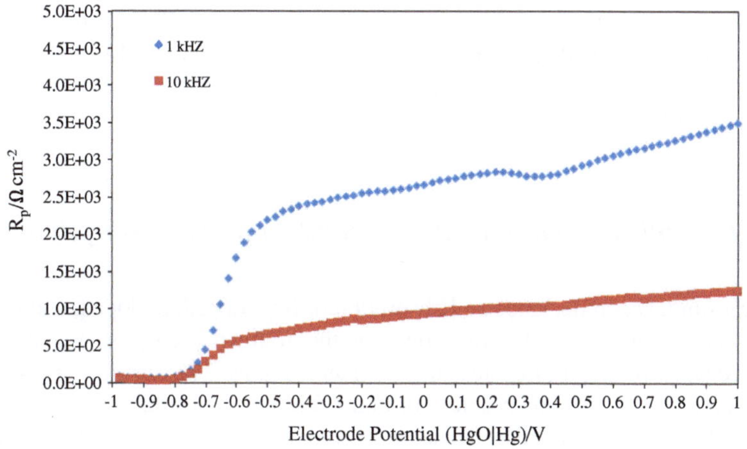

Fig. 6.23 Resistance of a reduced rutile (110) crystal. Data is shown for 1 and 10 kHz

second batch of Nb^{V}-doped rutile was 4×10^{-3} and 8×10^{-3} Ω cm^{-2}, respectively, whereas for the reduced material, it was 1×10^{-3} Ω cm^{-2}.

The resistance was clearly lower for the reduced electrode, despite having a lower charge carrier concentration according to the capacitance. The inconsistencies in resistance with frequency between the batches of Nb^{V}-doped material (same at 1 kHz, different at other frequencies) suggested that either the structure of the material differs, or that the data fitting was inadequate. It was also surprising that the resistance appeared to increase between potentials of −0.7 and −0.5 V. This is the potential range over which the capacitance measurements suggested that the charge carrier concentration increased.

It is difficult to reconcile these two observations: an increase in charge carrier concentration coupled with increasing resistance (decreasing conductivity). The conductivity is a product of the charge carrier concentration and the mobility of charge carriers. The mobility is determined by the scattering rate, which is in turn partially determined by the concentration of defects. Therefore, a higher concentration of dopants could (but is unlikely to) result in higher resistance.

A more likely reason for the increased resistance is that as the bands are bent, an electric field develops that opposes the flow of electrons from the semiconductor to the electrolyte. The measured resistance is likely to be a combination of the retarding effect of the electric field at the interface, and the increased charge carrier density.

6.10 Photocurrent Measurements

The characteristics of the space charge region, as estimated by Mott-Schottky analysis, are now used to attempt to model the measured photocurrent using the equations derived in Chap. 2 (Principles of Photoelectrochemical Cells). First, the optical characteristics, i.e. the band gap and absorption coefficient of the Nb^{V}-doped and reduced rutile (110) single crystals were determined.

6.10.1 Optical Characteristics—Band Gap and Absorption

The transmittance, reflectance and absorption of Nb^{V}-doped, undoped, and reduced rutile (110) single crystals are shown in the following chapter, together with measurements of samples coated in gold nanoparticles. The absorption coefficient of the bare crystals is shown below in Fig. 6.24. The measured transmittance was corrected by the measured reflectance to obtain the transmission losses due to absorption. The absorption coefficient was calculated from the Beer-Lambert law (described in the chapter, Experimental Methods).

The absorption coefficient measurements indicated a band gap between 400 and 420 nm. The absorption peaked at 15,000 m^{-1} for the undoped sample and at 18,000 m^{-1} for the doped and reduced samples. The difference was due to higher reflectance of the doped and reduced samples. This could be attributed to the higher concentration of free electrons, the effect of which was also evident in the absorption in the visible and infrared part of the spectrum. Cronemeyer [10] reported absorption coefficients of 10,000 to 20,000 m^{-1} around 400 nm for rutile crystals over a range of temperatures up to room temperature; such values are of the same order of magnitude as the measurement reported here. The absorption

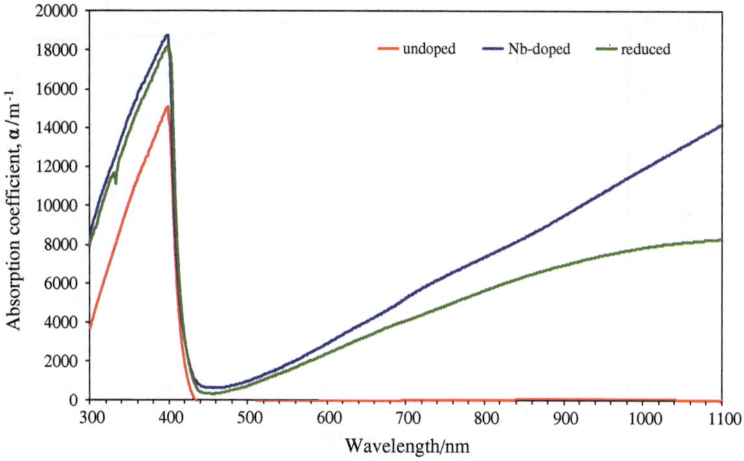

Fig. 6.24 Absorption coefficient of rutile (110) crystals measured by UV-V is spectroscopy. Data is shown for undoped (*red line*), Nb-doped (*blue line*), and reduced (*green line*) samples. The coefficient was calculated from the Beer-Lambert law with the measured thickness of the crystals (0.2–0.5 mm)

coefficient data also showed that the reduction and doping results in a slight shift in the absorption edge to lower energies in the order of tens of meV.

Absorption coefficients of 15,000–18,000 m^{-1} are extremely small compared to typical solar cell materials such as Si and GaAs, which have absorption coefficients in the range 10^8–10^{10} m^{-1}. The low absorption was also demonstrated by the presence of significant secondary reflections from the back of the sample that were the convolution of the transmittance and reflectance spectra. Such a low absorption coefficient is generally indicative of a material with an indirect bandgap. A proper determination of the band gap is possible from a Tauc plot of $(\alpha h\nu)^n$ against $h\nu$, the photon energy, where n = 0.5 for an indirect transition and n = 2 for a direct transition. The Tauc plots in Fig. 6.25a, b suggest that there was a direct band gap at about 3.03 eV for undoped rutile (110), which shifted to 3.02 eV for reduced rutile. The indirect Tauc plot does not have an obviously linear region. A survey of the theoretically and experimentally determined band gaps of rutile by Ekuma and Bagayoko [11] suggested that rutile does indeed have a direct band gap between 3.0 and 3.05 eV, which agrees with the measurement here. They also note the possibility of an indirect transition at 2.95 eV, which could almost be inferred from the indirect Tauc plot.

6.10.2 *Modelling the Photocurrent*

An attempt was made to fit the photocurrent data using the model outlined in Chap. 2.

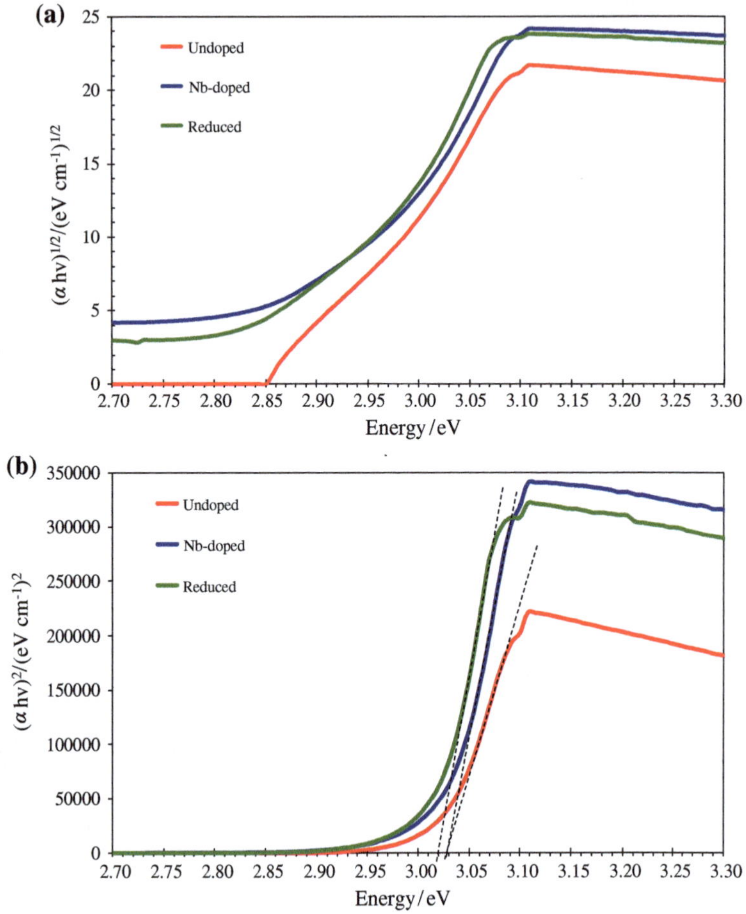

Fig. 6.25 Tauc plot of rutile (110) crystals. Data is shown for undoped (*red line*), Nb-doped (*blue line*), and reduced (*green line*) samples. The direct bandgap model fitted the data better

The flat band potential of rutile TiO_2 is reported typically to be slightly positive of water reduction at ca. −0.8 V (SHE). This is consistent with the Mott-Schottky analysis described in previous paragraphs. Based on visual inspection of the dark and total current density-potential data, the onset of photocurrent appeared to be in the region of −0.7 V (HgO|Hg), which is broadly commensurate with reported data [12] and the results of the Mott-Schottky analysis. A fit using a Gärtner type model demonstrates that it is possible to model the potential range −0.4 to +1.5 V and capture the gradual increase of photocurrent with band bending (Fig. 6.26). The fit took the following form and values:

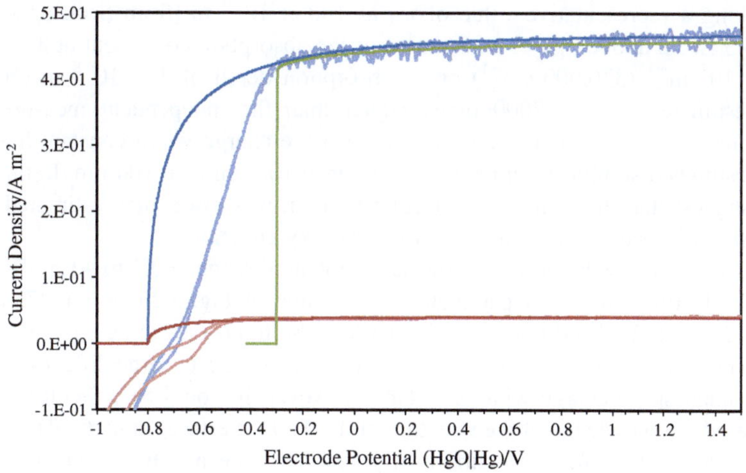

Fig. 6.26 Fit of the photocurrent of Nb-doped rutile (110) using the Butler Gärtner model outlined in Chap. 2. Fit parameters described in main text

$$j_{TOT} = A\left[1 - Be^{-C\sqrt{\Delta\phi_{SCR}}}\right] \quad (6.1)$$

- Flat band potential of −0.8 V (HgO|Hg) to calculate $\Delta\phi_{SCR}$ from electrode potential
- A = 0.47, B = 1, C = 3.3

It is noted that B should take values of 1 or less to be physically realistic.

Given that $B = 1/(1+\alpha L_p)$ and $C = \alpha w_0$, this indicates an electrode with a space charge width constant w_0 larger than the absorption depth α^{-1}, and a small diffusion length. However, the model fits extremely poorly for the first 0.4 V of band bending. Attempts to fit the onset current between −0.8 and −0.4 V by changing the fit parameters were unsuccessful. It was not possible to obtain any reasonable fit of the model in this region, even ignoring the fit from −0.4 to +1.5 V. The conclusion was that the shape of the total current density-potential data in this range of potential does not follow the form of the fit Eq. (3.79). It is suggested that in addition to the photo-electrochemical oxidation of water, simultaneously there is a reduction process occurring that changes the shape of the current density-potential data, preventing analysis by fitting. The fit of the data recorded with a UV filter in place also had the same problems. The smaller photocurrent of the measurement made with a UV filter showed that there was indeed a cathodic dark current occurring simultaneously with the photocurrent between −0.8 and −0.6 V (HgO|Hg). As discussed previously this may be the reduction of photo generated O_2.

Taking a representative value of w_0 as 100×10^{-9} m (from Sect. 6.9.4), and C = 3.3 (from above), gives an estimate of the absorption coefficient at 375 nm of 3.3×10^7 m^{-1} (330,000 cm^{-1}) or an absorption depth of 3×10^{-8} m (30 nm). This estimate is about 2000 times higher than the independent measurements presented above. This may suggest that the space charge width constant has been underestimated significantly; i.e. it was closer to 100 μm than 100 nm. In turn, this may suggest that the charge carrier concentration was lower than estimated using analysis of impedance data by the Mott-Schottky equation.

In fact, it is possible to model the same potential range (−0.4 to +1.5 V) with a completely different set of parameters (green line of Fig. 6.26, A = 0.47, B = 8, C = 1.2, and a flat band potential of −0.3 V). The fit appears to be equally good if only a portion of the data is considered. This would suggest a large diffusion current and a small space charge width constant. However, it would be difficult to justify this based on knowledge of the structure of the TiO_2 interface. In particular, there is no way to explain the photocurrent at potentials more negative than of −0.3 V. While the original fit was also poor at such potentials, the deviation from the model has a possible explanation. Furthermore, taking B = 8 is physically unrealistic.

The incident light was also partially reflected internally, and was further absorbed in the space charge region, adding to the photocurrent, and modifying the shape of the current density-potential data. A model for the additional photocurrent due to reflections has not been attempted here.

The charge carrier concentration and space charge width constant varies continuously from the flat band potential at about −0.8 V to about −0.4 V, due to the ionization of impurities and dopants. This is demonstrated in Sect. 6.9. A model applying a variable charge carrier concentration should be developed.

Overall, this inability to fit the Gärtner model to the data precludes the use of the Butler-Gärtner approach to estimating the flat band potential from the photocurrent. More complex models, including numerical rather than analytical, have been reported [13]. However, these do not take into account 'dark' processes happening at the electrode surface, and it is re-iterated that in the photocurrent data there appears to be a reduction occurring over the same potential range as the onset of photocurrent, precluding fitting the data with a model of only the photocurrent. To make a meaningful analysis of the total current, it is necessary either to obtain the photocurrent alone (by removing the dark current), or to model both dark and light processes in the same model, de-convoluting the peaks. The de-convolution of peaks in semiconductor electrochemistry is not a standard technique and has not been attempted here.

The photocurrent data obtained with mechanical chopping of the light and a lock-in amplifier also demonstrated a similar shape close to the onset (i.e. not in accordance with the Gärtner model). Even though the dark current was removed, the cathodic process was linked to the photocurrent: the photocurrent generates oxygen, which was then reduced. As such, it was not possible to remove the reduction process from the data.

6.11 Impedance Spectroscopy—Equivalent Circuits

The initial impedance spectroscopic measurements used a simple assumed equivalent circuit to determine the space charge layer capacitance, from which to estimate the flat band potential and the charge carrier concentration. However, as discussed in Sect. 6.9.5, the resistance data derived from the impedance was incompatible with the capacitance data. It was assumed that this was due to the use of an unrepresentative equivalent circuit.

To test this hypothesis, additional impedance spectroscopic measurements were carried out. The potential was fixed during each measurement, and the frequency was varied from 100 kHz down to 0.1 Hz. The measurements were repeated for a range of potentials from −1.0 V (HgO|Hg) to +1.0 V. The measurement generated a Nyquist plot of Z' versus $-Z''$ at each potential. The fitting routine of the NOVA software was used to fit a range of equivalent circuits to the data, of increasing complexity. Initially, only capacitive and resistive elements were used. However, it was necessary to introduce constant phase elements in place of capacitive elements to obtain a good fit (as measured by the χ^2 value calculated for each fit by NOVA). Constant phase elements and equivalent circuits are discussed further in Chap. 4 (Experimental Methods).

Klahr et al. [8] used this technique to good effect to characterize the impedance of hematite electrodes.

6.11.1 Nb^V-Doped Rutile (110)−Sample 2, Batch 2

Impedance data for the Nb^V-doped rutile (110) crystal (Sample 2, Batch 2) were measured from −1.0 to +1.0 V (HgO|Hg) in increments of 0.1 V. At potentials from −0.2 to +0.5 V, the gathered impedance data was unreliable. Instead of the characteristic 'arcs' with |Z| increasing with frequency, the measurement would reach a certain frequency and the impedance would return to a much lower value of |Z|. From +0.6 to +1.0 V, the measurements returned to the expected behaviour. When the measurements were repeated, the same effect was seen.

A selection of data for potentials in the range −1.0 to −0.3 V is shown below in a Nyquist plot in Fig. 6.27 and for potentials +0.6 to +1.0 V in Fig. 6.28. The plots show that the impedance decreased as the potential was increased from −1.0 to −0.8 V, at which it reached a minimum, before increasing from −0.7 V and above. As the potential approached −0.3 V (HgO|Hg), the imaginary component of the impedance increased dramatically and the impedance approached purely capacitive behaviour at lower frequencies. At potentials positive of +0.6 V, the imaginary component of the impedance begins to reduce again, and an arc becomes apparent.

The large imaginary component of much of the impedance data was attributed to a combination of a high resistance, a small capacitance, and no charge transfer processes, only charging of the space charge layer. As shown in the chapter on the

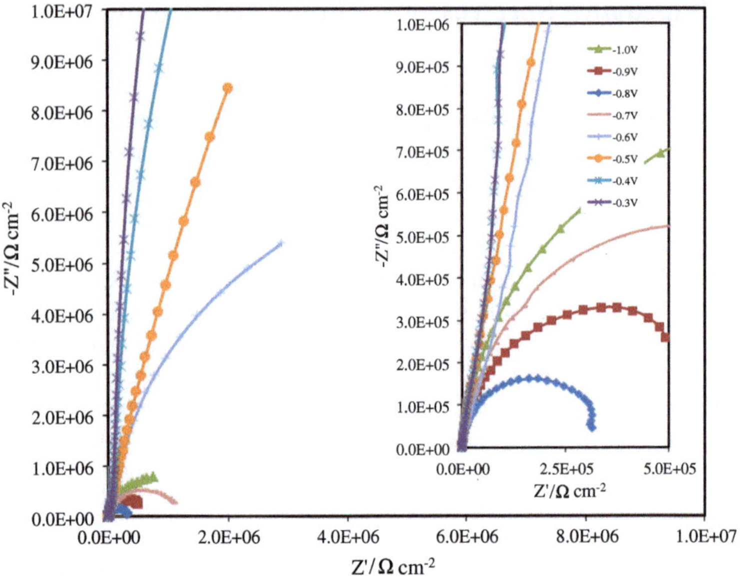

Fig. 6.27 Nyquist plot of the impedance of Nb^{V}-doped rutile (110) (Sample 2, Batch 2) over the potential range −0.7 to −0.3 V (HgO|Hg) in 1 M NaOH. The *inset* shows there was a (relatively) low impedance arc nested in the larger arc. The data was fit with the same equivalent circuit, with RC values varying steadily with potential

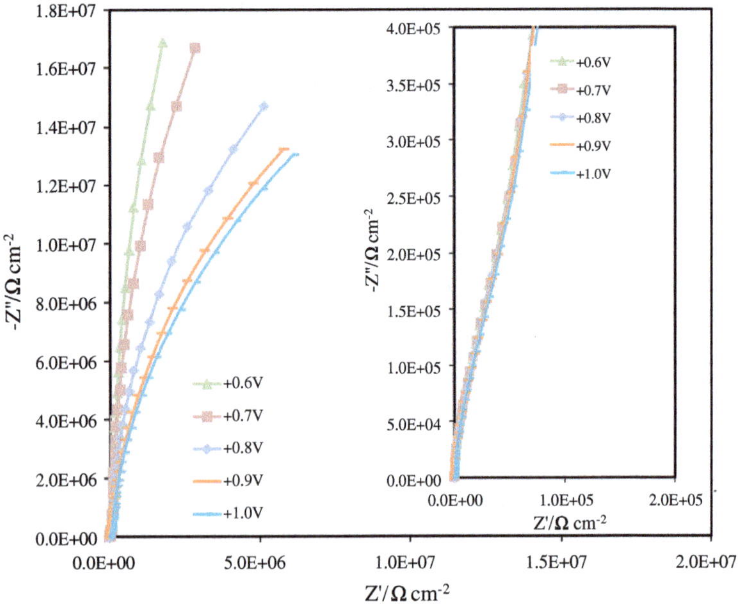

Fig. 6.28 Nyquist plot of the impedance of Nb^{V}-doped rutile (110) (Sample 2, Batch 2) in 1 M NaOH over the potential range 0.6 to 1.0 V (HgO|Hg). The *inset* shows there was a (relatively) low impedance arc nested in the larger arc. As the potential approached +1.0 V, the behaviour of the impedance became less purely capacitive

principles of photo-electrochemical cells, the capacitance decreases with increased band bending.

It is suggested that at −1.0 V, the electrode had an accumulation layer at the interface, and readily passes current (due to the lower impedance). There was also a capacitance associated with the accumulation layer. As the potential was made more positive, the thickness of the accumulation layer decreased and the capacitance increased, decreasing the impedance. At potentials less negative than −0.8 V, the impedance increased again, presumably due to the formation of a depletion layer, the capacitance of which decreased at more positive potentials until ca. +0.6 V, above which the decreasing imaginary component of the impedance showed that a Faradaic component to the current appeared, as was evident in the voltammetry.

This indicates that the flat band potential lies between −0.9 and −0.7 V (HgO|Hg), and is likely to be close to −0.8 V, at which the maximum capacitance occurred. Additional impedance spectroscopy between −0.9 and −0.7 V would give a better indication.

On this basis, it seems that the highly capacitive nature of the current flow, coupled with the high resistance, caused the measurements between −0.3 and +0.6 V (HgO|Hg) to become unstable, perhaps due to exceeding the minimum current capabilities of the potentiostat (this could have been rectified by selecting a higher resistance for the potentiostat).

A voltammetric measurement of the electrode following impedance spectroscopy showed that the qualitative behaviour of the electrode had changed from the first set of measurements. The electrode behaved more like the first batch of Nb^{V}-doped rutile (110), showing almost purely capacitive behaviour at potentials less negative than −0.5 V (HgO|Hg). It is assumed that the electrode oxidized between measurements, removing any traces of reduced phases at the surface.

The Nyquist plots show two circuit elements, corresponding to two charge transfer processes. There was a relatively low impedance arc at high frequencies, and a higher impedance arc at low frequencies; the higher frequency arc was nested in the lower frequency arc. The two processes are more clearly seen in the Bode plots of Figs. 6.29 and 6.30, which show the phase of the complex impedance against the frequency. The plot in Fig. 6.30 shows two distinct process peaks at 1 and 1000 Hz. The plot is essentially the same for all potentials positive of −0.5 V inclusive. Based on the behaviour of the larger impedance arc, it would appear that this corresponded to the capacitance of the space charge layer. The smaller arc at high frequency may have corresponded to either: Faradaic charge transfer, charging of trap states, the impedance of the bulk material, or the electrode contacts. However, it is not yet clear exactly what this represented. Typically, it has been reported by others working with oxides for photo-electrochemical water splitting that this element is due to charging of trap states [14]. In fact, the equivalent circuit used to fit the impedance data (Fig. 6.31) is based on this physical interpretation.

The parameters used to fit the equivalent circuit to the impedance data are given in Table 6.1. It appeared that there was a constant uncompensated solution resistance at about 7 Ω between the electrode surface and the tip of the Luggin probe,

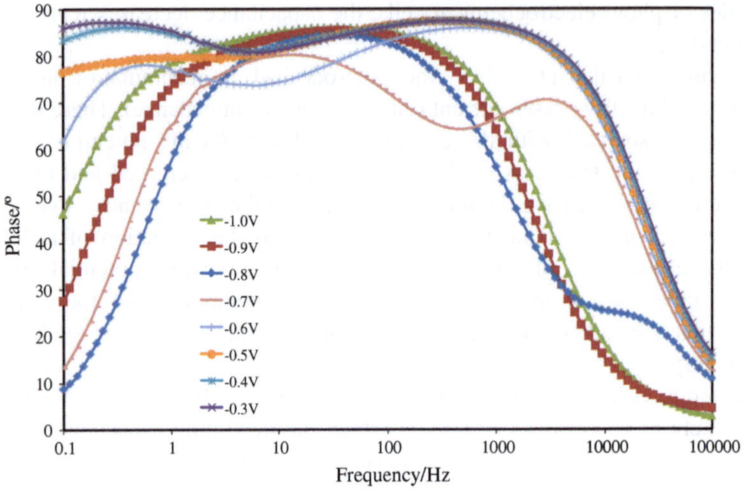

Fig. 6.29 Bode plot of the phase of the impedance of Nb^V-doped rutile (110) (Sample 2, Batch 2) in 1 M NaOH over the potential range −1.0 to −0.3 V (HgO|Hg)

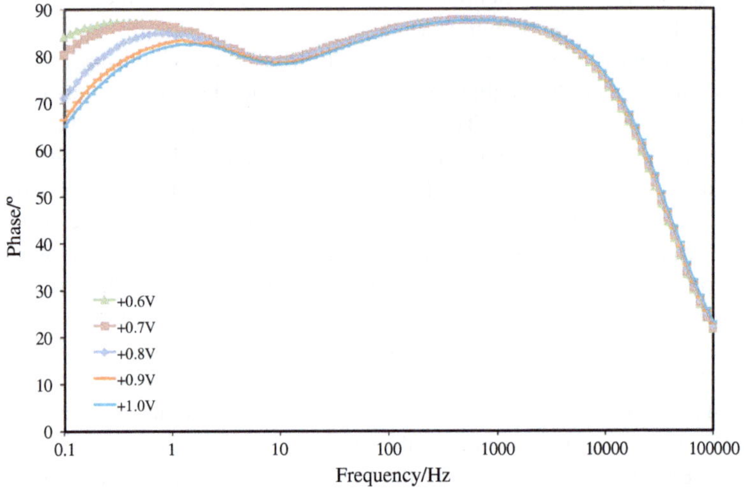

Fig. 6.30 Bode plot of the phase of the impedance of Nb^V-doped rutile (110) (Sample 2, Batch 2) in 1 M NaOH over the potential range +0.6 to +1.0 V (HgO|Hg)

together with ohmic potential losses in connections and contacts. There was a resistance R_1 corresponding to the space charge layer, with a minimum at ca. −0.7 V. The resistance R_2 had a minimum at ca. −0.8 V, increasing to over 200 MΩ cm^{-2}. It is suggested that this corresponded to the transfer of electrons into the conduction band of TiO_2 by tunnelling across the space charge layer; the wider the space charge layer becomes, the larger this value.

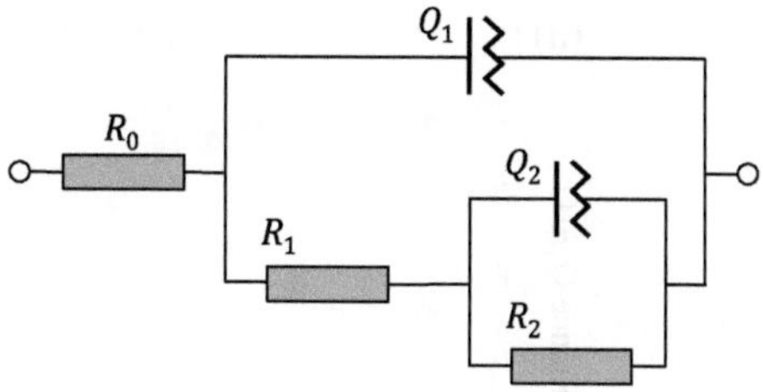

Fig. 6.31 Equivalent circuit used to fit the impedance of Nb^V-doped rutile (110) in 1 M NaOH; constant phase elements (CPEs) are denoted by 'Q'

Table 6.1 Values of elements of equivalent circuit for Nb^V-doped rutile (110) in 1 M NaOH

Potential (HgO|Hg) (V)	R_0 (Ω)	R_1 (kΩ)	Q_1 ($\mu\Omega^{-1}$)	Q_1/n	R_2 (kΩ)	Q_2 ($\mu\Omega^{-1}$)	Q_2/n
−1.0	6.95	40.4	9.3	0.974	139	4.34	0.621
−0.9	7.53	31.4	11.8	0.966	33.0	6.59	0.745
−0.8	6.94	11.4	2.53	0.947	30.9	6.99	0.971
−0.7	6.83	0.343	1.75	0.964	104	2.82	0.931
−0.6	6.75	34.7	1.48	0.968	1150	…	…
−0.5	6.95	41.4	1.11	0.989	29,700	0.879	0.749
−0.4	6.93	66.1	1.07	0.986	23,100	0.444	0.925
−0.3	6.92	66.3	1.01	0.986	155,000	0.417	0.943
−0.2 to +0.5	…	…	…	…	…	…	…
+0.6	6.94	61.4	0.702	0.988	283,000	0.340	0.968
+0.7	6.94	58.5	0.682	0.988	13,200	0.341	0.969
+0.8	6.95	49.9	0.659	0.989	53,700	0.379	0.935
+0.9	6.98	38.5	0.634	0.990	4130	0.443	0.888
+1.0	6.98	39.9	0.619	0.990	3750	0.453	0.884

It is noted that the circuit was modelled using constant phase elements (CPEs) rather than capacitive elements; this was required to obtain the best fit. The reasoning was that the exponent of the CPE would indicate how close to capacitive was the behaviour of the impedance. Circuits using capacitive elements in place of either of the CPEs did not fit the data well. The data in Table 6.1 shows that the exponent of the CPE Q_1 that corresponded to the space charge region capacitance was between 0.95 and 0.99. The data from the table is presented in Fig. 6.32. At positive potentials, n was close to unity, indicating almost ideal capacitive behaviour. The minimum value of n occurred close to the flat band potential at −0.8 V (HgO|Hg). This may indicate that when the width of the space charge layer is small, the effect of surface roughness results in less than ideally capacitive behaviour, i.e. there were spatial differences in the electrode. As the space charge layer increased in width, the charge depletion occurs in a plane in the bulk of the material, resulting in ideal behaviour.

In the case of the CPE Q_2, the departure from capacitive behaviour was more extreme, with n varying from 0.6 to 0.97. Interestingly, n was closest to capacitive behaviour around the flat band potential and at potentials over which current was at a

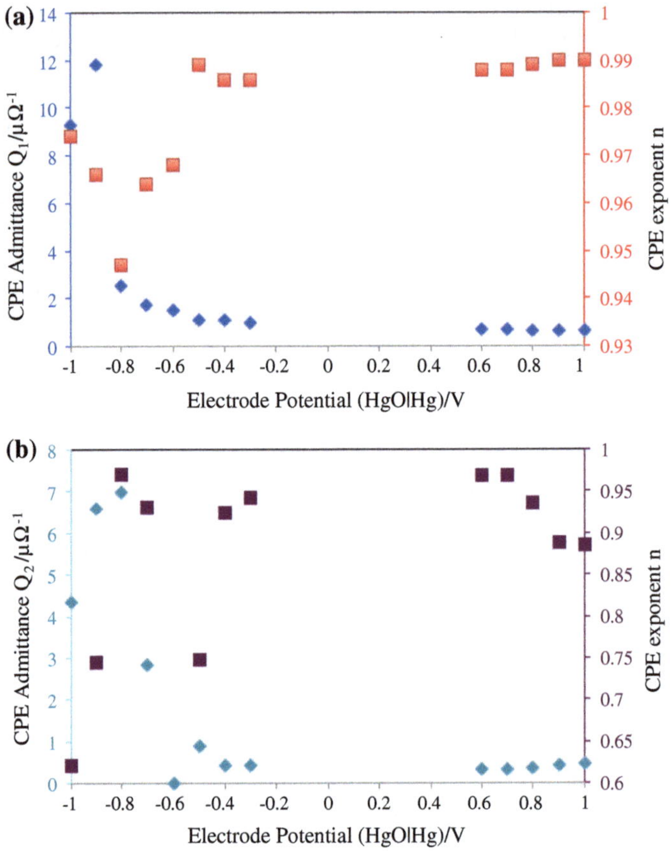

Fig. 6.32 Parameters used to fit the impedance of Nb^{V}-doped rutile (110) in 1 M NaOH. The circuit employed two constant phase elements (CPEs) denoted by Q1 and Q2, the magnitudes of which are plotted on the left hand axis with the exponent n plotted on the right hand axis

minimum. At potentials negative of the flat band potential, electrons were transferred from the conduction band of TiO_2 to the solution species. The magnitude of the CPE and its exponent were extremely dependent on frequency at these potentials, with an exponent as low as 0.621. This may indicate that charge transfer occurred unevenly over the surface of the electrode. For example, the charge transfer may have occurred preferentially at surface defects or crystallographic defect planes. At more positive potentials, at which oxidation current began to flow, the exponent again decreased, suggesting that electron transfer to TiO_2 from the solution species also occurred preferentially at certain spatially non-uniformly distributed sites.

It is suggested here that the sites at which charge transfer may have occurred were surface Ti^{3+} defects due to doping and partial reduction of the rutile crystal. Current may have flowed via the crystallographic shear planes that are expected to form in the Magnéli phases of rutile when reduced slightly.

6.11.2 Reduced Rutile (110)—Sample 1, Batch 1

The reduced undoped rutile (110) crystal (Sample 1, Batch 1) was also measured from −1.0 to +1.0 V (HgO|Hg), in increments of 0.1 V, at frequencies from 100 kHz to 1 Hz.

A selection of data for potentials in the range −1.0 to −0.4 V is shown below in a Nyquist plot in Fig. 6.33 and for potentials +0.2 to +1.0 V in Fig. 6.34.

It is immediately clear from a comparison of data in Figs. 6.27 and 6.28 to Figs. 6.33 and 6.34 that the impedances of reduced rutile (110) at any frequency were significantly smaller than those of Nb^{V}-doped rutile, not exceeding 20 kΩ cm^{-2} on the imaginary axis.

At potentials from −1.0 to −0.8 V, there was an initial low impedance arc (inset of Fig. 6.33). However, this did not appear to be present at potentials less negative than −0.7 V, at which only a single impedance arc was evident. However, the phase of the arc did not decrease at low frequency, but decreased and then continued to rise, giving the impression of a diffusive element such as the Warburg impedance (a 45° line in the complex plane). However, the data cannot be modelled as such. The data was initially fitted using the same equivalent circuit (Fig. 6.31) as for the Nb^{V}-doped rutile, but with different parameters. However, this circuit produced an adequate fit only at potentials −0.8 to −0.6 V. To fit the data well, an alternative circuit (Fig. 6.35) was required. This was the simplest circuit that could fit the data. Circuits using a capacitive element in place of the CPE did not fit the data well.

It is difficult to assign a physical interpretation of the additional CPE Q_3, but may have been due to the effect of the changing nature of the reduced phases at the

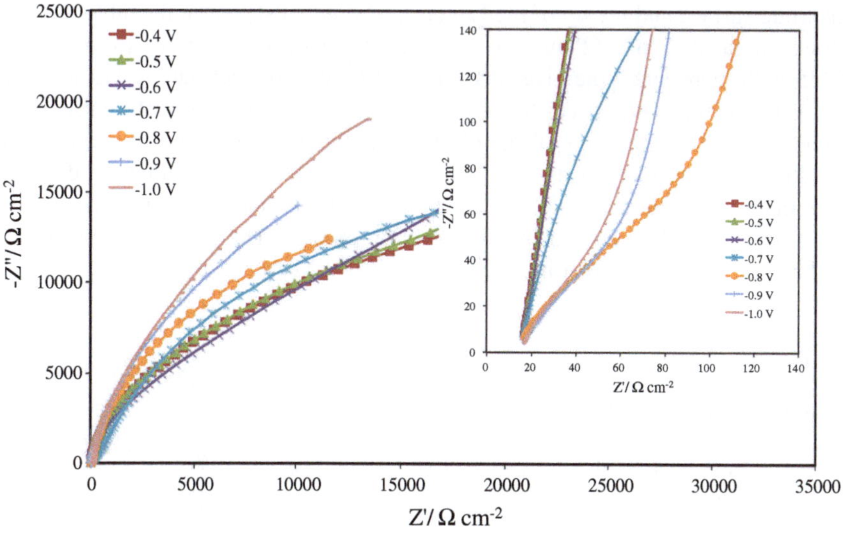

Fig. 6.33 Nyquist plots of the impedance of reduced rutile (110) (Sample 1, Batch 1) over the potential range −1.0 to −0.4 V (HgO|Hg). The *inset* shows the high frequency data points

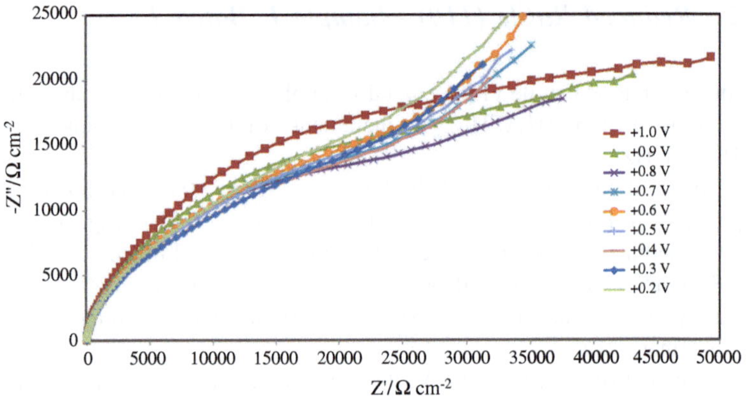

Fig. 6.34 Nyquist plot of the impedance of reduced rutile (110) (Sample 1, Batch 1) in 1 M NaOH over the potential range +0.2 to +1.0 V (HgO|Hg)

surface of the electrode. It may be possible to convert R_1 and Q_3 into an equivalent parallel pair, which could then be viewed as an alternative charge transfer pathway. At this point, it is considered sufficient to note that the equivalent circuit and fitted parameters suggest that the reduced rutile demonstrated even less ideal capacitive behaviour than the Nb^V-doped sample and that there were additional processes occurring, as evident in the voltammetry that exhibited a range of redox processes over the potential range −1.0 to +1.0 V (HgO|Hg).

The departure from ideal resistor-capacitor behaviour is clear to see from the fit parameters in Table 6.2. The uncompensated solution resistance R_0 between the electrode surface and the tip of the Luggin probe is not shown in the table, but was about estimated at ca. 5 Ω cm^{-2}. The exponent of CPE Q_2 increased from 0.767 to 0.935 with increasing potentials. As for the Nb^V-doped electrode, it was assumed that as the width of the space charge layer increased, the effect of surface roughness decreased, and the charging of the electrode capacitance occurred in a plane. The fact that the capacitance was less than ideal up to even +1.0 V may be due to a combination of a thinner space charge region, a higher charge carrier concentration, and a rougher surface caused by the reduction process. The potential dependences of the exponents are shown in Fig. 6.36.

Fig. 6.35 Equivalent circuit used to fit the impedance of reduced rutile (110) in 1 M NaOH; constant phase elements (CPEs) are denoted by 'Q'

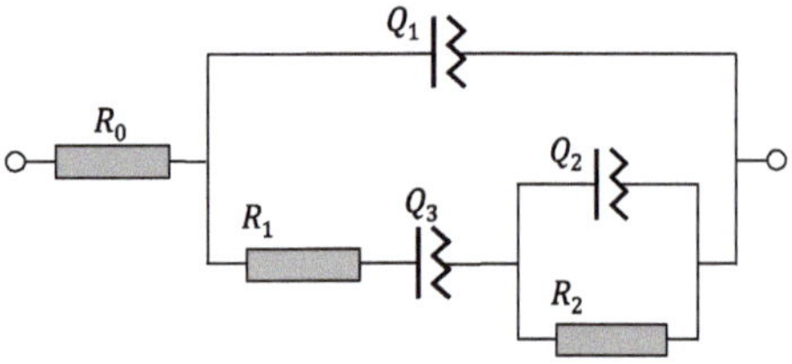

Table 6.2 Values of elements of equivalent circuit for reduced rutile (110) in 1 M NaOH

Potential (HgO|Hg) (V)	R_1 (Ω)	Q_1 (μΩ^{-1})	Q_1/n	R_2 (kΩ)	Q_2 (μΩ^{-1})	Q_2/n	Q_3 (μΩ^{-1})	Q_3/n
−1.0	Not fit							
−0.9	Not fit							
−0.8	48.7	19.1	0.767	10.2	10.5	0.953	…	…
−0.7	242	4.03	0.878	12.3	18.2	0.745	…	…
−0.6	1980	2.83	0.901	132	43.1	0.458	…	…
−0.5	3400	2.44	0.907	0.796	15.9	1.024	50.6	0.464
−0.4	3110	2.17	0.911	1.32	26.2	0.854	57.0	0.432
−0.3	2790	1.95	0.915	1.18	21.5	0.894	58.9	0.388
−0.2	2550	1.77	0.918	0.772	16.4	1.011	55.0	0.361
−0.1	3200	1.68	0.919	2.09	16.4	0.781	49.7	0.450
0	3400	1.58	0.920	6.45	12.9	0.637	36.4	0.703
+0.1	3160	1.50	0.921	8.67	12.7	0.611	37.0	0.795
+0.2	2830	1.43	0.922	9.83	14.0	0.592	38.6	0.819
+0.3	2610	1.38	0.923	8.88	16.5	0.579	49.5	0.783
+0.4	2460	1.31	0.925	11.0	16.7	0.552	56.7	0.862
+0.5	2550	1.25	0.927	10.4	14.2	0.565	46.1	0.831
+0.6	2550	1.20	0.928	9.38	12.4	0.574	39.1	0.792
+0.7	2470	1.15	0.929	7.56	12.0	0.576	46.9	0.669
+0.8	2730	1.11	0.930	7.35	10.4	0.601	55.0	0.630
+0.9	2700	1.06	0.932	8.06	9.77	0.597	55.0	0.540
+1.0	1800	0.992	0.935	9.74	12.9	0.540	66.0	0.360

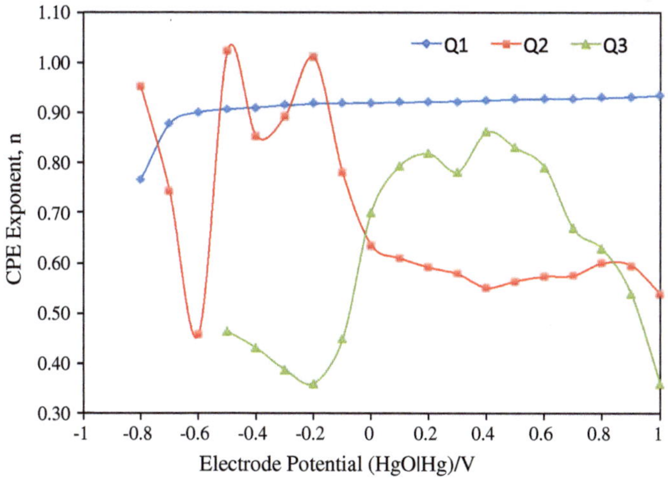

Fig. 6.36 Exponents of CPEs used to fit the impedance of reduced rutile (110) in 1 M NaOH. The CPE Q1 (associated with the space charge layer and Helmholtz layer) approaches capacitive behaviour with increasingly positive potential

References

1. Patel, M., Mallia, G., Liborio, L., Harrison, N.M.: Water adsorption on rutile TiO_2 (110) for applications in solar hydrogen production: A systematic hybrid-exchange density functional study. Phys. Rev. B **86**, 045302 (2012)
2. Liborio, L., Harrison, N.: Thermodynamics of oxygen defective Magnéli phases in rutile: a first-principles study. Phys. Rev. B **77**, 104104 EP (2008)
3. Kwon, D.-H., et al.: Atomic structure of conducting nanofilamentsin TiO. Nat. Nanotech **5**, 148–153 (2010)
4. Li, X., et al.: Magneli phase Ti_4O_7 electrode for oxygen reduction reaction and its implication for zinc-air rechargeable batteries. Electrochim. Acta **55**, 5891–5898 (2010)
5. Nelson, J.: The Physics of Solar Cells. Imperial College Press (2003)
6. Kennedy, J.H., Frese, K.W.: Flatband potentials and donor densities of polycrystalline α-Fe_2O_3 determined from Mott-Schottky plots. J. Electrochem. Soc. **125**, 723–726 (1978)
7. Hankin, A., Alexander, J.C., Kelsall, G.H.: Constraints to the flat band potential of hematite photo-electrodes. Phys. Chem. Chem. Phys. **16**, 16176 (2014)
8. Klahr, B., Gimenez, S., Fabregat-Santiago, F., Hamann, T., Bisquert, J.: Water oxidation at hematite photoelectrodes: the role of surface states. J. Am. Chem. Soc. **134**, 4294–4302 (2012)
9. Diebold, U.: The surface science of titanium dioxide. Surf. Sci. Rep. **48**, 53–229 (2003)
10. Cronemeyer, D.C.: Electrical and optical properties of rutile single crystals. Phys. Rev. **87**, 876 (1952)
11. Ekuma CE, Bagayoko D (2010) Ab-initio Electronic and Structural Properties of Rutile Titanium Dioxide. http://arXivcond-mat.mtrl-sci:
12. Cardon, F., Gomes, W.P.: On the determination of the flat-band potential of a semiconductor in contact with a metal or an electrolyte from the Mott-Schottky plot. J. Phys. D Appl. Phys. **11**, L63 (1978)
13. Albery, W.J., Bartlett, P.N., Hamnett, A., Dare-Edwards, M.P.: The transport and kinetics of minority carriers in illuminated semiconductor electrodes. J. Electrochem. Soc. **128**, 1492–1501 (1981)
14. Sivula, K., Le Formal, F., Grätzel, M.: Solar water splitting: progress using hematite (α-Fe_2O_3) photoelectrodes. ChemSusChem **4**, 432–449 (2011)

Chapter 7
TiO_2 Thin Films on Fused Silica

This chapter details an investigation of the growth of TiO_2 films on fused silica. The initial motivation to use an amorphous glass substrate was to reproduce anatase films with the (101) orientation as seen by other groups working on TiO_2 for use as a transparent conductive oxide [1].

Fused silica is an amorphous form of silicon dioxide produced by sintering of high-purity silica. It avoids the problem of diffusion of impurities from the substrate to the film, as is found with soda-lime glass [2]. It also has a higher softening temperature than both soda-lime and borosilicate glasses. As such it is suitable for deposition of films and thermal treatment at higher temperatures.

A summary of the results and suggested further work are given in Sects. 7.1 and 7.2. The general experimental conditions are described in Sects. 7.3 and 7.4, and in greater detail in Chap. 4. Initial experiments on borosilicate glass are described briefly in Sect. 7.5.

The experiments for deposition of both undoped and Nb-doped TiO_2 films on fused silica are presented in Sect. 7.6, which describes a series of depositions that were carried out looking at the effect of varying the following parameters:

- In situ deposition oxygen pressure
- Ex situ post-annealing temperature of films deposited at room temperature
- In situ substrate temperature during deposition
- Thickness (number of laser pulses).

The characteristics of both undoped and Nb-doped films were determined using:

- X-Ray Diffraction (XRD)—to assess phase and orientation of films
- Raman Spectroscopy—to confirm structure and increase sensitivity to secondary phases
- Atomic Force Microscopy (AFM)—to assess the impact of deposition conditions and thermal treatment on film morphology.

© Springer International Publishing Switzerland 2016

J.C. Alexander, *Surface Modifications and Growth of Titanium Dioxide for Photo-Electrochemical Water Splitting*, Springer Theses,
DOI 10.1007/978-3-319-34229-0_7

Following this, the results of UV-Vis spectroscopic measurements are presented in Sect. 7.7 to assess optical properties including the direct or indirect nature of the band gap. These measurements were also used to assess film thickness and refractive index from the interference fringes

The impact of the Nb-doping on the phase and orientation of the films is discussed in Sect. 7.8.

Scanning Tunnelling Microscopy (STM) was also attempted in order to assess the atomic configuration at the surface of the films including surface reorganization and defects. However, the initial films deposited from an undoped TiO_2 target proved to be insufficiently conductive to allow a proper measurement. As such, films with both 1 and 10 % Nb-doping were also produced. As discussed later, the 1 % Nb-doped films did not exhibit any significant increase in conductivity, despite the expectation of highly conductive films on the basis of work by Hitosugi et al. [3, 4]. As such, it was not possible to obtain any images of the surface by STM. The issues surrounding the low conductivity are discussed in Sect. 7.9.

7.1 Summary of Results and Conclusions

The results of this chapter present the effect of growth conditions on TiO_2 thin films deposited onto an amorphous substrate: fused silica. The effect of the temperature, pressure, and number of pulses (thickness) on the thin film morphology was investigated. It was found that varying the deposition conditions and thermal treatment (substrate heating versus post-annealing) could be used to modify the orientation of the film as well as the phase. The addition of Nb doping also had an effect.

The study of the effect of oxygen pressure—for films deposited at room temperature and post-annealed—found the range 10–30 mTorr gave the best crystalline films. The post-annealed films were polycrystalline anatase with a preferential orientation of (101). The presence of other reflections of (004), (200), (013), and (112), showed that the films were not epitaxial, although they may have had a predominantly (101) surface. Interestingly, one film deposited at 32 mTorr seemed to have only the (101) orientation present in the XRD spectrum. This suggests that a further investigation should be done to establish the precise growth conditions that can result in a film with a single orientation of grains on fused silica.

The post-annealed anatase (101) film demonstrated a higher than expected anatase-rutile phase transition in the range, 1000–1100 °C. At no point during the process of annealing to successively higher temperatures did the film display any significant changes (such as changing orientation). At 1100 °C the film transitioned to a rutile film with the (110) orientation. This suggests that anatase films produced by PLD could be used for practical applications up to 1000 °C.

The films fabricated with substrate heating during deposition displayed a markedly different behaviour with temperature compared to the post-annealed films. At deposition temperatures of 400–500 °C, the films were anatase with a preferential (101) orientation, but with the presence of (103) and (200) grains. As the temperature increased between 500 and 700 °C, a contribution to the pattern from (004) grains started to appear, becoming the dominant orientation of grains at 700 °C. And at 800 °C the films grew as predominantly rutile films with a small fraction of anatase (004) grains. The rutile film grew preferentially with the (110) orientation and some grains oriented as (011) and (200).

This suggests that it is possible to use the deposition temperature to control the growth of oriented anatase films with predominantly (001) surfaces on amorphous substrates. These films may be used to investigate the properties of (001) surfaces in detail.

The effect of substrate temperature during deposition on the crystal phase and orientation was found to be approximately identical with 1 % Nb doping of the PLD target.

A film deposited at 800 °C from a 10 % Nb-doped target exhibited different behaviour. At 800 °C the film was predominantly anatase with (004) oriented grains, and only a small fraction of rutile present. In fact, the XRD pattern of the 10 % Nb-doped film was qualitatively similar to that of the undoped and 1 % doped films deposited at 700 °C. This demonstrated clearly the reported phenomena of Nb doping acting to stabilise the metastable anatase phase as discussed in Sect. 7.8. The temperature increments used were 100 °C: it is possible that studying additional intermediate temperature steps—say every 50 °C—might show a shift in the phase transition temperature for the 1 % Nb-doped films relative to the undoped ones.

7.2 Further Work

A possibility that has not been examined is the stability of the films deposited with substrate heating to further post-annealing. For example, the undoped and 1 % Nb-doped films deposited at 700 °C are predominantly with a c-axis or (004) orientation as opposed to (101) at lower temperatures. At present it is unknown whether the film maintains the (004) orientation on further heating or has a different anatase-rutile transition temperature than the films deposited at room temperature.

The dependence of the orientation of the film on the substrate heating temperature may be the most interesting finding—and one that has not been reported previously. Considering that the anatase (101) surface has been computed to be the lowest energy surface [5] it is unsurprising that thin films grown or annealed at low temperature (room to 500/600 °C) crystallize with this orientation to minimise the energy at the interface with the amorphous substrate and the environment. That a predominantly c-axis oriented film grows when the substrate temperature is increased seems to suggest that, at higher temperature, the (001) surface is the lowest energy surface. Given that density functional theory (DFT) calculations of the crystal structure and surface energy are usually carried out at a temperature of absolute zero, this idea is compatible with theory. It would therefore be interesting to investigate theoretically, the effect of temperature on the equilibrium crystal structure, using DFT or a similar computational approach.

The different growth mechanisms for films deposited with substrate heating or post-annealing are discussed in Sect. 7.6. For depositions with substrate heating it was noted that without the oxygen-annealing step before sample cooling, the film failed to crystallize. The study into highly Nb-doped films found that the films were not conductive. A review of the literature suggested that this might have been due to a relatively high oxygen partial pressure during annealing. As such, it is suggested that the effective of the oxygen pressure during annealing (with deposition pressure a fixed parameter) is done. This would aim to establish both the required oxygen pressure for the formation of a crystalline film, and the ideal pressure for conductive films.

7.3 Target Production and Pulsed Laser Deposition Methods

Details of the methods used to fabricate the 25 mm TiO_2 PLD targets are given in Sect. 4.4.1 of Chap. 4. A brief overview of PLD is also given in Sect. 4.4.2.

The specific conditions used and varied in each series of depositions are presented below.

7.4 General Deposition Conditions

The following conditions were typically fixed during a deposition experiment.

- 520 mJ per pulse (not investigated as a varied parameter)
  - Spot area of 0.35 cm^2
  - Fluence of 0.63 J cm^{-2}
- 90 mm substrate-target separation on glass (minimum possible before July 2012)
- 50 mm substrate-target separation on fused silica (minimum possible after July 2012)
- 5000 pulses at 2 Hz repetition rate (first few series of depositions)
- 10000 pulses at 8 Hz repetition rate (the majority of depositions thereafter).

7.5 Deposition on Borosilicate Glass

Initially, borosilicate glass was used as the substrate of choice, based on its low cost (and its typical use in electronics). However, it was noted that when depositing films at 700 °C that the substrate warped. Additionally, depositions at lower temperatures (400–600 °C) did not seem to yield crystalline films based on XRD analysis. Although, as discussed below, this is likely due to the very low thicknesses (<50 nm) of the initial batches of film deposited before the conditions were modified (reduced working distance and more pulses).

The following deposition conditions were investigated on borosilicate glass substrates (cover slips).

- Oxygen pressure of 10 mTorr; Temperatures of 400, 500, 600, and 700 °C
- Oxygen pressure of 100 mTorr; Temperatures of 400, 500, 600, and 700 °C

The deposition was done with 5000 laser pulses at 2 Hz. The XRD analysis did not identify any crystalline phase, just the amorphous background due to the glass substrate. Warping of the substrates was also encountered at 700 °C. The decision was made to move to fused silica, which has a higher softening temperature of 1683 °C compared to 820 °C for borosilicate glass.[1]

7.6 Deposition on Fused Silica

Before the first batch of deposition of TiO_2 films on fused silica, several films had been deposited on single crystal substrates of strontium titanate (STO) with the (100) orientation. With identical deposition conditions (400–600 °C, 10 mTorr, 5000 pulses at 2 Hz repetition rate, 50 mm substrate-target distance, same laser fluence) the films on STO were fully crystalline and epitaxial, whereas the films on both borosilicate glass and fused silica appeared to be amorphous.

Initially two films were deposited on fused silica at 400 and 700 °C, both at a pressure of 10 mTorr, with 5000 pulses and a higher repetition rate of 8 Hz. The film deposited at 700 °C showed a small peak in the XRD pattern corresponding to the rutile (110) reflection. In addition to using a different substrate, the substrate-target separation distance was decreased from 90 to 50 mm resulting in an increased deposition rate per pulse, giving thicker films that were easier to analyse by XRD. Increasing the number of laser pulses to 10,000 resulted in a crystalline film on fused silica as determined by a standard coupled 2θ XRD scan. This confirmed that the thicker TiO_2 films where required on an amorphous compared to a crystalline substrate, for detection of film peaks by XRD. Subsequently, a pulse number of 10,000 was taken as standard.

[1]See http://www.camglassblowing.co.uk/gproperties.htm for range of properties of glasses.

7.6.1 Effect of Deposition Pressure on Crystallization

Films were deposited over a wide range of oxygen pressures to identify the optimum range. The pressures used were 1000, 100, and 32 mTorr, and 20 μTorr. Ideally pressures in the range 1 mTorr down to 100 μTorr would also be investigated. However, the oxygen flow valves used in the PLD chamber could not accurately control the pressure in this region. The films were deposited at room temperature (RT) and subsequently post-annealed in air to successively higher temperatures. This allowed a wide parameter space of pressure and temperature to be investigated. The initial anneal was to 300 °C for 2 h followed by increments of 100 K in the dwell temperature up to 400, 500, and 600 °C, with an XRD pattern of each sample taken between each annealing step.

Of the four films deposited, only that deposited at 32 mTorr demonstrated the formation of a crystalline TiO_2 phase. The XRD measurement that followed annealing at 300 °C showed the film to be anatase, with only the (101) reflection present in the pattern, suggesting a textured film with preferential orientation.

A second study was then done to look at the effect of pressure at a finer resolution. Films were deposited at 10, 30, 40, and 50 mTorr, and also at room temperature. Figure 7.1 below shows the XRD patterns of the combined pressure studies, for films post-annealed to 300 °C. It was noted that annealing to even higher temperatures (of up to 700 °C) did not result in either the crystallization of previously amorphous films, or a change in phase or orientation films that crystalized at 300 °C.

It is apparent that the intensity of the anatase (101) peak decreases as the pressure increases from 10 to 40 mTorr, disappearing by a pressure of 50 mTorr. This indicated a pressure dependency when forming an anatase film that was most likely due to the reduced mean free path between collisions of ejected ions from the target and the oxygen in the chamber. It is well known that higher oxygen pressure results in a plume with lower kinetic energy on reaching the substrate [6], which reduces the growth rate. There seemed to be a slight shift of the peak position with oxygen pressure, which was probably due to a slight tilt in the sample as mounted on the goniometer of the XRD instrument. Typically, for a film on a known single crystal substrate, the substrate peaks act as reference points to allow data correction and real shifts in lattice parameter to be identified. However, on an amorphous substrate like fused silica, no such reference peaks are present. In this situation it should be possible to identify structural changes in the film by comparing the relative positions of film peaks. To do this, the dominant film peak—in this case (101)—is shifted to the theoretical 2θ position of 25.281 as given by the ICSD star pattern for anatase (Ref: 00-021-1272). The other major peaks should then appear at 37.801 (004), 48.050 (200), 53.891 (105), and 55.062 (211).

An interesting feature of the XRD pattern for the film deposited at 32 mTorr is that only a single peak is present, that of anatase (101). It is noted that due to technical issues during that particular deposition, that the pressure for this deposition was set manually. As the goal of the deposition on an amorphous substrate

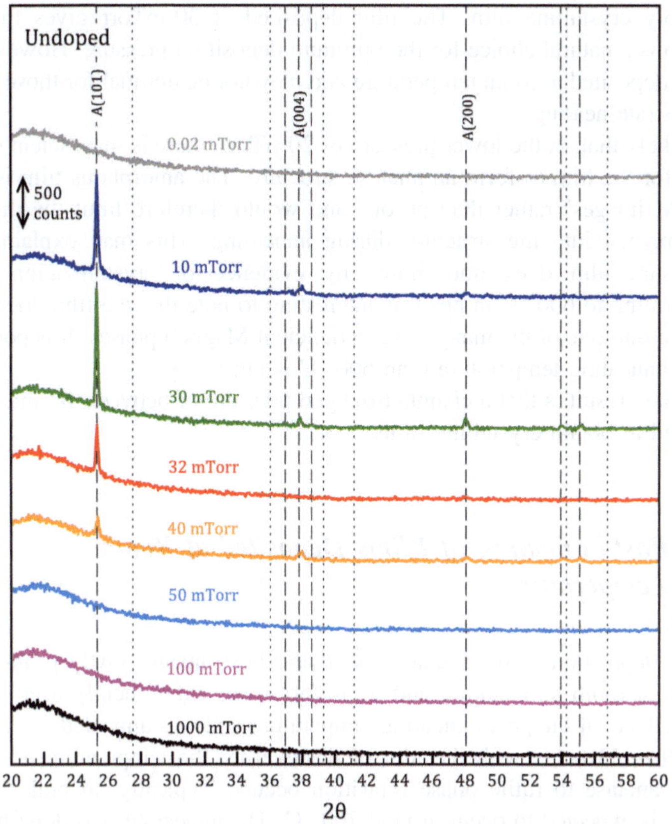

Fig. 7.1 XRD patterns of TiO_2 thin films deposited at a range of deposition pressures. All films deposited at room temperature and post-annealed at 300 °C for 2 h. Films formed by 10 k pulses at 8 Hz with substrate-target distance of 50 mm

was to achieve a film with only the (101) orientation of grains, this film could be described as the most successful. It seems likely that slight differences in the actual post-annealing conditions, rather than those intended and programmed, has resulted in differences in crystallite nucleation and growth. It may be worth looking systematically at cooling and heating rates, and position within the furnace to optimize the conditions for the growth of films with only the (101) orientation.

The films deposited at 10, 30, and 40 mTorr also have grains oriented (004) and (200) present at around 38° and 48° respectively. There is also the possible presence of a small fraction of (105) and (211) oriented grains. These are the most likely high intensity peaks according to the star-quality ICSD pattern for anatase TiO_2 (reference code 00-021-1272).

By comparing the intensities of the anatase (101) peaks in the patterns of films deposited at various pressures at room temperature and post-annealed in air, it appears that a deposition pressure between 10 and 40 mTorr is required to obtain a measurably crystalline film. The film deposited at 30 mTorr gives the clearest signal, so is a natural choice for the optimum deposition pressure. However, this is for films deposited at room temperature and may not be optimal for those deposited with substrate heating.

It is likely that, at the lower pressure of 20 μTorr, there is insufficient oxygen in the film for the film to form an anatase structure. The amorphous film is compact (see AFM images) rather than porous and would therefore limit the diffusion of further oxygen into the structure during annealing. This may explain why the low-pressure film does not show any evidence of crystallization even on post-annealing to 600 °C in air. It is interesting to note that the film does not even crystallize into one of the many oxygen deficient Magnéli phases. It is possible that a higher annealing temperature than 600 °C is necessary.

The main result is that a chamber oxygen pressure of between 10 and 40 mTorr is required to obtain crystalline films.

7.6.2 Post-annealing of Films Deposited at Room Temperature

The film deposition settings that showed the best anatase (101) peak—the film deposited at room temperature and 32 mTorr—was subsequently used to investigate the effects of the post-annealing temperature and was annealed in 100 K steps from 300 to 1100 °C with the aim of discovering the temperature at which the expected anatase to rutile phase transition occurs. Typically, in bulk, this phase transition is expected to occur around 700 °C. The annealing was done by heating the sample at a rate of 10 °C/min, holding it at the defined annealing temperature for 2 h, and then cooling at 10 °C/min. As shown in Fig. 7.2, the film crystallizes between 200 and 300 °C with the anatase crystal structure and a preferential orientation of (101). There was no trace of anatase detected in the XRD scan following post-annealing at 200 C (the scan is not shown). Surprisingly, the film remains in the anatase phase even after a post-annealing temperature of 1000 °C. Only when the film is annealed at 1100 °C is there a phase transition to the stable rutile crystal structure, the stable phase of TiO_2. The rutile film is preferentially oriented with the (110) planes parallel to the fused silica substrate. The XRD patterns taken after each annealing step are shown in Fig. 7.2 below, and the (101) peak of anatase can still be discerned. This is a surprising result, as the phase transition occurs between 1000 and 1100 °C, which is 300–400 °C higher than the conventionally accepted value. This suggests that anatase films deposited on fused silica might be suitable as substrates for materials and devices that are fabricated or operated at temperatures in the region of 1000 °C.

What is particularly interesting is that the fused silica substrate is amorphous. There is no substrate lattice encouraging the film to grow epitaxially (unlike with films deposited on STO, as presented in Chap. 8). In fact, this is the principal reason

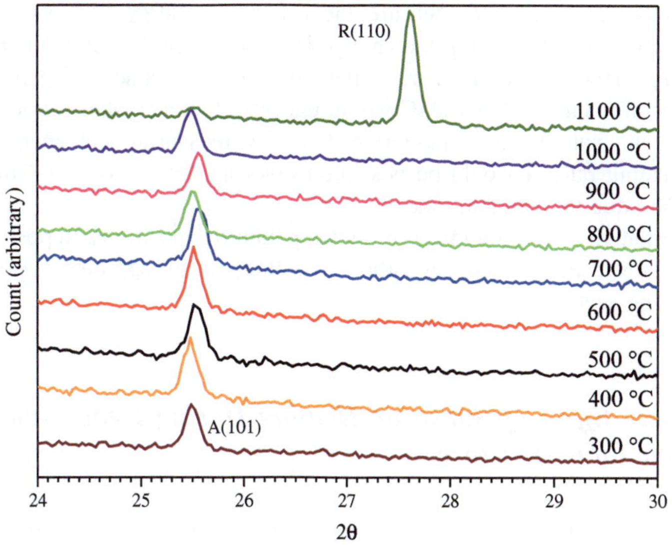

Fig. 7.2 XRD patterns showing the effect of post-annealing temperature on the structure of a TiO_2 thin film deposited on amorphous fused silica. The deposition was done at room temperature and a pressure of 32 mTorr. XRD patterns are ex situ

why an amorphous substrate is ideal for the growth of films with the (101) orientation. Firstly, there is experimental evidence showing the predominance of the (101) orientation in films deposited on glass [1]. Secondly, calculations performed at Imperial by project collaborators using hybrid Density Functional Theory (h-DFT) show that the (101) plane of anatase has the lowest surface formation energy [5]. As such, in the absence of additional constraints—such as a crystal lattice—one would expect the film to grow so as to minimise the surface energy at the interface with the substrate i.e. with the (101) orientation. As there is no obvious constraint imposed by the substrate, it is relatively unexpected that it should be particularly resistant to changing phase on annealing. That it forms a rutile film with the lowest energy (110) surface is however expected as this is calculated to be the lowest energy surface of rutile [7].

The reasons why the film remains in the anatase phase up to ca. 1000 °C are now discussed. The phase transition of anatase to rutile grows by surface nucleation i.e. a rutile particle will grow on the surface of an anatase particle or film. Given that a PLD film is compact, there is little free surface (within the film) at which the rutile can grow (compared to a powder or a porous bulk ceramic). To form a new crystal structure with a different orientation would require the breaking of Ti and O bonds within the film, with a re-arrangement. This is limited by the compactness of the film. Only the top surface is relatively unconstrained, although this is constrained by the film below. The entire film (or large grains) would then have to transition simultaneously resulting in higher activation energy for the phase transition.

It should be possible to measure the activation energy for this process by annealing films and measuring the change in the enthalpy by differential scanning calorimetry (DSC). This may also demonstrate a thickness dependence. An amorphous film annealed in a DSC would undergo; first a crystallization to anatase and then; an anatase-rutile phase transition. By measuring the energy loss for various heating rates it would be possible to plot an Arrhenius plot. This has not been to date due to available time.

An alternative growth regime to room temperature deposition and post-annealing is to heat the substrate during deposition. The effect of this approach is presented in the following section.

7.6.3 The Effect of Substrate Heating During Deposition on Film Structure

Films were also deposited with heating of the substrate during deposition. The substrates were heated to 400, 500, 600, 700, and 800 °C using a 1028 nm solid-state laser capable of approximately 80 W continuous heating. Ideally, the range from room temperature to 300 °C would also have been examined to identify the minimum temperature for crystallization with substrate heating. However, the temperature control of the laser as implemented was not steady below 350 °C. The heating and cooling rate was 10 °C min^{-1}. Typically the deposition pressure used was 30 mTorr, although 10 mTorr was also used in some cases. Following deposition at low oxygen pressure, the chamber was backfilled with oxygen to 500 Torr (just below ambient pressure) and the films were allowed to anneal as they cooled. This step was found to be necessary for causing the film to crystallize, but had a negative impact on conductivity (see Sect. 7.9).

The effect of substrate heating on the film structure is discussed below and shown in Fig. 7.4 for undoped TiO_2 and in Fig. 7.4 for 1 % Nb-doped films. The temperature dependence of the orientation and phase appears to be unaffected by the 1 % doping. The results discussed below are applicable to both the undoped and 1 % Nb-doped films. Results for films deposited at 400 and 700 °C with 10 % Nb-doping are also shown for comparison in Fig. 7.6, showing the effect of doping on the structure more clearly than the 1 % doping level.

The XRD patterns were analysed by comparison to high-quality (Star) powder diffraction files (pdf) maintained by the International Centre for Diffraction Data (ICDD). For anatase and rutile, the main peaks are given below in Table 7.1. While the results were also checked against patterns for the brookite phase, the patterns are not included here as only anatase and rutile were seen to be present. As the angle of the peak is related to the d-spacing by the wavelength of the exciting x-rays, the 2θ values given are correct only for Cu K_α x-rays.

Both anatase and rutile are structures in the tetragonal system. Anatase has a space group of I41/amd with lattice parameters a = b = 3.7852 Å and c = 9.5139 Å.

Table 7.1 XRD reference patterns for anatase and rutile

Anatase (pdf 00-021-1272)				Rutile (pdf 00-021-1276)			
(hkl)	2θ (deg)	d [Å]	I (%)	(hkl)	2θ (deg)	d [Å]	I (%)
(101)	25.281	3.520	100	(110)	27.447	3.247	100
(103)	36.947	2.431	10	(101)	36.086	2.487	50
(004)	37.801	2.378	20	(200)	39.188	2.297	8
(112)	38.576	2.332	10	(111)	41.226	2.188	25
(200)	48.050	1.892	35	(210)	44.052	2.054	10
(105)	53.891	1.700	20	(211)	54.323	1.687	60
(211)	55.062	1.667	20	(220)	56.642	1.624	20
(213)	62.121	1.493	4	(002)	62.742	1.480	10
(204)	62.690	1.481	14	(310)	64.040	1.453	10

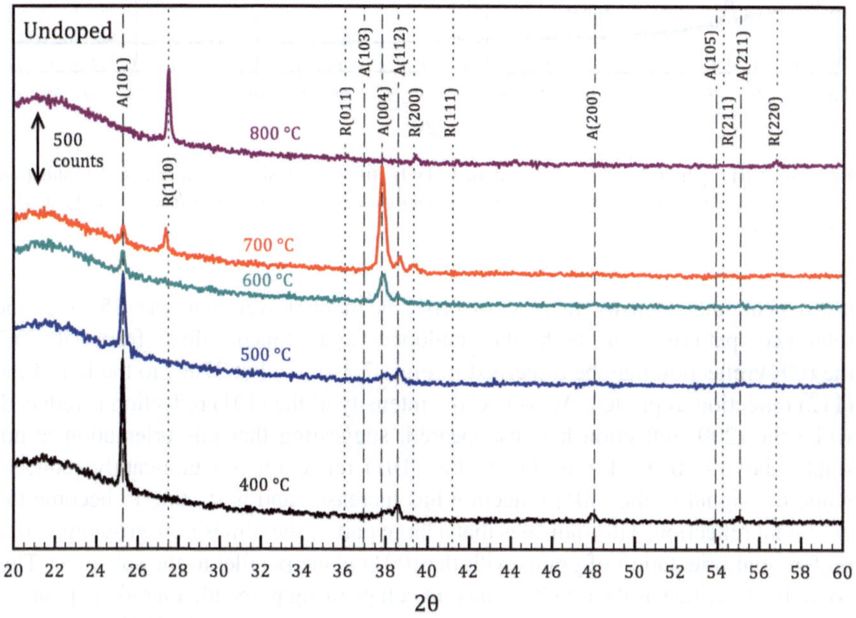

Fig. 7.3 XRD patterns of undoped TiO_2 films deposited at a range of substrate temperatures. The XRD peaks for anatase (*dashes*) and rutile (*dots*) are indicated by *vertical lines*

Rutile has a different space group P42/mnm and lattice parameters a = b = 4.5933 Å and c = 2.9592 Å.

At a substrate temperature of 400 °C the film forms with the anatase structure and a preferential orientation of grains with the (101) plane parallel to the substrate, as shown by the peak at 25.3° of Figs. 7.3 and 7.4. The reflections of the (112) and (200) planes at 38.6° and 48.0° are also present in the patterns. In the case of the

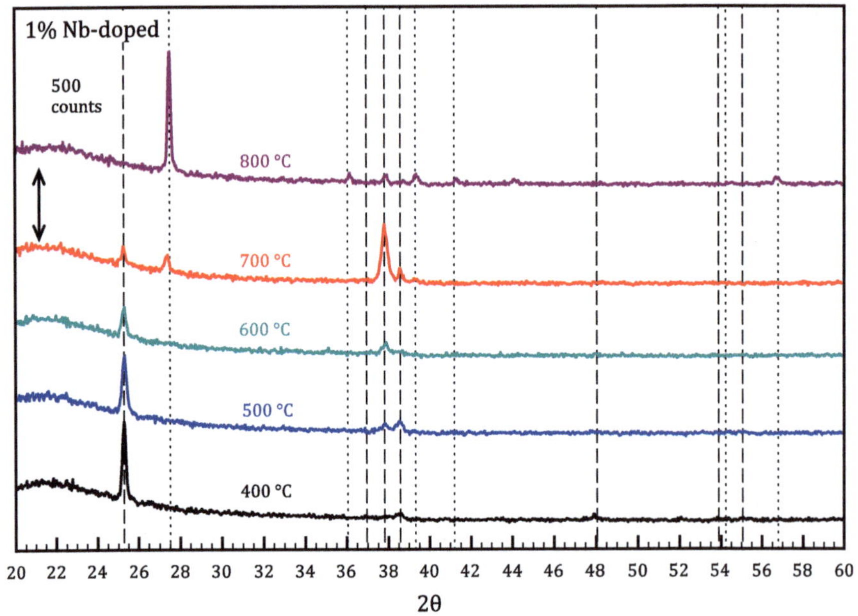

Fig. 7.4 XRD patterns of 1 % Nb-doped TiO_2 films deposited at a range of substrate temperatures. The XRD reference peaks for anatase (*dashes*) and rutile (*dots*) are indicated by *vertical lines*

(200) peak the intensity relative to (101) is much lower than the 35 % of the reference pattern. For both the undoped and doped films from 500 °C, the (004) reflection may be discerned in Figs. 7.3 and 7.4 at 37.8°, to the left of the (112) reflection as plotted. At 500 °C the intensity of the (101) reflection is reduced, while the (200) reflection has disappeared, suggesting that this orientation is not stable above 400 °C. From 600 °C the (101) reflection is significantly reduced, while the signal of the (004) reflection has increased and is starting to become the dominant reflection. This indicates that a large part of the film is now growing with a preferential orientation of grains with the (001) plane parallel to the substrate. This could be described as the c-axis of the unit cell growing perpendicular to the plane of the substrate. At this temperature, there is no rutile observed in the XRD pattern. At 700 °C the (004) reflection increases significantly compared to the (101) reflection, suggesting that this film is now comprised largely of grains with the c-axis perpendicular to the substrate. There is also a signal that can be attributed to the rutile (110) and (200) reflections at approximately 27.5° and 39.3° respectively. Increasing the substrate temperature during deposition to 800 °C results in a film that is rutile with grains oriented preferentially with the (110) planes parallel to the substrate, but with some rutile (200) oriented grains present. There is still a small peak anatase (004) peak in the XRD pattern at 800 °C, suggesting that a higher deposition pressure is required for a single-phase rutile film.

These results suggest that it is possible to use substrate heating during deposition to control the orientation—as well as the phase—of a TiO_2 film grown on an amorphous substrate such as fused silica. The use of a STO buffer layer to encourage the growth of c-axis oriented anatase films on Si substrates has been shown previously [8]. To the author's knowledge, control over orientation of films deposited directly onto an amorphous substrate has not been demonstrated before.

This suggests a further study into optimizing the growth of anatase c-axis oriented films. It may be possible to investigate in more detail the growth of films in the 600–700 °C temperature range, with the aim of finding the temperature that maximises the proportion of c-axis oriented grains while minimizing the presence of rutile.

This is of interest for photo-electrochemical and photocatalytic applications as the (001) surface of anatase has been reported to be more active than the (101) surface [9]. The ability to grow this surface on a relatively cheap amorphous substrate is an advantage compared to using expensive single crystal substrates such as $SrTiO_3$.

While the broad results discussed above are valid for both the undoped and 1 % Nb-doped films, there are some small differences. While the (004) reflection is present at 400 °C for the undoped film (Fig. 7.3), it does not appear until 500 °C in the doped film (Fig. 7.4). For the doped films the rutile (011) reflection also appears at approximately 36.0° and there is still visible a peak that corresponds to anatase (004). This seems to indicate that the Nb-doping slightly modifies the relative stabilities of each surface and stabilizes the anatase phase to a slightly higher temperature. However, the number of samples is statistically insignificant so the differences may be attributed to slight differences in the preparation of the substrates or growth conditions (such as how good the thermal contact between the substrate and the heater was).

The effect of substrate temperature during deposition on the film structure was also investigated using Raman spectroscopy. The Raman spectra of the undoped films are presented in Fig. 7.5 for films deposited with a substrate heating temperature of 400–800 °C. They confirm—on the whole—the findings from the XRD analysis, showing the transition from anatase to rutile above 600 °C. At 400 and 500 °C the film is comprised solely of anatase. As with the XRD patterns, the Raman spectra for 700 °C shows a mixed film of anatase and rutile. The main difference with the XRD is seen at a substrate heating temperature of 600 °C where in addition to the standard anatase spectra, the presence of rutile is suggested by a small peak starting to form around 440 cm^{-1}. This is reasonable as Raman is a slightly more sensitive technique than XRD. However, at 800 °C there is no indication of anatase, unlike in the XRD pattern, which shows some anatase (004) grains.

It is clear that the film growth changes with higher substrate temperature. At lower temperatures of 400 and 500 °C, the anatase (101) surface is favoured. At 600–700 °C the anatase (001) surface becomes more favourable than (101), and finally the rutile (110) surface becomes dominant.

To further investigate how Nb doping affected the film growth and characteristics, a series of films were also deposited with a higher doping concentration of

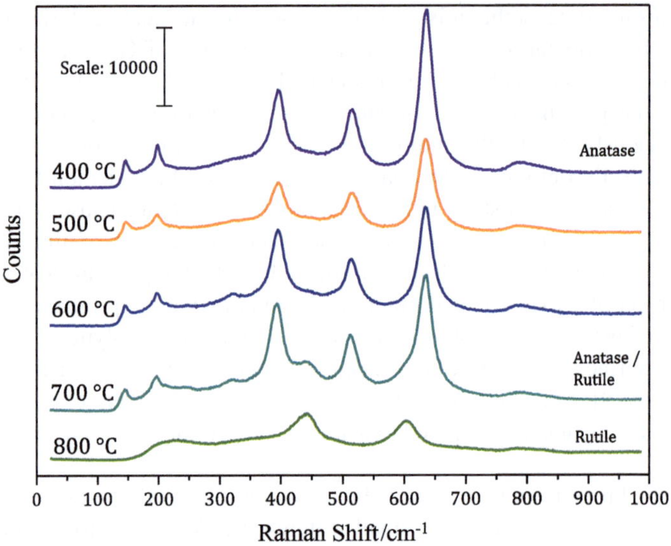

Fig. 7.5 Raman Spectra of 1 % Nb-doped films on fused silica for films deposited with substrates heated to 400–800 °C. Spectra measured with a 514 nm laser

10 % Nb. The primary purpose was to investigate conductivity in these films. This is discussed in detail in Sect. 7.9. The XRD patterns of samples deposited at 400 and 800 °C are shown in Fig. 7.6. All other deposition parameters were set to be the same as for the undoped and 1 % Nb-doped films. At 400 °C there seems to be little difference in the structure of the deposited films. The predominant peak is anatase (101) with smaller (112) and (200) peaks. But at 800 °C—unlike the undoped and 1 % doped films—the film is clearly still largely comprised of anatase with a predominant (004) or c-axis orientation. There is a small rutile (110) peak. Qualitatively, the XRD pattern of the 10 % Nb-doped film at 800 °C is similar to the undoped and 1 % doped films at 700 °C i.e. the dominant peak is anatase (001), with anatase (101) and rutile (110) also apparent. This suggests that the niobium doping is stabilising the anatase phase with respect to rutile. This is discussed at the end of the chapter in Sect. 7.8.

7.6.4 Discussion of Substrate Heating Versus Post-annealing

Initially it was presumed that the films grown with substrate heating must have been growing with the temperature dependent orientation from the start of the deposition. During most depositions, the chamber was filled with oxygen to a pressure of 500 Torr following the deposition step, but prior to cooling from deposition temperature to room temperature. However, a deposition was made that omitted the

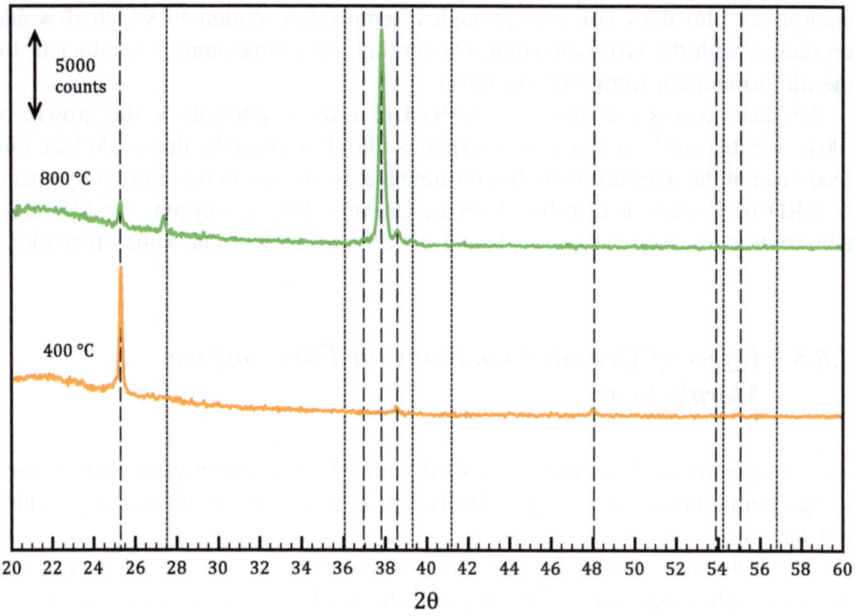

Fig. 7.6 XRD patterns of films with 10 % Nb deposited on fused silica with substrate heating at 400 and 800 °C. The anatase (*dashes*) and rutile (*dots*) XRD reference pattern peaks are indicated by *vertical lines*

final oxygen-annealing step before cool down. The film did not crystallise (there were no peaks detectable in the XRD patterns). This suggested that the film does not crystallise during the deposition step, but rather during the short period of oxygen annealing whilst cooling (from 30 min to 2 h depending on the deposition temperature). As such, it was concluded that the as-deposited film is predominantly amorphous, until the introduction of the oxygen. It should be noted that even without the introduction of oxygen, that the film is still annealed, just at a lower oxygen pressure. This suggests that the oxygen pressure is critical in the formation of anatase i.e. during deposition there is insufficient oxygen incorporated into the film to form the anatase structure. The necessity of oxygen annealing in the formation of rutile has not been investigated. The films deposited at room temperature are amorphous until post-annealing in air. When annealed to the same temperatures of 600—700 °C there is no reorientation. This may be because the films have to pass through a temperature range in which the can reorganize but in which the (101) surface is lowest energy. So the films crystallize between 200 and 300 °C in the (101) orientation. On heating to 600–700 °C the film does not have sufficient energy to change from one solid crystalline orientation to another. In the case of the films deposited with substrate heating, the amorphous film is formed directly at 700 °C for instance. With the introduction of oxygen, the film is able to crystallize, and does so at a temperature at which (004) rather than (101) is the lowest energy surface. So the transition is directly from amorphous to a (004) crystal grain. The amorphous film does not pass through a temperature region in which it would crystallize with the (101) direction. On cooling, the temperature is insufficient for the film to reorient from (004) to (101).

Another process that may be involved or fully responsible is the growth of (004) oriented seeds in a largely amorphous film. For example, this could be a thin seed layer at the interface with the substrate that is too thin to be detected by means of XRD or Raman or distributed seeds. On annealing in oxygen, the film crystallisation is nucleated at the seeds and the film grows with the same orientation.

7.6.5 *Effect of Growth Conditions on Film Surface Morphology*

The surfaces of the TiO_2 films deposited with 1 % Nb-doping were characterized using Atomic Force Microscopy (AFM). The experimental methods are presented in Chap. 4.

The following figures present the measurements made on both the 1 % Nb-doped films deposited with varying substrate-heating temperature (Figs. 7.7, 7.8, 7.9 and 7.10), and the undoped films deposited at room temperature with

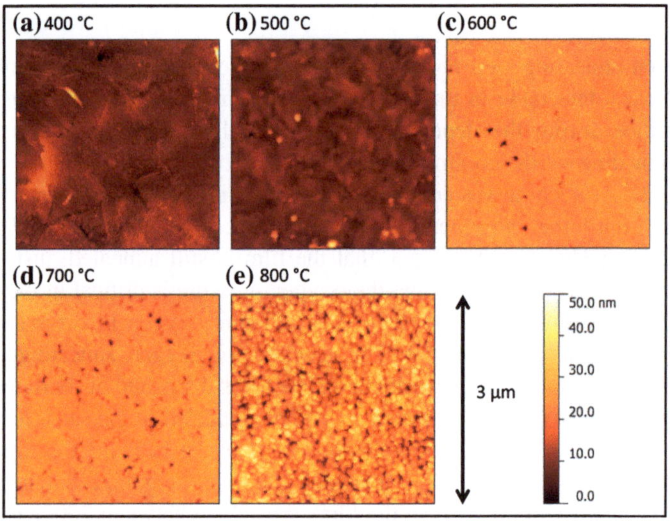

Fig. 7.7 AFM images of 1 % Nb-doped films on fused silica. Deposition pressure was 30 mTorr, with 10,000 pulses at 8 Hz. The colour map of each image has been scaled to 50 nm to allow visual comparison of the morphology. Each image is 256 × 256 pixels covering an area of 3 × 3 µm

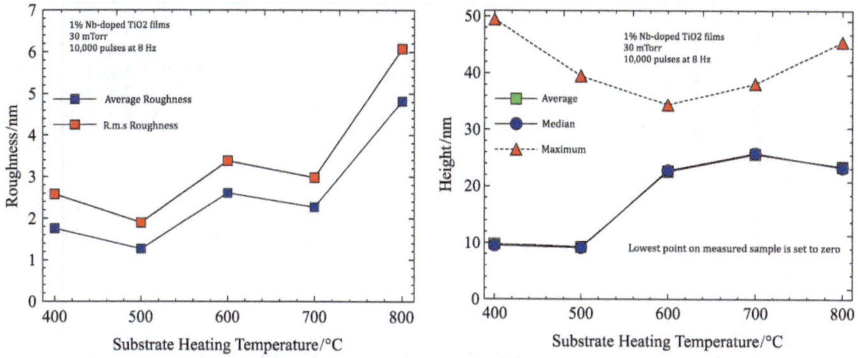

Fig. 7.8 Statistics from AFM images of 1 % Nb-doped films on fused silica. Deposition pressure was 30 mTorr, with 10,000 pulses at 8 Hz. The minimum measured height is set to zero. The maximum (*red triangle*), average (*green square*) and median (*blue circle*) values relative to the minimum are given in the *left hand* panel. The *right hand* panel shows the surface average (*blue square*) and root-mean-square (*red square*) roughness of the films

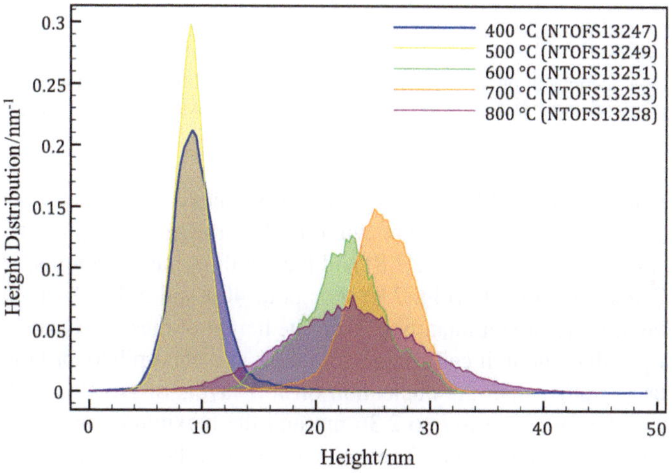

Fig. 7.9 Height distribution statistics from AFM images of 1 % Nb-doped films on fused silica. The minimum measured height is set to zero. The figure shows the distribution of heights of points on the film in the AFM images for each deposition temperature

varying pressure and subsequently post-annealed (Figs. 7.11, 7.12, 7.13 and 7.14). A comparison was also made between films deposited with substrate heating and those with post-annealing.

Looking at Fig. 7.7 through Fig. 7.10, the film grown with a substrate temperature of 400 °C has lines on the 1 micron length-scale denoting grain boundaries with a fine texture on the grains. The film has an average roughness of 1.76 nm as a whole, and of 1.62 nm excluding the occasional island-like features (these show up as bright or dark spots). At 500 °C the film still demonstrates the island-like

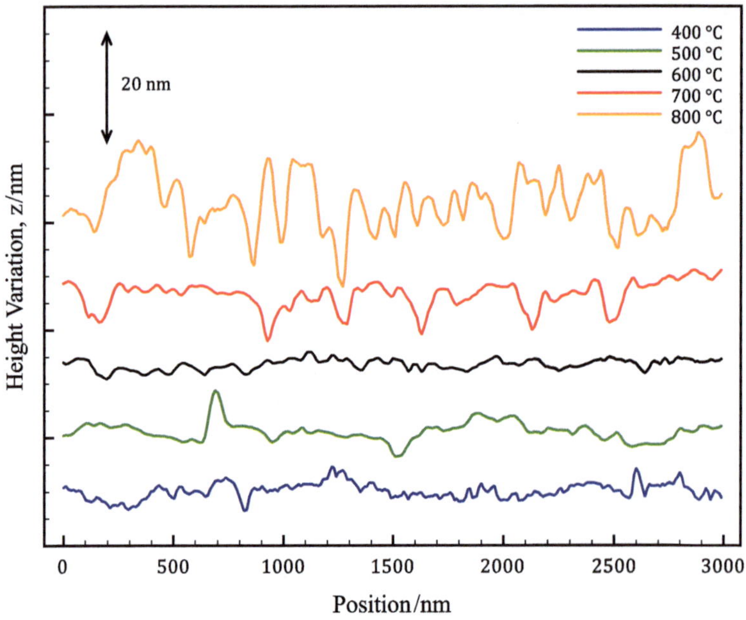

Fig. 7.10 Line Sections through AFM images of 1 % Nb-doped films deposited on fused silica at temperatures of 400–800 °C. Each line is a section through the AFM image showing the variation in the z-profile

features, with a smooth film between. The roughness reduces to only 1.27 nm including the features, and 1.17 nm without. The range of heights of peaks and valleys of the film decreases (Fig. 7.8) and the width of the distribution is narrower for 500 °C than for 400 °C (Fig. 7.9). For both 400 and 500 °C, the statistically outlying features are predominantly peaks. The height of these features (10–40 nm) is much larger than the unit cell, suggesting these are nano-pillars or large particles from the target rather than the nucleation of new layers of growth. At 600 °C the roughness of the film increases to 2.36 nm and the maximum range of heights on the sampled area drops to 37 nm. The average height (relative to the lowest point on the film) moves closer to the maximum. This can be seen as a shift in the height distribution of points imaged in the scan (Fig. 7.9). The outlying features are predominantly pits 10–20 nm deep, while the majority of the film is smooth. The film deposited at 700 °C also shows pitting. The roughness decreases slightly, to 2.27 nm, and the pits appear to be slightly deeper due to the shift in the average height towards the maximum. At a deposition temperature of 800 °C there is a large increase in film roughness to 4.82 nm and an increase in the range of heights across the film. The average height occurs close to the centre of the range, and it is clear from the image of the scan that there are both peaks and pits present.

From 400 to 500 °C the decreased roughness could be attributed to the higher temperature annealing defects in the surface, as there is no major change in the XRD patterns or Raman spectra. There is a slight decrease in the intensity of the

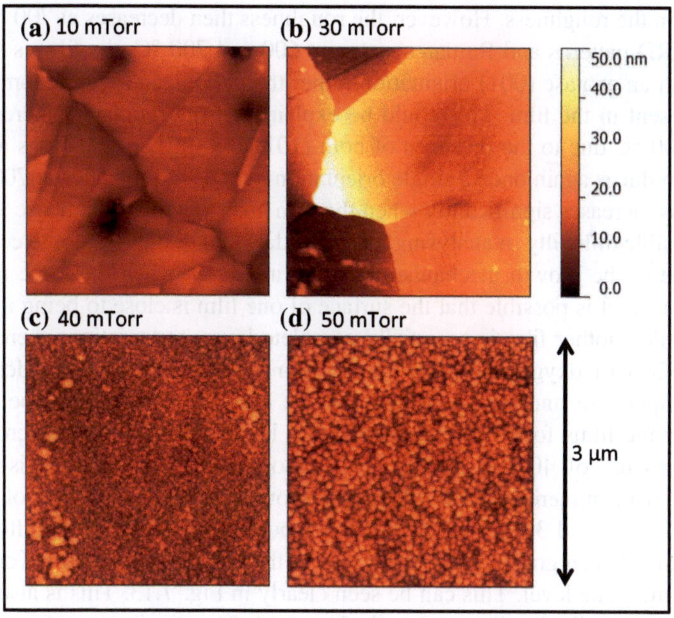

Fig. 7.11 AFM images of undoped TiO_2 films deposited on fused silica at room temperature with 10,000 pulses at 8 Hz. The oxygen pressure was varied to identify the optimum pressure for film crystallization. The pressures used were 10, 30, 40, and 50 mTorr. Following deposition the films were post-annealed to 300 °C in air. The colour map of each image has been scaled to 50 nm to allow visual comparison of the morphology. Each image is 256 × 256 pixels covering an area of 3 × 3 µm

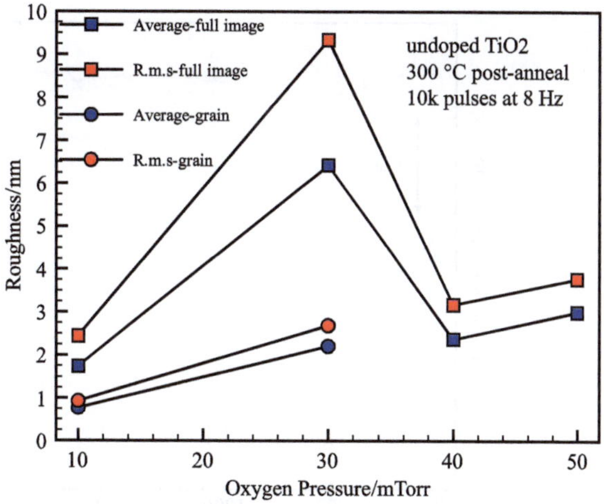

Fig. 7.12 Roughness of undoped TiO_2 films deposited at room temperature with varying oxygen pressure. The statistics are presented for the whole area of the sample (*squares*) and for a single grain (*circles*) where possible. While the films are rough due to the presence of large grains, the grains themselves are extremely smooth

(101) peak, which could indicate a less crystalline material. But this is counter-intuitive, as higher temperatures should lead to better crystallization. From 600 °C the film starts to develop a preferential c-axis orientation, and we see an

increase in the roughness. However, the roughness then decreases at 700 °C. Based on the XRD patterns and Raman spectra, at 600 and 700 °C, the film is starting to grow with an anatase (001) orientation rather than (101), and with a component of rutile present in the film. This could be explained if the roughness increases from 500 to 600 °C due to the presence of both (101) and (001) orientations at 600 °C, but then reduces again once a single orientation—(001)—dominates at 700 °C. The roughness increases significantly when the film grows as the rutile phase at 800 °C.

A possible difficulty in analysing the AFM data and drawing definite conclusions arises due to the growth mechanism of the films. Even when the same number of pulses is used it is possible that the surface of one film is close to being a complete layer, while another film is a partially complete layer, or variations thereof.

The effect of oxygen pressure on the film topography, for films deposited at room temperature and post-annealed, can be seen in Fig. 7.11. At both 10 and 30 mTorr the films forms large grains with a length scale of ca. 1 micron. At the higher pressure of 40 and 50 mTorr, the morphology of the films is distinctly different, with numerous small islands of about 100 nm. The grains of the films deposited at 10 and 30 mTorr exhibit sharp boundaries with height differences of 10–20 nm between grains, while the grains of the films at 40 and 50 mTorr seem to grow on the same level. This can be seen clearly in Fig. 7.13. This is also apparent from the height distributions shown in Fig. 7.14. The distributions for films at 40 and 50 mTorr are single populations described by a Gaussian (albeit slightly skewed). The distribution for the 10 mTorr film falls largely onto a single curve, but with a shoulder at low height that describes the pitting. For the film at 30 mTorr the height distribution is composed of multiple gaussians offset along the x-axis (height) that each describe a grain.

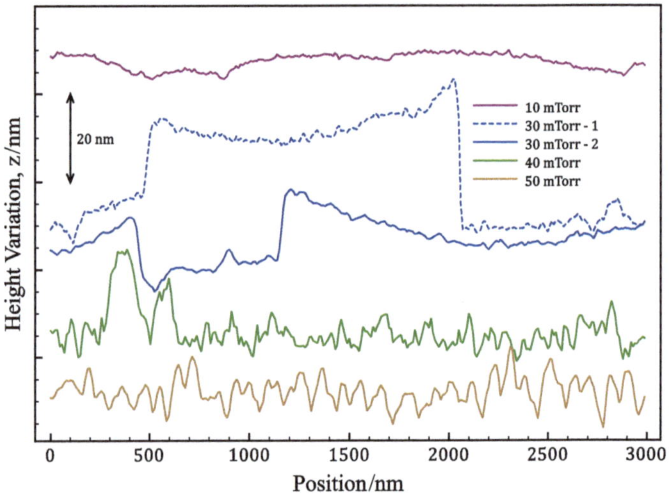

Fig. 7.13 Line profiles of undoped TiO_2 films deposited at room temperature with varying oxygen pressure. The profiles are offset from each other on the y-axis for clarity. The y-axis scale is shown as 20 nm. Two profiles are shown for the film deposited at 30 mTorr

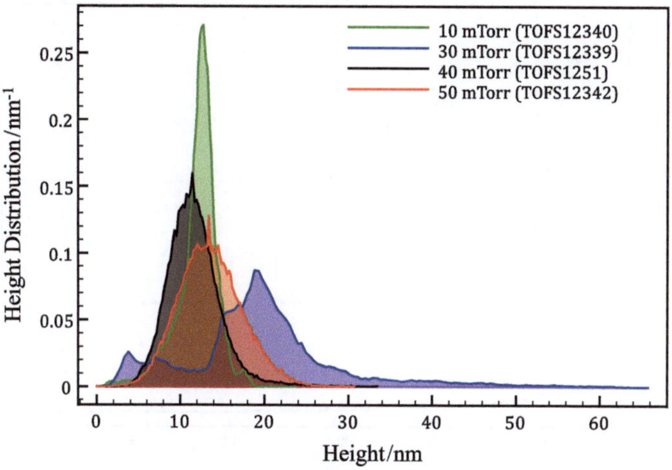

Fig. 7.14 Height distribution statistics from AFM images of undoped films on fused silica. The minimum measured height is set to zero. The figure shows the distribution of heights of pixels in the AFM images for each deposition pressure

The thickness of the films deposited at 10, 30, 40, and 50 mTorr are approximately 300 nm as determined by optical methods described in Sect. 7.7.

The effect of temperature on roughness can be seen in Fig. 7.12. Based on the roughness calculated for the whole image, the average roughness of the films increases from 1.74 nm at 10 mTorr to 6.42 nm at 30 mTorr. However, if the roughness is calculated for a single grain (such as seen in the upper-right corner of Fig. 7.11b) rather than for the entire image, then grain roughness increases from 0.768 nm to 2.21 nm (Fig. 7.12b). At 40 mTorr the roughness increases to 2.37 nm and then to 3.00 nm. This is close to being a linear increase in roughness with oxygen pressure from 10 to 50 mTorr.

By comparison to the films deposited with substrate heating at 10–30 mTorr, the films deposited at room temperature and post-annealed are similarly rough in the range 1.5–2.5 nm. Due to the growth of grains at different heights (presumably due to different growth rates for different crystal facets) the maximum range of heights of the film deposited at 30 mTorr is 65 nm over a 3 $\times$ 3 micron area. This is about 20 % of the thickness of the whole film. The film at 10 mTorr also suffers from pitting. This could form due to high-energy fragments from the material plume damaging the surface of the film. Overall, this suggests that films should be deposited with substrate heating if a uniformly smooth film is preferable.

7.6.6 *Effect of Film Thickness*

A series of depositions were made to calibrate the film thickness against the number of laser pulses for 5, 10, 20, and 30 thousand pulses. The calibration curve is

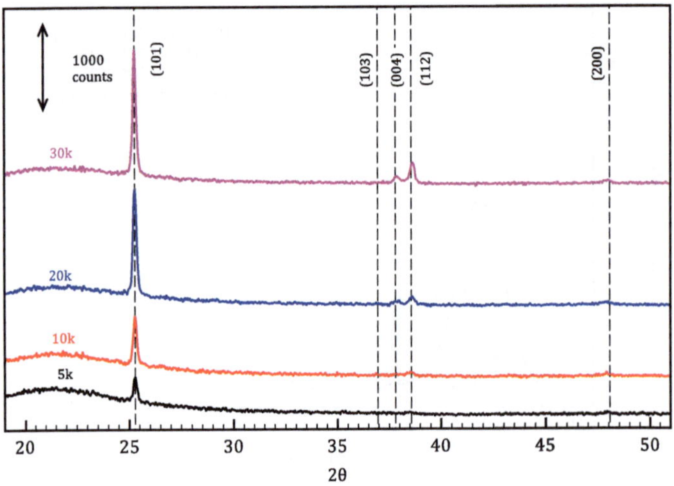

Fig. 7.15 XRD patterns showing effect of film thickness on peak intensity and presence of orientations. Films deposited with 5, 10, 20, and 30 thousand pulses

presented in the Chap. 4, on Experimental methods. The effects of thickness on the structure and properties are discussed below.

The XRD patterns (Fig. 7.15) show that the predominant peak is the anatase (101) peak. The intensity of the peak increases proportionally with the thickness. The (112) and (004) reflections are also apparent and the intensity increases with thickness. The (200) reflection is barely visible and the intensity doesn't increase significantly. The stronger presence of the (112) and (004) peaks for the film deposited with 30 k pulses and the scaling of the (101) peak intensity suggest that (112) and (004) oriented grains may also be present in the thinner films. In that case, the XRD scan speed used for the 10 k pulse films was too fast to give an adequate signal to noise ratio for resolving the peaks.

The effect of thickness on the surface morphology of the films is shown in Fig. 7.16. The film of approximately 150 nm in thickness, deposited with 5 k

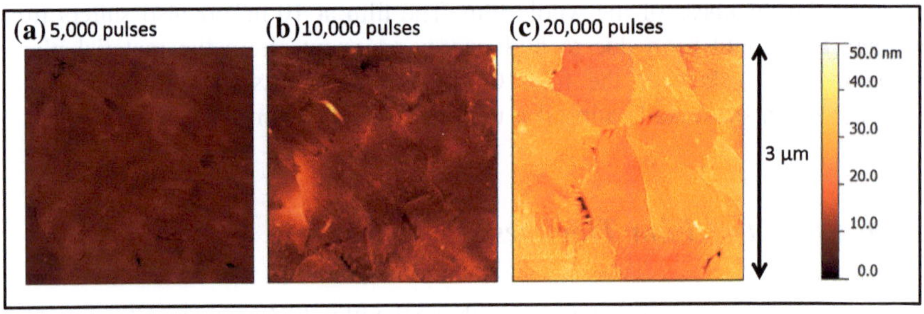

Fig. 7.16 AFM image showing effect of film thickness on morphology of films. Films deposited with 5, 10, 20 thousand pulses. The film deposited with 30 thousand pulses was not imaged

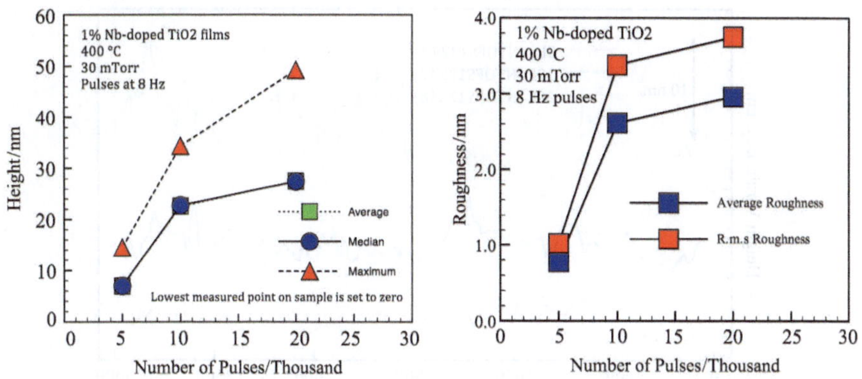

Fig. 7.17 Statistics from AFM images of films deposited with a range of thicknesses. Deposition pressure was 30 mTorr and the temperature 400 °C. The minimum measured height is set to zero

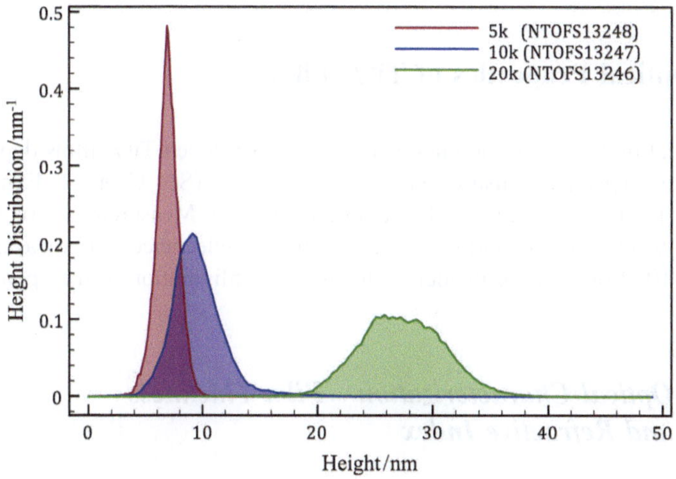

Fig. 7.18 Height distribution statistics from AFM images of undoped films on fused silica deposited at a range of thicknesses. The minimum measured height is set to zero. The figure shows the distribution of heights of pixels in the AFM images for each deposition pressure. Films deposited with 5, 10, 20 thousand pulses

pulses, appears to be uniform. At ca. 300 nm thickness (10 k pulses), grain boundaries start to become apparent. Otherwise, the grains in the film are still of equal height. The grain boundaries become even more apparent at ca. 600 nm (20 k pulses) and there is a clear difference in height of adjacent grains of 5–10 nm, as seen in the line profile of Fig. 7.19.

The statistics of the film show that both the roughness (Fig. 7.17) and the range of heights (Fig. 7.18) increase with thickness at a less than linear rate (but it is hard to be certain with only three measurements). The average roughness is 0.78 nm at 5 k, 1.76 nm at 10 k, and 2.96 nm at 20 k.

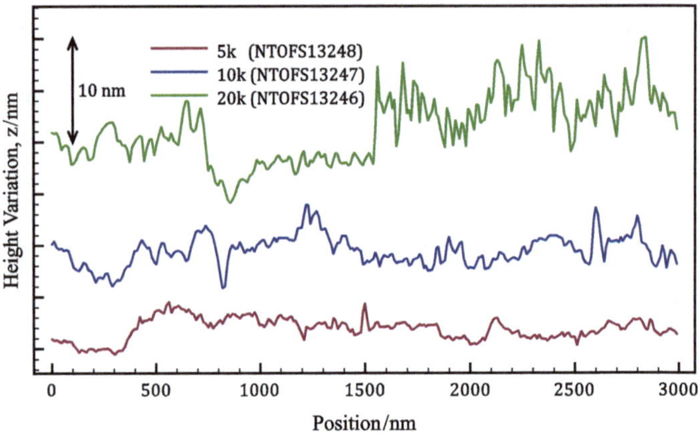

Fig. 7.19 AFM line profiles of films deposited with a range of thicknesses demonstrating increase of roughness with thickness

7.7 Optical Properties of TiO_2 Films

The optical properties of the undoped and 1 % Nb-doped TiO_2 films deposited on glass were investigated using UV-Vis spectroscopy (See Chap. 4, Experimental Methods for further detail on the setup and theory). Measurements of the transmittance were made at normal incidence, while reflectance was measured at an angle of 10° from normal incidence due to the configuration of the apparatus.

7.7.1 Optical Characterization—Film Thickness and Refractive Index

The thin film interference fringes in the UV-Vis spectra were used to calculate film thickness and refractive index using Swanepoel's method [10]. The transmittance, reflectance, calculated refractive index and thickness for each thin film of interest are given in Appendix C. Figure 7.20 shows a selection of transmittance spectra comparing undoped and Nb-doped TiO_2 films. The spectra clearly show the thin film interference fringes that were used to estimate the film thickness and other optical properties. The spectra for the film deposited at 800 °C has fewer fringes due to the higher refractive index of rutile compared to anatase. The effect of deposition conditions on the optical properties is presented below for refractive index, absorption, and band gap.

The measured reflectance data was used to correct the measured transmittance for reflections to allow the calculation of the absorption in the film. In hindsight, it would have been preferable to measure the transmittance at 10° from normal incidence as this would have made the correction of the transmittance by the

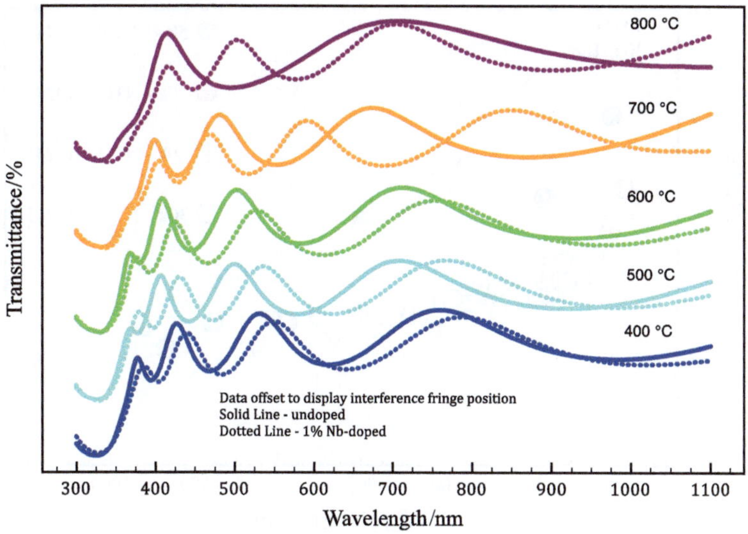

Fig. 7.20 Transmittance of undoped versus 1 % Nb-doped films. The spectra are offset for clarity. The undoped films (*solid lines*) are slightly thinner than the Nb-doped films (*dotted lines*) deposited at the same conditions. This is demonstrated by the blue shift of the peaks and in some cases fewer peaks

reflectance data more accurate. As will be seen below, the presence of the interference fringes (coupled with incomplete correction) makes it difficult to estimate the exact position of the band edge.

To examine whether the thickness of the film had any impact on the refractive index, films with 5, 10, 20, and 30 thousand pulses were deposited at 400 °C and 30 mTorr. In theory, the refractive index should be independent of thickness, however, slight changes in the crystal structure could result from thickness differences. For example, resulting from a strain gradient with the substrate or different orientations of grains in the films. The calculated film thicknesses are given in Table 7.2.

The estimated thicknesses given by the Swanepoel method are shown with a range of one standard deviation in the range of values calculated at each interference fringe. Apart from the film of 114 nm in thickness (deposited with 5000 pulses), the refractive index could be modelled by fitting a quadratic to the reciprocal of the wavelength (essentially a Cauchy model) with R^2 values between 0.988

Table 7.2 Thicknesses of Nb-doped TiO_2 films estimated by the Swanepoel method

Pulses	Thickness (nm)
5,000	114 ± 7
10,000	300 ± 3
20,000	651 ± 5
30,000	921 ± 2

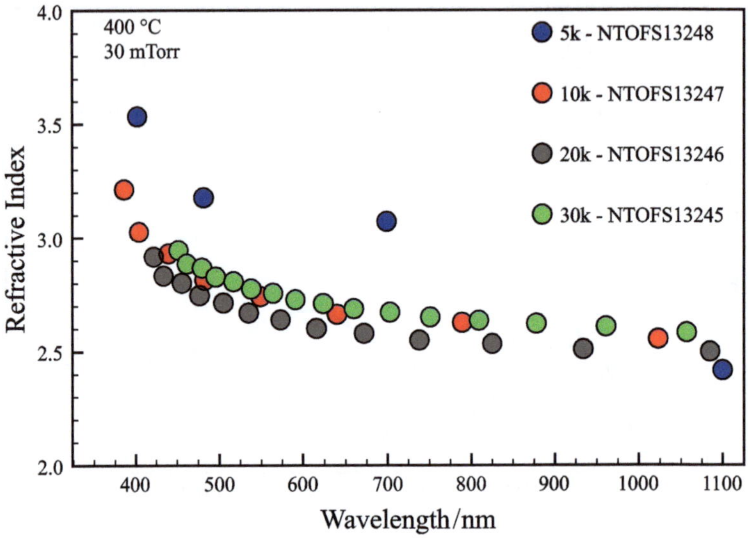

Fig. 7.21 Fitted refractive index of 1 % Nb-doped TiO_2 films deposited at a range of thicknesses. Apart from the number of pulses, all other conditions were held constant. The sample with 5000 pulses could not be modelled accurately due to insufficient interference fringes

and 0.997. That the thickness and refractive index can be modelled accurately suggests that there may be a slight variation in refractive index with thickness. The modelled refractive index of each sample is shown in Fig. 7.21. As the refractive index of the 300 nm film is between that of the 651 and 921 nm films there does not appear to be a trend. Further films deposited with a range of intermediate pulses would be required to identify a trend.

The effect of temperature on the refractive index is shown in Fig. 7.22 for undoped films and in Fig. 7.23 for 1 % Nb-doped films. The refractive index increases slightly from 2.9 at 400 nm for the 400 °C film to 3.0 at 400 nm for the 500 °C film. There is no change from 500 to 600 °C. At 700 °C there is an increase in refractive index to 3.2 at 400 nm as the ratio of rutile to anatase increases, reaching about 3.6 at 400 nm for the rutile film deposited at 800 °C. The refractive index of the Nb-doped films does not follow any distinct trend and does not follow the expected increase in refractive index as the proportion of rutile increases relative to anatase. This suggests that either the fitting of the data may be insufficiently good, or that the Nb-doping had an effect on refractive index.

7.7.2 *Optical Characterization—Absorption*

The transmittance and reflectance spectra were used to obtain the absorption of each film. The estimated film thickness was then used to calculate the absorption coefficient

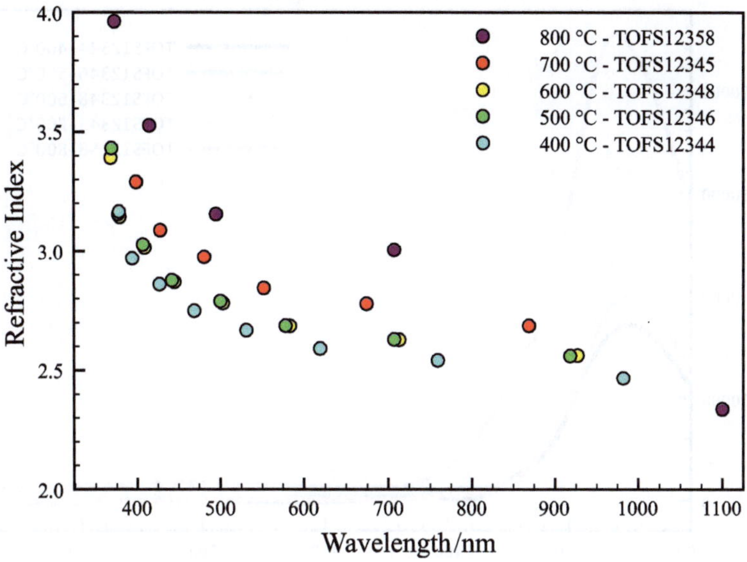

Fig. 7.22 Fitted refractive index of undoped TiO_2 films deposited at a range of substrate temperatures

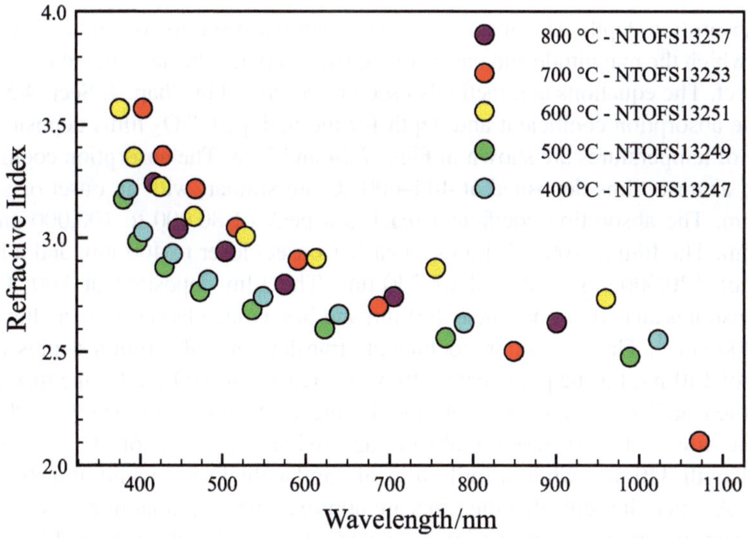

Fig. 7.23 Refractive index of 1 % Nb-doped TiO_2 films deposited at a range of substrate temperatures

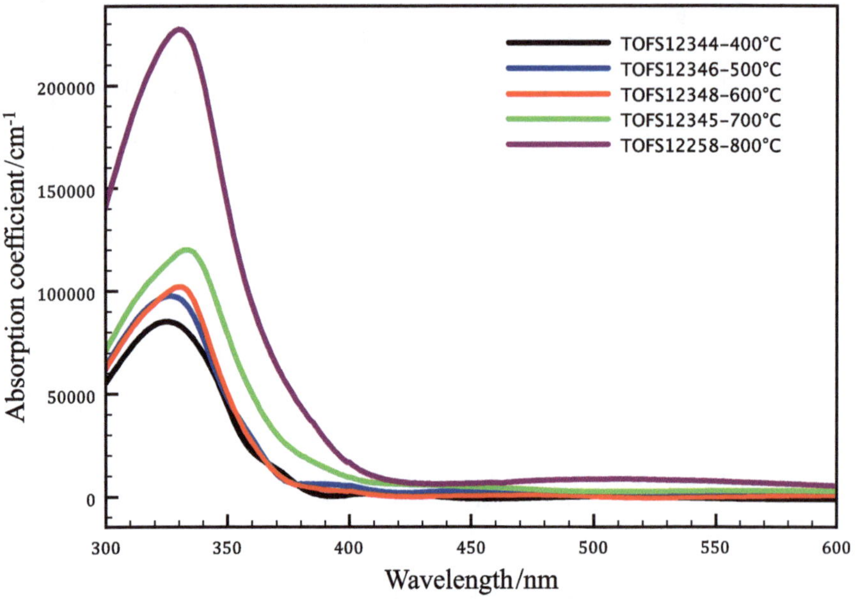

Fig. 7.24 Absorption coefficient of undoped TiO_2 films deposited at temperatures from 400 to 800 °C

and absorption depth. The absorption coefficient was used to construct a Tauc plot from which the magnitude and nature of the band gap may be determined i.e. indirect or direct. The equations and methods used are described in Chap. 4, Sect. 4.5.7.

The absorption coefficient and depth for the undoped TiO_2 films deposited at a range of temperatures are shown in Figs. 7.24 and 7.25. The absorption coefficients of the anatase films deposited at 400–600 °C are similar, with an onset of around 380 nm. The absorption coefficient reaches a peak of 80,000 to 100,000 cm^{-1} at 330 nm. The film at 700 °C shows an earlier onset closer to 400 nm, and a higher peak of 120,000 cm^{-1}, shifted to 340 nm. The film deposited at 800 °C also demonstrates an earlier onset near 400 nm and has a much larger peak coefficient of 230,000 cm^{-1}. The absorption coefficients translate into absorption depths on the order of 150 nm for the pure anatase films, decreasing to 100 nm for the mixed film deposited at 700 °C, and only 50 nm for the rutile film. This suggests that the anatase films of 250–300 nm absorb approximately 90 % of the photons of wavelength 330 nm, while a rutile film of similar thickness would absorb around 99 %. As such, it seems that the films are approximately optimal thickness in terms of maximum absorption from minimal material (at least for absorption of UV light).

To assess the onset of absorption (the magnitude of the band gap) more accurately, Tauc plots were constructed and are shown in Fig. 7.26 assuming an indirect band gap and in Fig. 7.27 assuming a direct band gap. In a material with an indirect band gap, the absorption of a photon requires the simultaneous creation or annihilation of a phonon, in order to conserve momentum as well as energy. The Tauc

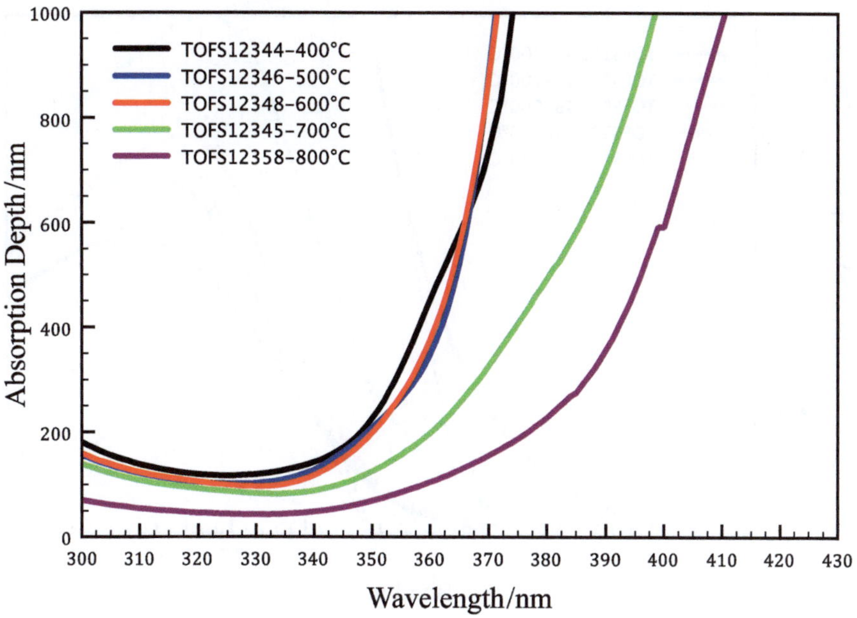

Fig. 7.25 Absorption depth of undoped TiO_2 films deposited at temperatures from 400 to 800 °C

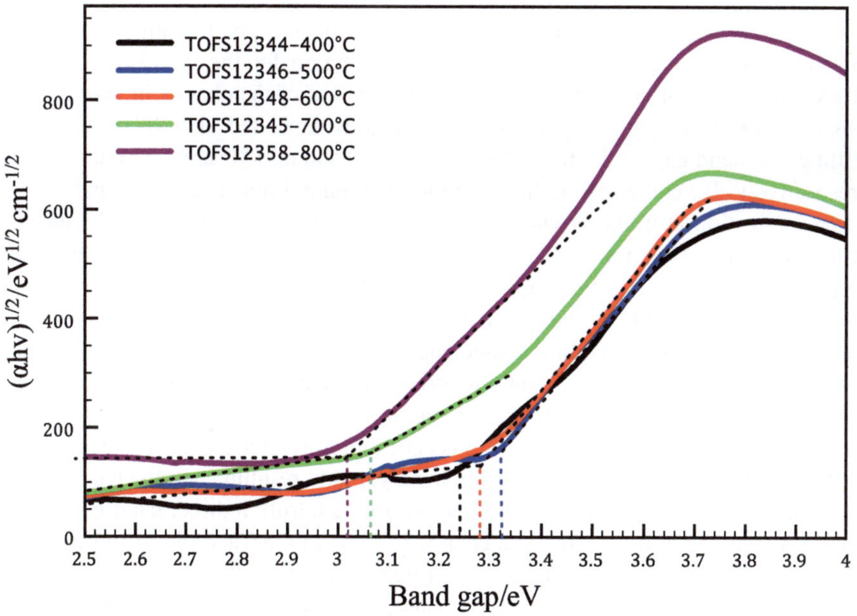

Fig. 7.26 Tauc plot for indirect band gap of undoped TiO_2 films deposited at a range of temperatures

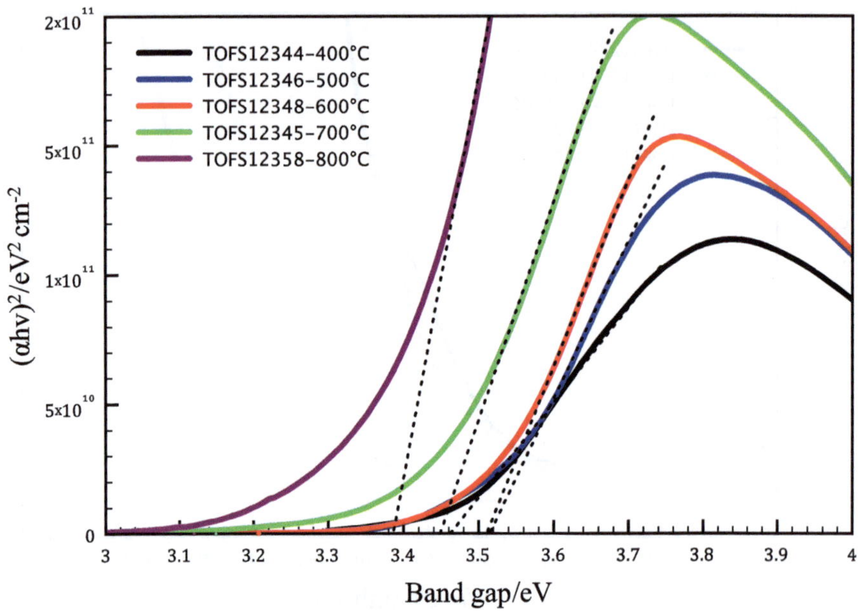

Fig. 7.27 Tauc plot for direct band gap of undoped TiO_2 films deposited at a range of temperatures

plot assuming an indirect band gap shows that the data can be fit with a straight line in the region 3 to 3.5 eV. The incomplete correction for the interference fringes makes it somewhat difficult to determine the position of the change in gradient that signifies the band gap, in particular for the film deposited at 400 °C. From the fitting, the band gap of the films deposited at 500 and 600 °C are 3.32 and 3.28 eV respectively. By comparison, the Tauc plot assuming a direct band gap indicates a direct transition at a higher energy of 3.5 eV. So the band gap E_g of the anatase films appears to be indirect and ca. 3.3 eV, slightly higher than the typically reported 3.2 eV[20]. The band gap of the predominantly rutile film deposited at 800 °C is estimated to be around 3.02 eV and indirect, while the mixed anatase/rutile film has an indirect band gap of 3.08 eV. It is noted that the Tauc plot for the rutile (110) single crystals reported in Chap. 6, Rutile Electrochemistry, showed the band gap to be direct. The confidence in that measurement is higher as it was for a bulk single crystal. The effects of thin film interference are masking the nature of the absorption and hence the band gap and resulting in misidentification. Alternatively, the thin films may be demonstrating a truly indirect band gap due to strain in the lattice. Interestingly, a change in slope of the Tauc plot is observed for the anatase/rutile film at around 3.35 eV (green line, Fig. 7.26), suggesting that this is the indirect band gap of the anatase phase, and hence that the 3.08 eV is for the rutile grains. The absorption coefficient and absorption depth for 1% Nb-doped films are presented in Figs. 7.28 and 7.29.

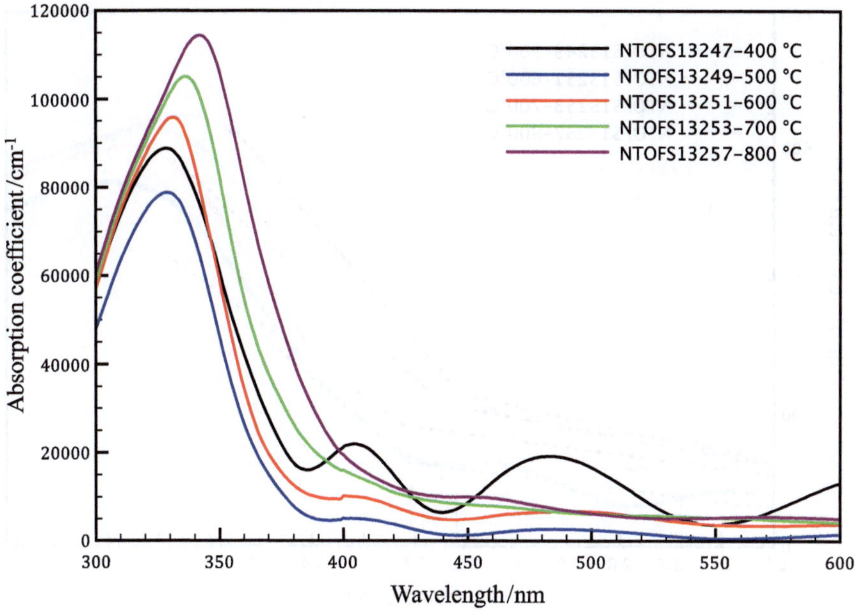

Fig. 7.28 Absorption coefficient of 1 % Nb-doped TiO_2 films deposited at a range of temperatures

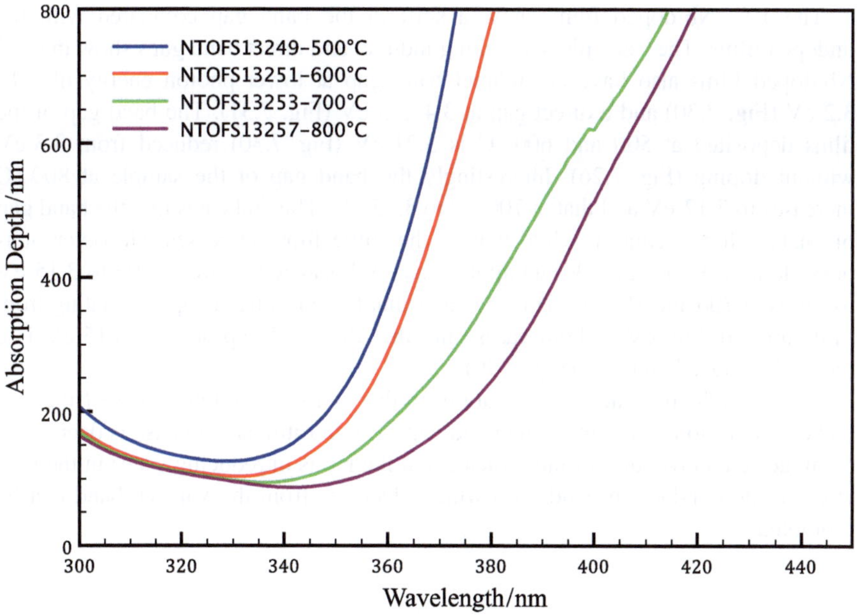

Fig. 7.29 Absorption Depth of 1 % Nb-doped TiO_2 films deposited at a range of temperatures

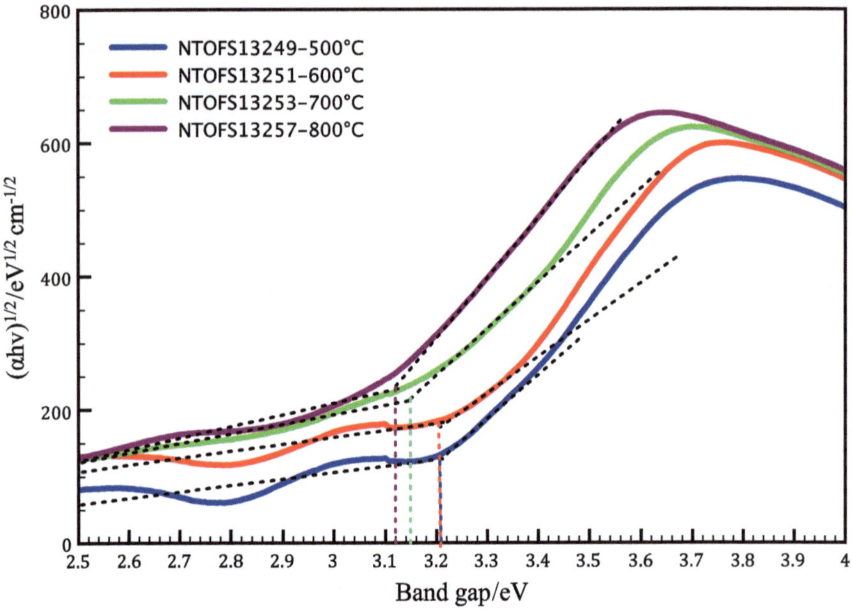

Fig. 7.30 Tauc plot for indirect band gap of 1 % Nb-doped TiO_2 films deposited at a range of temperatures

The 1 % Nb-doped films show a shift in the band gap compared with the undoped films. The Tauc plots assuming indirect and direct band gaps show that the Nb-doped films also have an indirect band gap at lower photon energy of 3.1–3.2 eV (Fig. 7.30) and a direct gap at 3.4–3.5 eV (Fig. 7.31). The band gap of the films deposited at 500 and 600 °C is 3.21 eV (Fig. 7.30) reduced from 3.3 eV without doping (Fig. 7.26). Interestingly the band gap of the sample at 800 °C increases to 3.12 eV and that at 700 °C to 3.15 eV. This indicates that the band gap of anatase films reduces, while that of the rutile films increases. However, it is possible to see (in Fig. 7.30) an approximately linear region from 2.95 to 3.15 eV for films at 700 and 800 °C. This indicates that there is a band gap originating from rutile around 2.95 eV in both cases, and that the band gap at 3.12–3.15 eV is a reduced anatase band gap (Fig. 7.30).

However, the presence of undulations in the data—due to interference fringes—makes it hard to be certain. If the band gap of both rutile and anatase is decreased, as suggested above, this would indicate that the Nb is introducing states in the gap, close to the conduction band, into which electrons from the valence band can be promoted.

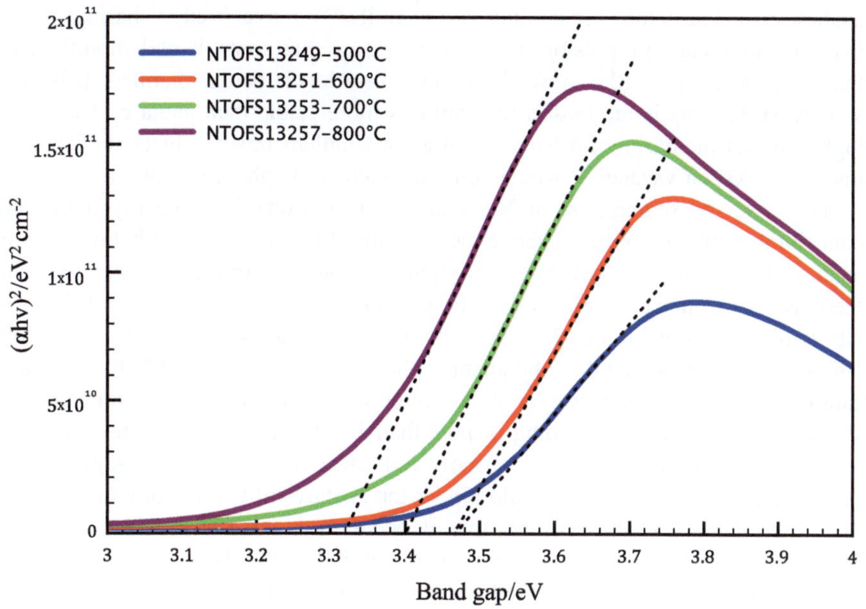

Fig. 7.31 Tauc plot for direct band gap of 1 % Nb-doped TiO_2 films deposited at a range of temperatures

7.8 Discussion—Stabilisation of Anatase with Nb

The stabilisation of the anatase phase to higher temperatures was seen in this work where the addition of 10 % Nb to the PLD target results in anatase films growing at higher temperatures than for undoped and 1 % Nb doped films. In particular, where an undoped or 1 % doped film grown at 800 °C is found to consist only of rutile; the 10 % doped film is still predominantly anatase. The precise anatase-rutile phase transition temperature for these highly doped films has not been determined. Deposition of films above 800 °C are required to establish this.

Arbiol et al. [11] studied the effect of Nb on the anatase-rutile transition by forming nano-powders doped with Nb from zero up to 24.5 at.%. The samples were annealed at 600–900 °C in steps of 50 °C. The ratio of anatase to rutile was determined from the intensities of the dominant peaks of the XRD patterns, anatase (101) and rutile (110). The temperature at which the 50 % anatase-rutile point occurred shifted to higher values with increasing Nb concentration. For the undoped TiO_2 the 50 % point occurred at about 700 °C, increasing to 850 °C for doping around 3 at.% and reducing towards 800 °C for the highest Nb doping levels of 10.9 and 24.5 %. The authors also note the formation of NbO at the anatase-rutile interface for the highly doped sample. It was suggested that the Nb segregates from the rutile as it forms from anatase. They suggest the reason for the higher tolerance of defects (Nb-doping) in anatase is due to the Ti-O octahedra sharing four edges in the anatase crystal structure versus only two in rutile. The greater edge sharing means that the cation-cation repulsion is higher in anatase than in rutile and should mean that a cation vacancy (the most likely ionic disorder) is better tolerated. It is also suggested that this is the reason that rutile is more stable than anatase. They also highlight that increasing the Nb concentration simultaneously reduces the concentration of oxygen vacancies, which could also affect the phase transition.

The observed segregation of Nb would seem reasonable especially if the Nb concentration in the sample is above the solubility limit for rutile but below that for anatase. Furthermore, if cation-cation repulsion leads to structural destabilisation, then cation vacancies could stabilise the structure. As the compensating defect for Nb doping in stoichiometric conditions is Ti vacancies it is consistent with Nb-doping acting to stabilise the anatase phase with respect to rutile by reducing cation-cation repulsion. However as Arbiol et al. report, the transition temperature seems to drop again when doping is more than 3–4 % indicating that the reduction of oxygen vacancies may become more important. Although, they suggest that decreasing oxygen vacancies would hinder the anatase-rutile transformation, while the experimental evidence shows that it becomes more easily achieved above 3–4 % of Nb. So the question remains as to why increasing the doping beyond this concentration results in a lower temperature phase transition?

Trenczek-Zajac and Rekas [12] found that TiO_2 doped with Nb had the rutile structure for up to 10 at.% of Nb, above which the $TiNb_2O_7$ phase was seen. This suggests that the solubility limit of Nb in rutile is approximately 10 at.%.

Interestingly, DeSario et al. [13] report the contrary for sputtered films: that Nb promotes the phase transition and that even higher solubility levels up to 40 % Nb are possible depending on the deposition technique. The P(O2) used was approximately 0.12 Pa or 0.9 mTorr or approximately 10^{-6} atm. DeSario et al. note that this pressure is considered high enough to form fully stoichiometric TiO_2 based on their previous publications. The Nb concentrations used were typically between 10 and 30 % Nb, but even the sample doped with 3 % Nb showed the rutile phase. The films were deposited at room temperature on glass, but used a substrate bias, which could have affected phase stability.

Given these studies from the literature, it is seems that for up to the 10 wt% of Nb-doping used in the PLD targets produced during this thesis, the effect of Nb is to stabilize the anatase phase. It seems likely that increasing the Nb concentration would then exceed the solubility limit and would result in the formation of both anatase and niobium oxide phases. It is also possible that above the solubility limit, the formation of rutile rather than anatase is favoured. This should be investigated.

7.9 Discussion—Conductivity

The principle reason for doping the films with Nb was to impart conductivity. Conductivity was desirable for a number of reasons. The following reasons are also applicable to films deposited on single crystal substrates, as discussed in Chap. 8.

- Scanning Tunnelling Microscopy (STM). The goal is to eventually be able to image the different surfaces of the anatase films to study the defects at the surface and to study the absorption of water and other molecules. STM can only be done on conductive samples. Most studies on oxides are done on highly reduced samples with a thin oxide layer i.e. they may be significantly different in behaviour to the stoichiometric oxides. Initial attempts at imaging the surfaces of the films on fused silica were unsuccessful due to the high resistivity of the samples. Achieving conductivity by doping was considered preferable to reduction, as it should have less impact on the structure (avoidance of Magnéli phases for example).
- Photoelectrochemistry and dark electrochemistry. The goal is to be able to compare the behaviour/performance of the different surfaces of anatase (and rutile) by in situ electrochemical techniques to characterize the interface with an aqueous electrolyte. The potential of the films needs to be controlled by a potentiostat, with current collected from the full area of the electrode. The films are deposited on non-conductive substrates (fused silica, MgO, STO) so it is not possible to collect the current at the back of the film. As such, the current would have to flow across the film. A low sheet-resistance would be required to collect current from the entire electrode.

On the basis of the work by Furubayshi [14] the introduction of 1 % Nb-doping is seen to increase the conductivity by several orders of magnitude compared to undoped anatase TiO_2. As such, a target with approximately 1 wt% Nb to TiO_2 was produced.

Despite the production of a 1 % Nb-doped target that should in principle result in conductivity similar to that of ITO or FTO, the films did not demonstrate a measurable degree of conductivity.

The lack of conductivity could be explained by the Nb not being incorporated into the film at a similar concentration as in the target. To investigate this both and undoped and 1 % Nb-doped films were examined using Secondary Ion Mass Spectroscopy (SIMS). The SIMS analysis confirmed the presence of Nb in the doped film only. It also identified the presence of a low-level concentration of aluminium impurities in both the undoped and 1 % doped films. Without a reference for sputtering yield of Nb and Al doping in TiO_2, it is difficult to quantify the concentration of dopant and impurity. However, the raw signals were of a similar magnitude (the SIMS data is not shown as it is not quantitative). Other techniques were considered including; XPS; ICP; and SEM. In TiO_2, the titanium cation is usually assumed to be in a 4+ state, so Nb acts as a 5+ donor and Al is a 3 + acceptor. As such, the presence of Al could counteract the n-type conductivity expected with Nb-doping. The most likely source of Al is from the Al_2O_3 crucible and furnace lining used during the sintering of the targets for PLD. But it is unlikely that the sintered ceramic would have incorporated a similar concentration of Al by chance as the intentional 1 % Nb-doping. It is also possible, but unlikely, that the stoichiometry of the target did not transfer to the substrate, resulting in similar concentrations of Nb and Al in the film. The absence of measurable conductivity (at room temperature as measured by multi-meter, four-point probe, Hall effect, and probe station) might suggest that there is sufficient Al to counteract intentional Nb-doping.

To try and rule out the effects of Al^{3+} compensation of the Nb-doping, a second doped target with a doping level of 10 % by weight was produced. The aim was to intentionally over dope the films to obtain a net n-type doping. This would indicate whether Nb-doping does indeed result in conductive anatase films. From this point it would be possible to decrease the doping to a lower concentration if necessary.

Conductivity measurements of an 8 mm pellet sintered at the same time as the PLD target demonstrated a high but measurable resistance. An XRD pattern of the pellet showed it to be purely rutile TiO_2. There was no indication that either anatase or Nb_2O_5 were present. As anatase should be more conductive than rutile—due to the fact that the donor levels are calculated to sit in or close to the conduction band, rather than 0.5–1.0 eV deeper in the band gap—it was expected that the anatase films deposited with 10 % Nb-doping would be conductive.

However, the 10 % Nb-doped films also proved to be too resistive to measure.

As such, it was decided to look in greater detail at the defect chemistry of TiO_2 to try to identify potential causes of the insulating behaviour of the highly doped films. The review and analysis of the defect chemistry of TiO_2 is presented in Chap. 3, and the conclusions in relation to this work are now summarized.

It is believed that the lack of conductivity is a matter of differences in how the films are grown in this study compared to those in the literature that report high conductivity. Here, the films have been deposited at raised temperature in a vacuum. Once the deposition has finished the chamber is backfilled with O_2 at 500 Torr (just below ambient pressure) and the films are then cooled in the O_2 atmosphere at 10 °C/min. As noted earlier in this discussion, the backfilling with oxygen appears to be required in order to cause the films to crystallize. The precise heating, cooling and backfilling procedures used by Furubayshi et al. [14] have not been reported in detail. From the literature review of the defect chemistry of TiO_2 it seems likely that when the film is formed at low pressure, oxygen can be lost from the lattice and a conductive film is formed.

At ambient $P(O_2)$ and in the region of stoichiometry, the expected compensating defect is ionic rather than electronic: the formation of Ti vacancies is favoured. The defect compensation mechanisms are typically given for equilibrium conditions, so it could be assumed that if the films are formed in similar conditions (in vacuum) and are then cooled before being brought back to ambient $P(O_2)$, that similar films and conductivities would be expected. If this is true, then the films being deposited by PLD are still in the stoichiometric region, and hence electronic conductivity would not be expected.

However, it is possible that the order of steps in the deposition process could lead to a situation in which one film is at equilibrium while the other is not. However, the non-equilibrium condition may be either metastable or only approach equilibrium slowly, resulting in conductivity.

For example, if the film is deposited in vacuum at high temperature it is likely to form electronic compensating defects. It is also easy for the lattice to eject a neutral

O_2 molecule into the vacuum with low oxygen activity. If the $P(O_2)$ is brought up to ambient, and the high temperature is maintained, it seems reasonable that the film would reincorporate oxygen into the lattice due to the preferred compensating defect at high oxygen activity. The high temperature would also aid diffusion of defects to the bulk or surface and the reorganization of the lattice. This would be expected to give an insulating film, and is the how the films have been grown in this study to-date. So the final film is in (or close to) equilibrium with the environmental oxygen pressure.

If the film was cooled to room temperature before increasing $P(O_2)$ to ambient, it might be possible to "lock-in" the electronic conductivity compensating the Nb^{5+} defect. Decreasing the temperature in vacuum should not result in a change of the compensation mechanism. With increasing $P(O_2)$ the film should be oxidized, with an associated change in the compensating defect. The lower temperature would act to prevent defect diffusion (such as interstitial Ti to the surface) and to the re-incorporation of oxygen into the lattice. It could be argued that the electronic conductivity might remain as a metastable defect, or that the film will slowly re-oxidize in ambient conditions. The final film would not be at equilibrium with the environment.

The corollary of the above being true is that conductive TiO_2 films can only be used at low temperature or in an oxygen deficient environment. In an oxygen environment, heating the film should result in the incorporation of oxygen into the lattice and the freezing out of the electron conductivity as the defect compensation mechanism changes. This process should be expected even at room temperature in an oxygen-rich atmosphere, as the equilibrium defect is probably interstitial Ti. However, it may occur too slowly to be of significance.

Based on the two different procedures of cooling and increasing $P(O_2)$ discussed above, it seems probable that the conductive films achieved in the existing literature have been cooled *in-vacuo* prior to being exposed to high oxygen activity. As such, it might be expected that the conductivity of these films might reduce over time as the film tries to establish equilibrium with the surroundings. The long-term performance of these electronically conductive anatase films is currently unreported in the literature.

It is noted that in addition to Nb-doped films, undoped TiO_2 films that have been annealed in reducing conditions have also demonstrated high conductivity [1, 15]. The conductivity of these films is likely due to intrinsic non-stoichiometry, and should also be unstable in the long term.

To test this hypothesis, a film was deposited at 30 mTorr and 400 °C as usual, then cooled back to room temperature at the usual rate, omitting the oxygen backfill step before cooling. As mentioned briefly earlier in this discussion in relation to the effect of deposition conditions on phase and orientation, the film failed to crystallise. It was also noted that Furubayashi et al. deposited films at a much lower oxygen pressures of 50 μTorr. A film deposited close to this pressure also failed to crystallise.

It appears that the oxygen-annealing step is required to form a crystalline film, but the act of introducing the oxygen causes the film to form ionic rather than electronic defects, resulting in highly insulating films. On this basis, it seems that a reasonable approach would be to form the films in vacuum, anneal them in oxygen to form a crystalline material, and then subsequently re-anneal in a reducing atmosphere to form a conductive film. Alternatively, the oxygen backfill pressure could be optimised for both crystallisation and conductivity.

This suggests that the next stage of the work should involve a systematic study of the following:

- Effect of changing the oxygen backfill pressure for annealing (from 500 Torr down to deposition pressure). For example at 500, 50, 5, 0.5, and 0.05 Torr.
- Effect of annealing in a reducing atmosphere; time, temperature, and pressure

It should be noted that the defects discussed in this section and the literature review are based on data for rutile rather than for anatase. This appears to be because most measurements are carried out at higher temperatures, which provides higher and therefore more easily measurable conductivity. The high temperature means that only the rutile phase is stable. The higher temperature may also improve the diffusion of Ti and O in the lattice and so allow quasi-equilibrium conditions to be reached more rapidly, and provide a more accurate picture of the defect chemistry in a reasonable measurement timescale.

It is proposed here, that the anatase thin films may prove useful as a means of investigating the defect chemistry of anatase TiO_2, which has not been examined in as much detail as for rutile. The stability up to at least 1000 °C of the polycrystalline anatase films, deposited at room temperature and post-annealed without undergoing a phase transition, suggests that measurements of conductivity at varying $P(O_2)$ may be possible up to 1000 °C. In addition, the apparent stabilisation of the anatase phase by Nb-doping might extend the range of experimental temperatures possible before the anatase transitions to rutile.

A full study of the conductivity of anatase over a range of temperatures, oxygen partial pressures, and doping concentrations would be a useful addition to the literature.

References

1. Hitosugi, T., et al.: Fabrication of TiO_2-based transparent conducting oxide films on glass by pulsed laser deposition. Jpn. J. Appl. Phys. **46**, L86–L88 (2007)
2. Osaka, A., Takao, S., Oda, K., Takada, J., Miura, Y.: The diffusion of sodium ions into tin oxide thin films from glass substrates. Mem. Fac. Eng. Okayama Univ. **24**, 53–61 (1989)
3. Hitosugi, T., et al.: Electronic band structure of transparent conductor: Nb-doped anatase TiO_2. Appl. Phys. Express **1**, 111203 (2008)
4. Hitosugi, T., et al.: Transparent conducting properties of anatase $Ti_{0.94}Nb_{0.06}O_2$ polycrystalline films on glass substrate. Thin Solid Films **516**, 5750–5753 (2008)
5. Sanches, F.F., Mallia, G., Liborio, L., Diebold, U., Harrison, N.M.: Hybrid exchange density functional study of vicinal anatase TiO_2 surfaces. Phys. Rev. B **89**, 245309 (2014)
6. Franklin, J.B., Downing, J.M., Giuliani, F., Ryan, M.P., McLachlan, M.A.: Building on soft foundations: New possibilities for controlling hybrid photovoltaic architectures. Adv. Energy Mater. **2**, 528–531 (2012)

7. Perron, H., et al.: Optimisation of accurate rutile TiO_2 (110), (100), (101) and (001) surface models from periodic DFT calculations. Theor. Chem Acc. **117**, 565–574 (2007)
8. Yamamoto, S., Sumita, T., Miyashita, A.: Preparation of TiO_2-anatase film on Si (001) substrate with TiN and $SrTiO_3$ as buffer layers. J. Phys. Condens. Matter **13**, 2875 (2001)
9. Yang, H.G., et al.: Solvothermal synthesis and photoreactivity of anatase TiO2 nanosheets with dominant 001 facets. J. Am. Chem. Soc. **131**, 4078–4083 (2009)
10. Swanepoel, R.: Determination of the thickness and optical-constants of amorphous-silicon. J. Phys. E Sci. Instrum. **16**, 1214–1222 (1983)
11. Arbiol, J., et al.: Effects of Nb doping on the TiO_2 anatase-to-rutile phase transition. J. Appl. Phys. **92**, 853 (2002)
12. Trenczek-Zając, A., Rekas, M.: Electrical properties of Nb-doped titanium dioxide TiO_2 at room temperature. Mater. Sci. Pol. **24** (2006)
13. DeSario, P.A., Graham, M.E., Gelfand, R.M., Gray, K.A.: The effect of Nb substitution on synthesis and photo-response of TiO_2 thin films prepared by direct current magnetron sputtering. Thin Solid Films **519**, 3562–3568 (2011)
14. Furubayashi, Y., et al.: Novel transparent conducting oxide: Anatase TiNbO. Thin Solid Films **496**, 157–159 (2006)
15. Furubayashi, Y., et al.: A transparent metal: Nb-doped anatase TiO_2. Appl. Phys. Lett. **86**, 252101 (2005)

References

Chapter 8
Epitaxial TiO_2 Thin Films on Single Crystal Substrates

This chapter describes the deposition of thin films of TiO_2 on single crystal substrates by pulsed laser deposition (PLD). The aim was to grow epitaxial films with different orientations and hence with different facets at the surface. This would allow the characteristics of each crystallographic face to be studied independently, for example, the adsorption of molecules or the photo-electrochemical performance. In particular, the low index (001) and (100) surfaces of anatase were of interest.

The results are summarized in Sect. 8.1 and the suggested next steps are identified in Sect. 8.2. The choice of substrates is outlined briefly in Sect. 8.3. The results of depositions on $SrTiO_3$ (STO) single crystals with the (100) orientation are presented in Sect. 8.4, focussing on analysis by both standard X-ray diffraction (XRD) and texture analysis. Optical measurements made on the films are presented in Sect. 8.4.5. The XRD analysis of films deposited on MgO single crystals with the (100) orientation is presented in Sect. 8.5 and optical measurements in Sect. 8.5.5. Finally, Sect. 8.5.6 reports the results of initial Raman spectroscopic measurements made on films deposited on both STO and MgO.

8.1 Summary of Results

The deposition conditions to grow epitaxial anatase films with the (001) orientation on STO, and anatase films with the (100) orientation on MgO have been identified. The epitaxial relations between the films and the substrates have been established.

The effect of varying the deposition conditions on the phase, orientation, and crystallinity of the films has been assessed. Practically, this gives a partial map of the parameter space of the specific deposition system used. In turn this will allow control of properties such as the film thickness, ratio of rutile and anatase, and the lattice parameter.

© Springer International Publishing Switzerland 2016

J.C. Alexander, *Surface Modifications and Growth of Titanium Dioxide for Photo-Electrochemical Water Splitting*, Springer Theses,
DOI 10.1007/978-3-319-34229-0_8

The work to date provides the foundation for future investigations into how the different surfaces of anatase and rutile TiO_2 interact with adsorbed molecules such as water.

To be able to image the surfaces at the resolution of individual atoms (by scanning tunnelling microscopy, for example) highly conductive films are required. The work undertaken to investigate the growth of conductive TiO_2 films is discussed in Chap. 7. As discussed in Chap. 7, the deposition process did not yield conductive films, and as such, no atomic resolution imaging of the surface was possible.

The phase of films deposited on STO was found to be relatively insensitive to the substrate temperature during deposition. For substrate temperatures between 400 and 800 °C, the film always grew as (001) oriented anatase. The temperature does appear to affect the strain in the films, in that higher temperatures seem to result in more relaxed films. The films deposited on MgO, however, were found to be sensitive to substrate temperature. At 400–500 °C, the films were (100) oriented anatase. At 600 °C a mixed anatase/rutile film resulted, and at 700 °C a film with the (110) rutile orientation was achieved. Both the anatase and rutile films were seen to grow epitaxially on MgO.

The effect of other deposition parameters such as substrate-target distance, pulse number, pulse rate, and oxygen pressure seem to have little effect on the films deposited on STO except to change the thickness of the film. For the films grown on MgO, it appears that the substrate-target distance and the oxygen pressure also affect the phase of the film in addition to temperature. On this basis, the growth of the anatase (100) orientation appears to be the least stable film and require the most precise control of growth conditions. For example, it is possible to grow crystalline rutile at temperatures as low as 400 °C depending on the working distance and the pressure. Typically, an anatase film would be expected at this temperature.

Comparing the films deposited on single crystals to those deposited on fused silica (discussed in Chap. 7) it is clear that there is a complex interplay between the growth conditions that affect the phase and orientation of the films. For example, the relatively small lattice mismatch of anatase and STO results in a film that is insensitive to the deposition temperature, while the poor match of anatase and MgO results in sensitivity of phase to not just temperature, but other parameters such as substrate-target distance. The films grown on amorphous fused silica (discussed in Chap. 7) show that the orientation as well as the phase can be influenced by temperature alone.

8.2 Further Work

The films deposited on fused silica showed that the films would crystallize as anatase at low temperature (ca. 300 °C) and remain stable until over 1000 °C, before undergoing a transition to rutile: a much higher temperature than the 700 °C observed with substrate heating during deposition. It would be informative to see whether the anatase films formed on STO remain stable to a higher temperature due to the constraint of the lattice of STO (by a clamping effect). Likewise, it would be of interest to know whether the anatase (100) films deposited with substrate heating may be subsequently heated to higher temperatures without a phase transition to rutile.

The effect of the substrate-target distance on the growth of anatase or rutile on MgO (100 would also merit further investigation, to provide additional insight into what determines whether a film grows as rutile (110) or anatase (100).

8.3 Substrate Overview

The choice of $SrTiO_3$ (STO) and MgO single crystals was based on the work of Silva et al. [1] who obtained the (001) orientation of anatase on STO (100) and the (100) orientation on MgO (100) single crystals. As described in the literature review, the use of STO (100) to grow anatase is common due to the reasonably well-matched lattice parameters. To the knowledge of the author the growth of anatase (100) on MgO (100) has not been reported widely (the work by Silva may be the only example). Interestingly, there is no obvious reason why epitaxial (100) i.e. a-plane films, would be expected to grow on MgO due to a lack of any lattice matching. This will be discussed later.

The chapter presents the growth conditions for epitaxial, oriented TiO_2 films using the pulsed laser deposition. The effects of modifying several parameters such as temperature and thickness are investigated.

The STO crystals were bought in (Crystal GmbH, Germany). The crystals were polished on both sides, with the edges parallel to the (010) and (001) directions.

The MgO crystals were also bought-in (Crystal GmbH, Germany). They had the (100) orientation with the edges parallel to (010) and (001). The original batch of substrates (samples TOM12111 to TOM12114) was polished on both sides. Subsequent depositions were made on a different batch of substrates with only one polished side. As will be discussed below, this could be a factor in the growth of anatase rather than rutile films.

Prior to deposition, each substrate was cleaned in an ultrasonic bath using acetone, isopropanol, and deionized water in turn. The substrate was then dried using nitrogen. No other pre-treatment—such as annealing—was used.

The initial deposition conditions were based on those reported in the work of Silva et al. The deposition pressure set was 10 mTorr; the laser repetition rate was set at 2 Hz to give time for material to diffuse on the surface of the substrate, and the maximum pulse energy of 520 mJ per pulse was set. An initial pulse number of 5000 was chosen as the starting point (and was to be modified depending on the results).

The initial substrate-target distance was set to 90 mm, which was at that time, the minimum possible for the PLD chamber. The chamber was later modified to allow a minimum 50 mm substrate-target distance. The substrate-target distance used is highlighted were relevant.

Deposition series were carried out looking at the effect of varying the following parameters:

- In-situ deposition temperature
- In-situ deposition pressure
- Thickness (number of pulses)
- Distance between substrate and target
- Ex-situ post-annealing temperature of films deposited at room temperature

8.4 TiO_2 Films on STO (100)

Based on the existing literature, films deposited on STO (100) were expected to be anatase (001).

8.4.1 Effect of Substrate Temperature

Films were deposited at a range of substrate heating temperatures between 400 and 800 °C. All other conditions remained identical between depositions (as far as possible).

The full XRD patterns are shown in Fig. 8.1 for films deposited with substrate heating temperatures of 400, 500, 600, 700, and 800 °C. The deposition oxygen pressure was 10 mTorr, and the substrate-target distance set to 90 mm. The films were formed with 5000 pulses at 2 Hz. Figure 8.2 shows the main film peak in detail.

The only film peaks present in the XRD scan are those of the anatase (004) and (008) peaks in the range of 37.5° to 38° and 80° to 81° respectively. These correspond to reflections from planes parallel to the (001) plane. All other peaks are from the substrate, such as the Cu K_β peak at 41° to 42°. The patterns are shown on a logarithmic y-axis to allow both film and substrate peaks to be seen together. The presence of reflections from planes parallel to (001) indicates that the c-axis of the anatase film is perpendicular to the substrate. The films grow with at least preferential orientation (texture) due to the effect of the substrate lattice. To test whether the film is epitaxial requires the application of additional XRD techniques (texture analysis), and is discussed later.

A closer view of the (004) peak for each film is shown in Fig. 8.2. It is clear that as the temperature is increased, the angle of the peak maximum shifts to lower values of 2θ. This corresponds to an increase in the c-axis lattice parameter of the anatase film. The peak maximums and corresponding d-spacing are given in Table 8.1.

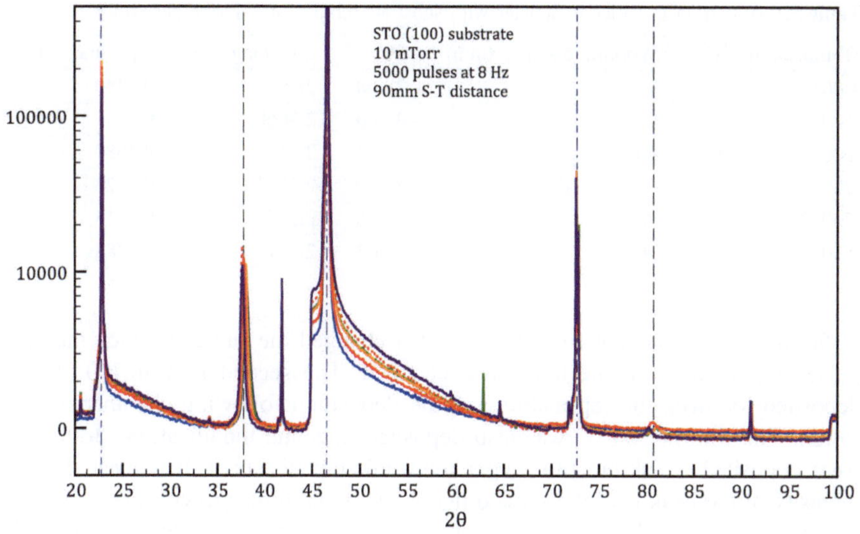

Fig. 8.1 XRD patterns of undoped TiO_2 films deposited on STO (100) with substrate heating temperatures of 400–800 °C

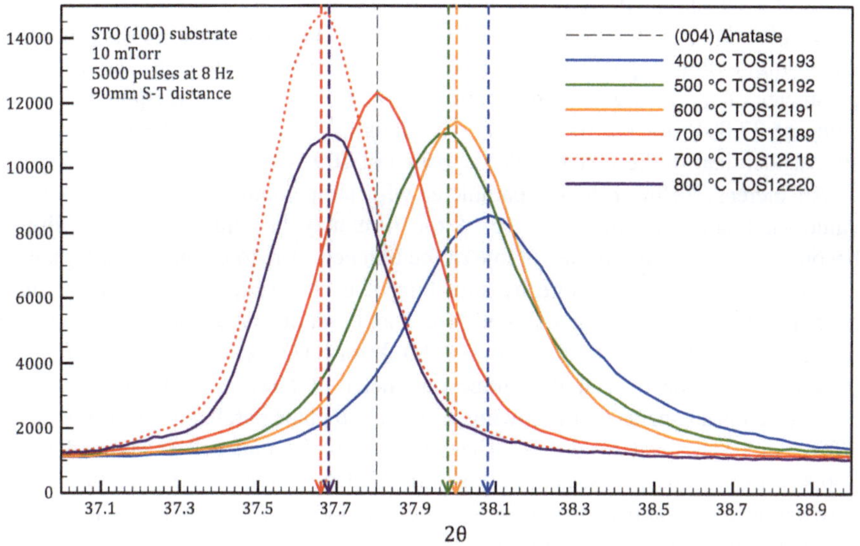

Fig. 8.2 View of anatase (004) XRD peak of undoped TiO_2 films deposited on STO (100) with substrate heating temperatures of 400–800 °C. A peak shift to lower angles is apparent at higher temperatures

A second film (700–2, sample TOS12218 in Fig. 8.2) was deposited at 700 °C following the modification of the PLD chamber to allow a shorter working distance

Table 8.1 Variation in lattice parameter with substrate temperature during deposition

Temperature (°C)	Working distance (mm)	2θ	d-spacing (Å)	c parameter (Å)
400	90	38.08	2.3612	9.4449
500	90	38.00	2.3660	9.4641
600	90	37.98	2.3672	9.4689
700–1	90	37.80	2.3781	9.5123
700–2	90	37.68	2.3854	9.5415
800	90	37.65	2.3872	9.5488

of 50 mm. The modification to the chamber changed the angle at which the laser pulse was incident on the target had changed. The second film at 700 °C was deposited to check the repeatability of the deposition following chamber modification. The film at 800 °C was also deposited after the modification. Both films were deposited at 90 mm (no longer the minimum distance). It is clear that the intensity of the peak is of the same order as the initial films grown, suggesting that the film thickness and crystallinity is similar despite the change in the incident angle of the laser on the target. However, there is clearly a shift in the lattice parameter between samples deposited at 700 °C, which could indicate that incident angle is having an effect with a similar magnitude as temperature. This may have been a result of a changing spot size and hence fluence.

Despite the chamber modification, there does appear to be a general trend of increasing *c* lattice parameter with temperature. By comparison, the *c* lattice parameter of the ICDD reference pattern is 9.5139 Å which corresponds to a 2θ value of 37.801° for the (004) peak. This falls in the middle of the range, and is almost identical to the first sample deposited at 700 °C.

An increase in the *c* lattice parameter with temperature—towards the powder value—indicates that the structure of the film may be relaxing i.e. the film is becoming less strained. If the volume of the unit cell were to be maintained, then an in-plane strain of the structure by stretching it in the *a-b* plane might result in a decrease in the height of the unit cell. The a and b lattice parameters of the anatase unit cell are 3.7852 Å, compared to 3.9050 Å for STO (based on ICDD reference patterns). This means that the lattice parameter of STO is 1.032 times that of anatase. Thus an anatase film growing with the a-b plane parallel to the a-b plane of STO will be strained by 3.2 %. The a-b face of the unit cell is strained to be 106.4 % of the reference structure. Assuming that the volume is constant this would correspond to a *c* lattice parameter 94 % of the ideal. This would give a *c* lattice parameter of 8.939 Å at an angle of 2θ = 40.32°, which is significantly outside of the measured values in the range 9.50 ± 0.04 Å. The smallest measured value of 9.4449 Å for the film at 400 °C is still 99 % of the reference value. Assuming that the *a-b* plane of the thin film matches that of the substrate, the measured values of the *c* parameter would indicate that the volume of the unit cell is increasing relative to the reference at lower temperature. This calculation assumes that the *a-b* lattice parameters of the film are strained to match the substrate throughout the thickness of the film. It ignores the likely presence of strain gradients in the film, which will be discussed below.

The peaks of the films deposited at 400–600 °C display an asymmetry: there is a right-hand shoulder on each peak (higher values of 2θ). The (004) peak of films deposited at 700 and 800 °C are symmetrical. The symmetry (or lack thereof) of the peaks was determined by fitting the data using the HighScore Plus software produced by PANalytical. Peaks were fit with Voigt functions (see Experimental Methods) with components for both $K_{\alpha 1}$ and $K_{\alpha 2}$ X-rays.

The peaks for films deposited at 700 and 800 °C were fit closely with a single set of Voigt functions. Any perceived asymmetry in the peaks is due to the presence of the $K_{\alpha 2}$ X-rays only. The films deposited at 400–600 °C required the addition of a second pair of Voigt functions to match the data at higher values of 2θ. This asymmetry is most easily seen in Fig. 8.2 for the sample deposited at 400 °C (the right-hand side of the peak is wider than the left-hand side). The presence of a second peak at a higher value of 2θ (smaller d-spacing) is an indication of a strain gradient or layers in the thin film. To give an indication of the magnitude of the peaks, the film deposited at 400 °C has a $K_{\alpha 1}$ peak intensity of 7626 counts ($K_{\alpha 2}$ has 3813 counts) for the low angle peak fit. The high angle peak fit has an intensity of 3902 counts ($K_{\alpha 2}$ has 1951 counts). The widths of the low angle peak fits are 0.3636 (Full Width Half Maximum) while the widths of the high angle curves are 0.5231. The lower intensity and greater width of the high angle peak fit suggests that there is a thinner, highly strained layer (or layers) in the film with a smaller *c* lattice parameter. The majority of the film is less strained as evidenced by the higher intensity and narrower peak with a larger *c-axis* lattice parameter.

For the film at deposited at 400 °C, the intensity of the $K_{\alpha 1}$ peak (3902 counts) of the strained layer is almost equal to the intensity of the $K_{\alpha 2}$ peak of the relaxed layer (3813 counts). At 500 °C the intensity of the $K_{\alpha 1}$ peak (1795 counts) of the strained layer is much less than the intensity of the $K_{\alpha 2}$ peak of the relaxed layer (5033 counts). The values are similar for 500 and 600 °C. This suggests that the ratio of the thickness of the relaxed layer to the strained layer is increasing with temperature. The higher the temperature of the substrate that the film is deposited onto, the more the film is able to relax. This might be possible if the higher temperature acts to distribute the strain into the substrate as well as the film. Alternatively, the higher temperature might encourage the growth of an intermediate phase at the interface that is too thin to detect by XRD. It is also possible that the films deposited at higher temperature spent longer at high temperature due to the time taken to cool down following deposition. This would have given films more time to reach equilibrium between phases. To properly study the effect of temperature on the interface, and to confirm the existence of a strain gradient or layers, it would be preferable to look at a cross-section of the film using a high-resolution transmission electron microscope (HR-TEM). This was not possible to do in the timeframe of this work.

It is interesting to note that all films from 400 to 800 °C were anatase and oriented with the c-axis perpendicular to the substrate i.e. (001) films. There was no indication in this temperature range that any rutile was present in the film, unlike with films deposited on fused silica (Chap. 7) or on MgO (as described later in this chapter). Neither did the film show any change in the preferred orientation (as with the films deposited on fused silica).

8.4.2 *Effect of Post-annealing*

The effect of depositing a film at room temperature, followed by post-annealing is shown in Fig. 8.3. The film was post-annealed for 1 h in air at successively higher temperatures from 100 to 600 °C, in 100 K steps. The (200) peak of the STO substrate in each scan was aligned with the reference pattern at ca. 37.8°. The as-deposited film was amorphous. With annealing of the film in air at 100 and 200 °C, the film remains un-crystallised. At 300 °C a peak corresponding to the (004) reflection of anatase becomes evident at about 37.8°. The intensity of the anatase (004) peak increases with increasing temperature up to 600 °C. At 400–600 °C an additional peak is evident at $2\theta = 29.5°$ to 30°. The peak does not correspond to any of the peaks of anatase, rutile or $SrTiO_3$ recorded in reference patterns. Interestingly, the peak drops from 400 to 500 °C, before increasing at 600 °C. It seems likely that this peak may be due to contamination of the sample from the furnace e.g. powder deposits. At 600 °C a peak between 27.5° and 28°

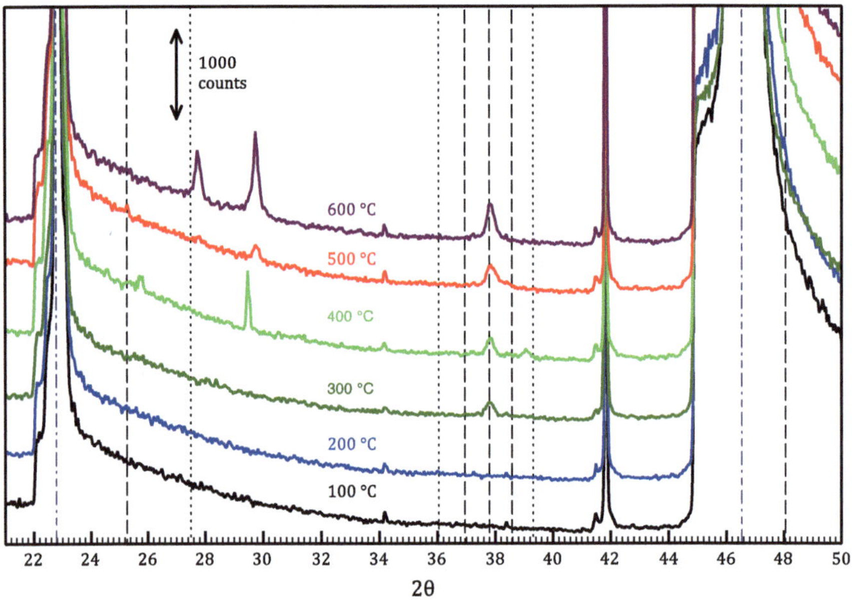

Fig. 8.3 XRD patterns of an undoped film deposited at room temperature and subsequently post-annealed at 100–600 °C in 100 K increments. The reference patterns for anatase (*dashed line*) and rutile (*dotted line*) are indicated for comparison

appears. This peak may be that of the rutile (110) reflection, although it is at slightly too high an angle (too small d-spacing). This could be due to the lattice of the STO substrate causing strain in the film.

8.4.3 *Effect of Other Deposition Parameters*

Following the deposition of films at a range of temperature, the remaining PLD parameters were varied to establish the effect on film growth if any. A substrate heating temperature during deposition of 400 °C was used in all cases.

An initial study on the effect of oxygen pressure during deposition was done at 10 and 100 mTorr. The intensity of the (004) peak decreased significantly above 100 mTorr. This is due to a reduced growth rate resulting in a thinner film. From this, a deposition pressure of 10 mTorr was selected as being preferred (it was not possible to go to 1 mTorr due to limitations of the system).

Following the modification of the minimum substrate-target distance the effect of pulse rate and pulse number was examined briefly. The full XRD patterns for the results are given in Fig. 8.4, and a detailed view of the anatase (004) peak in Fig. 8.5. The patterns have been aligned with the STO (200) reference peak.

The XRD patterns of Fig. 8.5 show how the deposition parameters affect the film growth. The initial film deposited at 400 °C (TOS12193, mid-blue line) was made with a substrate-target distance of 90 mm, and was formed by 5k pulses at a

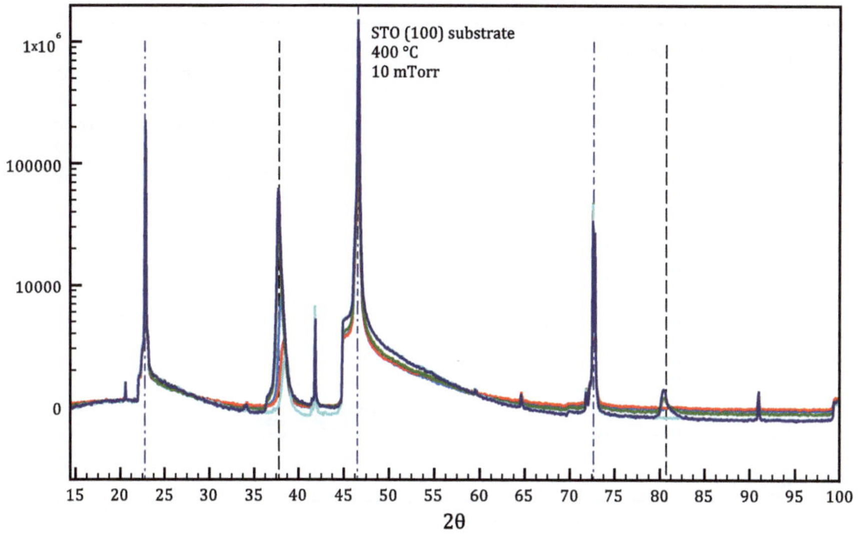

Fig. 8.4 XRD patterns of undoped TiO_2 films deposited at a range of thicknesses by altering pulse number, rate, and substrate-target distance. The scans are shown on a logarithmic y-axis as the intensities of the film peaks vary by an order of magnitude

repetition rate of 2 Hz. All further films were deposited with a shorter 50 mm substrate-target distance. In comparing this film (TOS12193, mid-blue line) to the film (TOS13060, red line) deposited at the shorter distance, but with the same repetition rate and number of pulses, it seems that the intensity of the (004) peak decreases significantly. This is somewhat surprising as the plume should be denser with a shorter distance, which should result in a thicker film, and in this instance should have given a higher intensity. That it did not suggests that the modifications to the chamber may have affected the formation of the plume (for example the angle of incidence of the laser pulse).

It is more straightforward to compare only the films deposited with a 50 mm substrate-target distance (Fig. 8.5). The films deposited at a rate of 8 Hz with 10k and 5k pulses show very similar peaks. In fact the (004) reflection of the film formed with 10k pulses (TOS12354) has almost twice the maximum intensity as the film with 5k pulses (TOS12355) as can be seen in Fig. 8.6, which uses a linear y-axis. This would suggest that the film is close to double the thickness with double the number of pulses, which is expected. From profilometry measurements, the thickness of sample TOS12354 (10,000 pulses) is ca. 100 nm.

As with the films deposited with temperatures in the range 400–600 °C (refer to Fig. 8.2), the (004) film peak is asymmetric, with a right-hand shoulder at higher angles that cannot be attributed to $K_{\alpha 2}$ alone. The film peaks were fit with Voigt functions and the peak maximum of the films occurs at 37.74° in the cases of the low angle peak. Additional Voigt functions were used to fit the right-hand shoulder. However, the film deposited with only 2500 pulses (TOS13053, light blue line)

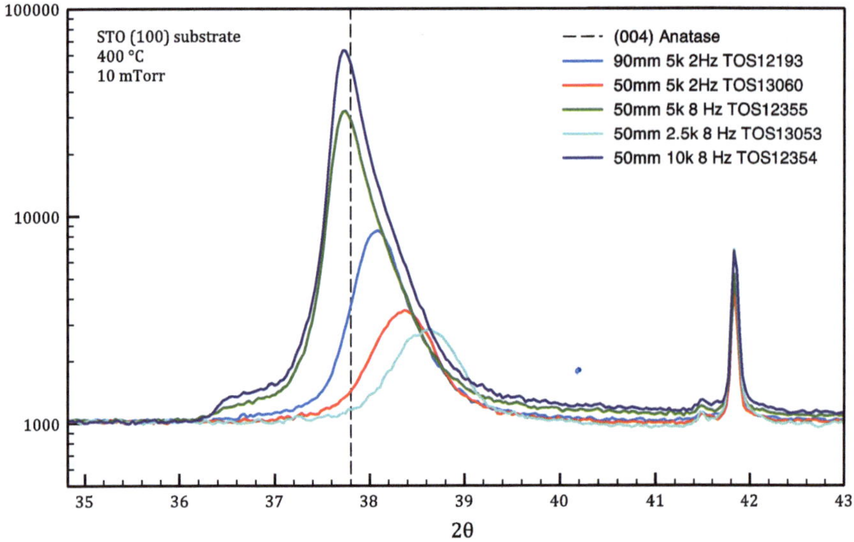

Fig. 8.5 View of anatase (004) XRD peak of undoped TiO_2 films deposited at a range of thicknesses by altering pulse number, rate, and substrate-target distance

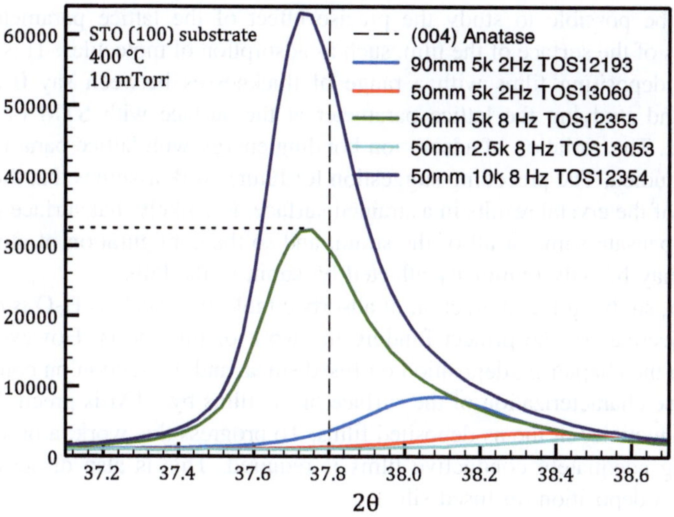

Fig. 8.6 Anatase (004) XRD peak on linear scale. Effect of thickness on the strain in the thin film

does not follow this trend, which is clear when the scan is viewed on a linear intensity scale (see Fig. 8.6).

The additional high angle peak occurs at about 38.5°, and corresponds to a smaller d-spacing between the (004) planes. The peak is also more symmetric than those of the thicker films as it can be fit with one pair of Voigt functions. This suggests that the film is similarly strained throughout its thickness. The ratio of the width of the peak of 0.6586° relative to its height of 1934 counts is increased 3.4×10^{-4}. The thicker films with 5k and 10k pulses can be fit with three pairs of Voigt functions, but there is some variation in the intensities and widths due to the arbitrary nature of the fitting process (and as such the exact quantitative data is not meaningful).

Films show a shift in the position of the peak intensity to lower angles with increasing thickness, giving an increase in the *c* lattice parameter as the strain reduces away from the substrate. For the increase of thickness from the 5000 to the 10,000 pulses sample, there is no discernible shift, indicating that the upper layers of the film are fully relaxed and are no longer strained by the substrate. That the thicker films have a shoulder at high angles, and need to be fitted with extra Voigt functions, shows that the strained layer or layers is still present at the interface with the substrate. It is not clear whether the strain varies continuously from the substrate to the point at which the crystal structure of the film is relaxed, or whether there are discrete strained layers i.e. discontinuous strain. To assess this properly High Resolution Transmission Electron Microscopy (HR-TEM) could be used to examine a section of the thin film and substrate interface.

As the *c* lattice parameter relaxes (increases), the in-plane *a-b* parameters are also relaxing (decreasing) towards the reference value. Practically, this suggests that it might be possible to study the precise effect of the lattice parameter on the properties of the surface of the film, such as adsorption of molecules. This would be done by depositing films with a range of thicknesses between say 0 and 5000 pulses, and studying the lattice parameter at the surface with STM or a similar technique. The variation of adsorption bonding energy with lattice parameter could then be studied. The preceding suggestion for future work assumes that the strain in the bulk of the crystal results in a strained surface. It is likely that surface relaxation will compensate some or all of the strain, and so the configuration of atoms at the surface may be only minimally affected by strain in the bulk.

In fact, studying the interaction of adsorbed molecules such as H_2O is one of the main objectives of the project funding the work of this thesis. However, as discussed in the chapter on deposition on fused silica, and the section on conductivity, the precise characterization of the surface of the films by STM is precluded by the poor conductivity of the as-deposited films. To progress this work, a procedure for producing adequately conductive films is required. This is also discussed in the chapter on deposition on fused silica.

8.4.4 X-ray Texture Analysis—Epitaxy of Films on STO (100)

To confirm that the thin films grown on STO were epitaxial rather than simply preferentially oriented, a number of pole figures were produced for several critical values of 2θ.

The pole figures are a plot of the intensity of the reflection for a set value of 2θ, and varying ψ and ϕ, which are plotting as the radial and angular coordinates in a polar contour plot. The value of 2θ is chosen either by a preceding 2θ line scan or by setting it to the reference value in a pdf pattern. Once a peak is identified in ψ and ϕ, the value of 2θ can be refined by fixing ψ and ϕ. A pole figure with $2\theta = 37.5833°$ is shown in Fig. 8.7 for an anatase film with a known (004) orientation (Sample TOS12218). This pole figure is of minimal interest other than that it shows an intense peak centred $\psi = \phi = 0$ that corresponds to the anatase (004) reflection. There are four small peaks at $\psi = 45°$ and separated by 90° in ϕ. At this value of 2θ, there are no obvious candidates for this peak. By repeating the scan with a blank substrate (Fig. 8.8), it is possible to see that these peaks are due to the combination of the optics and the highly crystalline nature of the substrate. As these poles occur at $\psi = 45°$ on an (100) oriented cubic substrate, they correspond to the {110} planes. It is more intuitive to write this as a (001) oriented substrate and poles from the {101} planes. The offset of the peaks from $\phi = 0$ shows that the sample is mounted with a slight rotation in ϕ of 2.5°. The resolution of the scan is 5° in both ψ and ϕ, so the offset is really between 0° and 5°.

To establish the epitaxial relation between the film and the substrate, a pole figure (Fig. 8.9) was produced with $2\theta = 25.4658°$ to identify the position of the

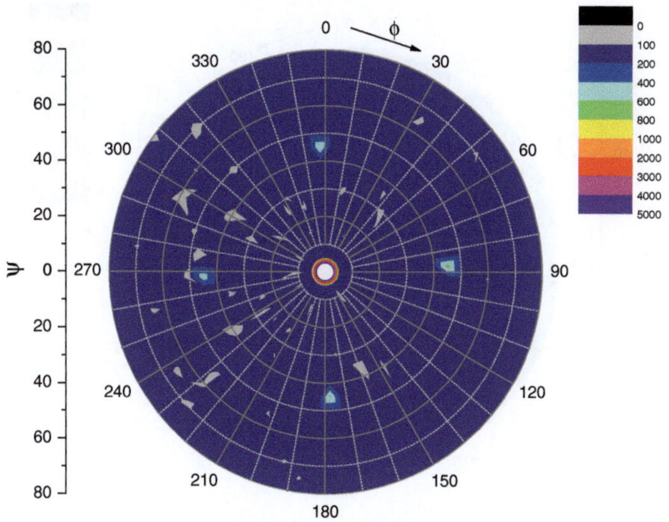

Fig. 8.7 Pole Figure of Anatase (004) film on STO. Scan for peaks at $2\theta = 37.8°$ corresponding to anatase (004)

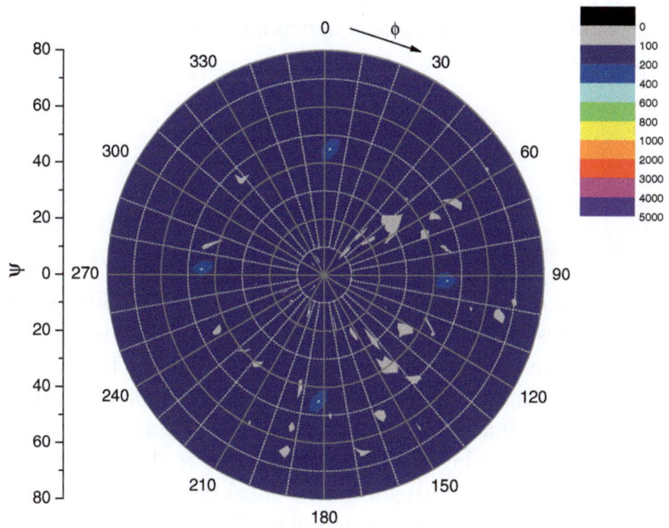

Fig. 8.8 Pole Figure of STO (100) substrate. Scan for peaks at $2\theta = 37.8°$ corresponding to anatase (004)

anatase (101) peaks relative to the substrate. The initial pole figure identified four poles at $\psi = 68°$ and at a spacing of 90° in ϕ. The ϕ positions of the (101) poles of anatase are identical to those of the substrate peaks attributed to the {101} planes of STO. As the film is oriented with the *c*-axis of the unit cell parallel to the

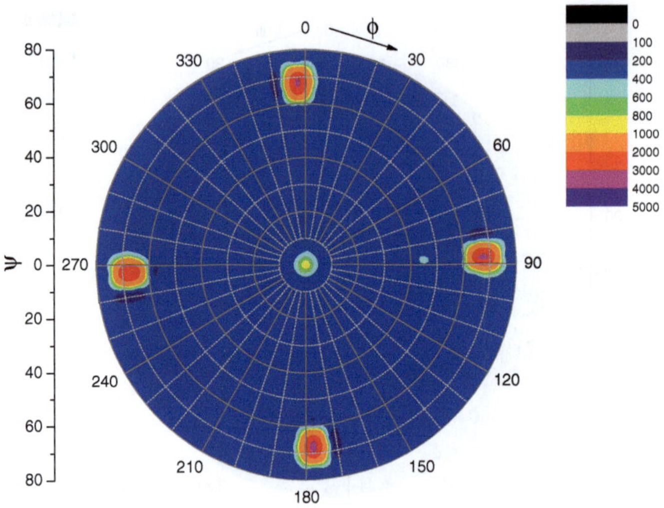

Fig. 8.9 Pole Figure of Anatase (004) film on STO. Scan for peaks at 2θ = 25.4658° for anatase (101)

(001) direction of the substrate, this shows that the *a* and *b*-axis directions of the film are also parallel to the *a* and *b*-axis directions of the STO substrate, giving the epitaxial relations:

$$[001]_{anatase}//[001]_{STO}$$

and

$$\langle 100\rangle_{anatase}//\langle 100\rangle_{STO}$$

This is a slightly different formulation from that given by Silva et al. [1]. Here the *a* and *c* directions of STO have been exchanged without loss of accuracy. It is important to note that because anatase is tetragonal while STO is cubic, that the [101] direction in anatase is not parallel to the [101] direction of STO.

The full width half maximum of the peaks in a ϕ-scan are $\Delta\phi = 0.90°$ suggesting good in-plane ordering (by comparison Silva et al. achieve $\Delta\phi = 1.3°$).

However, a ψ-scan (Fig. 8.10) shows less ideal behaviour. The *a* and *c* lattice parameters of the anatase reference pattern are 3.7582 and 9.5139 Å respectively. Ideally the (101) planes should be at an angle of $\tan\psi = c/a$, or $\psi = 68.44°$. This is approximately where the anatase (101) poles appear. However, the peak is wide ($\Delta\psi \sim 10°$), going from about 63° to 73° and has a flat top. This seems to suggest that there is a range of lattice parameters i.e. that both the *a* and *c* lattice parameters of the film have a variety of values throughout the film, so the (101) plane occurs at a range of angles with respect to the *c*-axis. This is in keeping with the asymmetry of the anatase (004) peak seen in the standard 2θ scans for the same sample

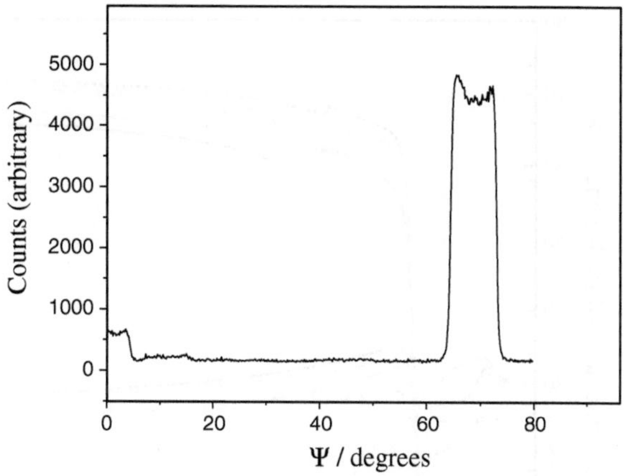

Fig. 8.10 ψ-scan of anatase (004) film on STO, examining the (101) film pole. 2θ = 25.4658° corresponding to anatase (101)

(TOS12218). At $\psi = 63°$ the ratio of c/a is 1.96, while at $\psi = 73°$ the ratio of c/a is 3.27. This corresponds to an extremely wide range of possible lattice parameters. The smaller values of c/a correspond to the more strained layers that are closer to the substrate. It is not known whether the unit cell of each grain is strained in both the *a* and *b* directions, or whether half the grains are strained along *a* and the other half along *b*.

8.4.5 Optical Measurements—Films on STO

The films were characterized using UV-Vis spectroscopy to try to evaluate the absorption coefficient and band gap. The transmittance and reflectance data shown in Fig. 8.11 is for anatase (001) films on STO (100) using a substrate-target distance of 90 mm. The substrate transmittance and reflectance is also shown for reference. It is clear that the band edge of the films is indistinguishable from that of the substrates. As such it was not possible to calculate the absorption coefficient for the films or obtain a Tauc plot from which to estimate the bandgap.

8.5 TiO_2 on MgO (100)

The aim of depositions on MgO (100) was to grow the (100) orientation of anatase, as achieved by Silva et al. [1]. To the knowledge of the author of this thesis, the work by Silva et al. is the only previous example of the growth of anatase films with the (100) orientation. As discussed in detail below, the initial set of films did not grow as expected. Following the depositions made on fused silica and modifications to the chamber, further successful films were deposited.

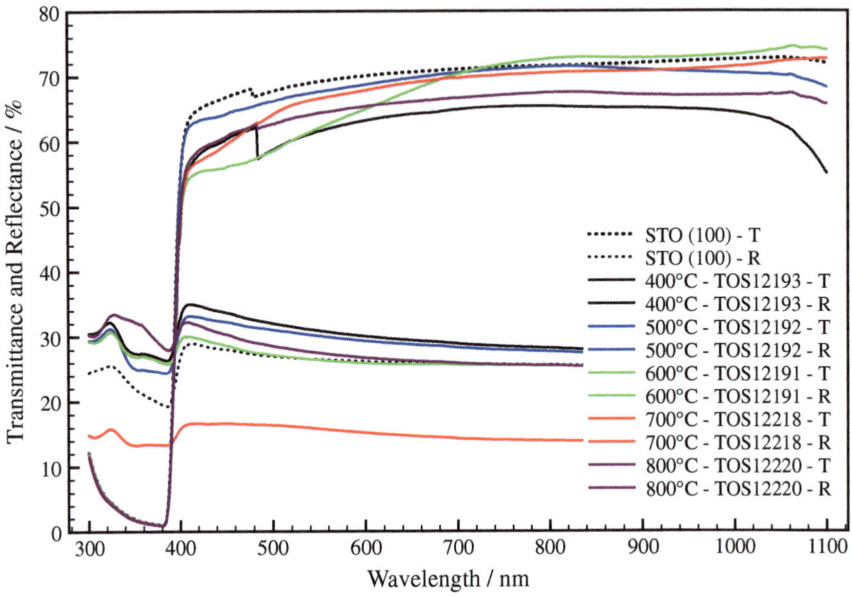

Fig. 8.11 Transmittance and Reflectance of anatase (001) films on STO (100) substrates

8.5.1 *Effect of Temperature—90 mm Substrate-Target Distance*

The following films were deposited at a range of substrate heating temperatures from 400 to 700 °C. The films were deposited at 10 mTorr with 5k laser pulses at 2 Hz. The effect of temperature on the structure of the films can be seen in Figs. 8.12 and 8.13. Figure 8.12 shows the patterns over a range $45° < 2\theta < 70°$. The reflection of the anatase (200) planes was expected to be present at $2\theta \sim 48°$. The (200) peak is the lowest order peak expected for the (100) orientation. It is clear that there is no such peak present. There is however an indication of rutile due to the presence of a small (220) peak at about 56.8°. The remaining features can all be attributed to the substrate (see pink line, Fig. 8.12). The patterns are not shown at $2\theta < 45°$ as only substrate peaks were present.

The low-angle scan presented in Fig. 8.13 shows the rutile (110) peak: it is clear that the film is composed of rutile alone. The plot indicates that the film is crystalline and grows as the rutile phase from as low as 400 °C. At 400 °C anatase is expected to form rather than rutile. There is no peak evident for even the low energy anatase (101) orientation. It seems like there is a slight decrease in the intensity of the (110) peak with temperature from about 1500 counts to 1000 counts. But this could be explained by slight differences in the thickness of the film. The features between 29° and 30° can also be seen in the substrate.

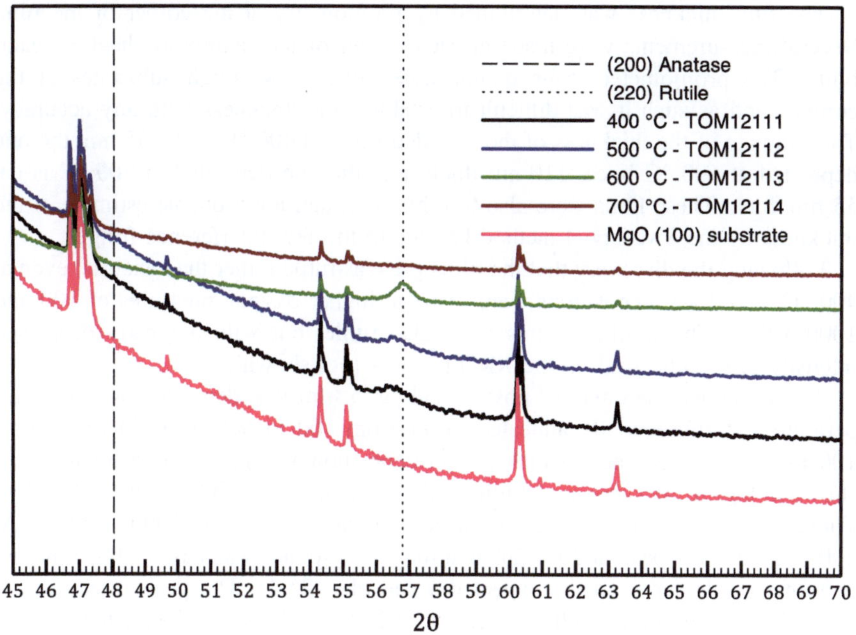

Fig. 8.12 XRD patterns of undoped TiO_2 films deposited on MgO (100) with substrate heating temperatures of 400–700 °C. The oxygen pressure was 10 mTorr; the pulse rate was 2 Hz for 5000 pulses

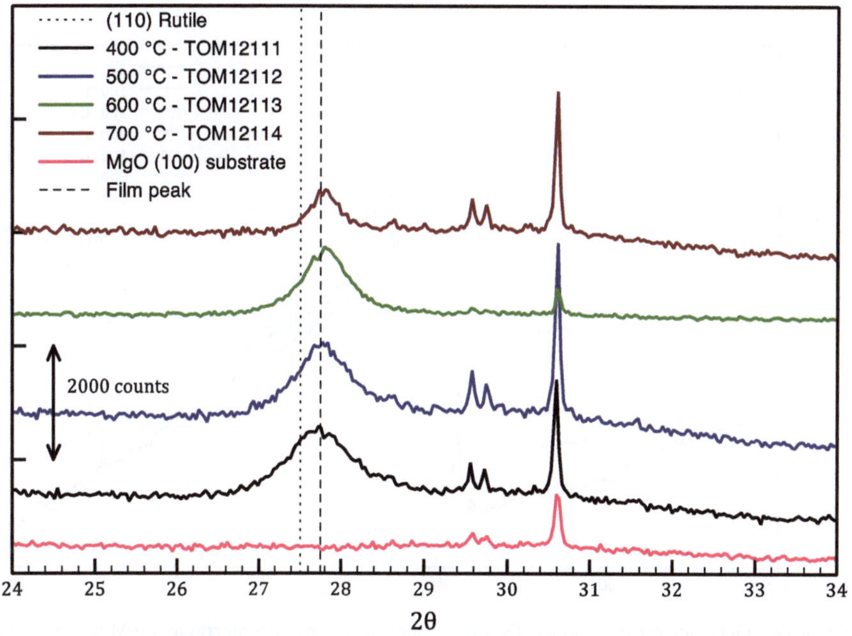

Fig. 8.13 Closer view of the rutile (110) XRD peak of undoped TiO_2 films deposited at a range of temperatures on MgO (100)

The film thickness was determined by profilometry at the corner of the film. Several measurements were made at each corner of the sample to check repeatability. The profilometer scans demonstrated extremely rough substrates at the corners, and as such it was difficult to establish the thickness with any accuracy. The estimate of the thickness of the film deposited at 400 °C is ca. 65 nm, the one deposited at 600 °C is ca. 110 nm thick, and the one deposited at 700 °C is ca. 55 nm thick. These films were also too thin to obtain a reasonable estimate of the thickness by the Swanepoel method [2] due to too few interference fringes.

Following the discovery that the films grew as rutile rather than anatase, even at 400 °C, additional depositions were made at higher oxygen pressures of 100 and 1000 mTorr. The resulting films were also rutile, but with a greatly decreased intensity of the film peak in the XRD patterns (not shown).

To investigate the effect of post-annealing, a film was deposited at room temperature and subsequently annealed in air from 100 to 600 °C in increments of 100 K. The full XRD patterns taken after deposition and then after each anneal are shown in Fig. 8.14. There are no film peaks apparent except in the range 25° to 30°. There are no film peaks visible for annealing temperatures from 100 to 300 °C. At 400 °C, there is a peak at $2\theta = 26°$ that may correspond to anatase (101), although it is reasonably far from the expected value of ~25.4°. The peak disappears at 500 °C, although there is a hint of a peak at 26.5°, which could be related to rutile (110). In summary, the method of annealing did not produce even the clearly (110) oriented rutile films that deposition with substrate heating did.

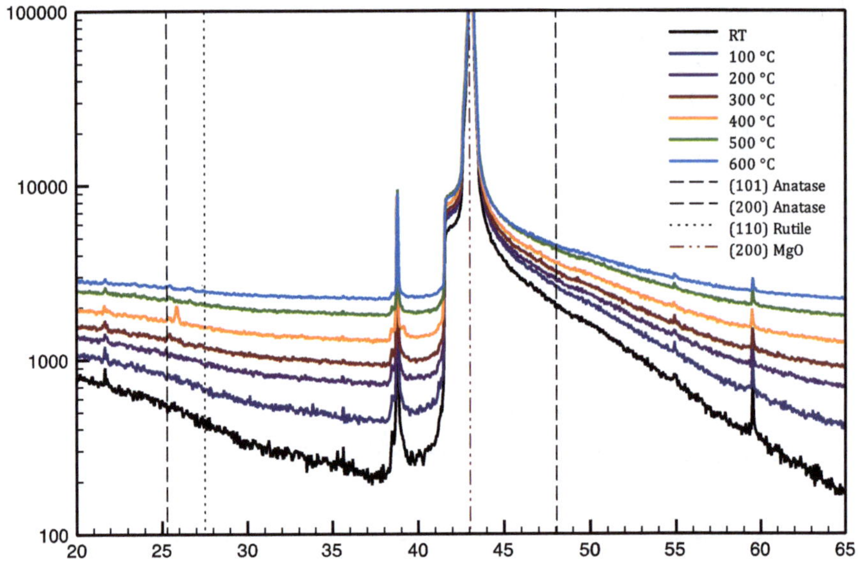

Fig. 8.14 XRD patterns of undoped TiO_2 films deposited at room temperature on MgO (100) and post-annealed at temperatures increasing from 100 to 600 °C

At this stage it appeared that anatase films oriented with the (100) planes parallel to the substrate could be achieved, even with similar deposition conditions to Silva et al. The deposition pressure of 10 mTorr was considered optimal (and as low as practicable in the PLD system). The substrate-target distance was at a minimum (90 mm at the time of the initial deposition), and the energy of the laser pulse was the maximum possible. The films produced were also relatively thin and difficult to measure.

Following modification of the substrate-target distance, and experience with deposition on fused silica, a further set of depositions was made and is discussed here now.

8.5.2 Effect of Temperature—50 mm Substrate-Target Distance

The substrate-target distance was reduced to 50 mm, the repetition rate increased to 8 Hz, and the number of pulses to 10k. Films were deposited at a range of substrate heating temperatures from 400 to 700 °C. The pressure was also increased from 10 to 30 mTorr based on the parallel work on deposition on fused silica that found 30 mTorr to give more intense film peaks in the XRD spectrum.

Additionally, the depositions were made on a new batch of single crystal MgO (100) substrates with only single side polishing. This precluded use of optical transmission measurements for both the properties of the film and to estimate the film thickness. Reflectance measurements were possible, but do not yield information on the absorption or the bandgap.

With the reduced working distance, the XRD patterns (Fig. 8.15) clearly show the presence of the anatase (200) reflection at $2\theta = 48°$, for the films deposited at 400, 500, and 600 °C. At 700 °C the (200) reflection has disappeared. There is now a rutile (220) peak apparent at 56.5°. Examining the low angle patterns in Fig. 8.16 gives more information about the anatase/rutile ratio of the film. At 700 °C the rutile (110) peak is clearly visible. At 600 °C the peak intensity has dropped to the same order as the anatase (200) peak. The film deposited at 600 °C should therefore be approximately equal parts anatase to rutile. At 500 °C there is still a rutile peak visible, however of much lower intensity than anatase (200) peak. This suggests that the film is largely anatase at 500 °C and at 400 °C there is only the anatase (200) peak present.

The film peaks were analysed by fitting a Voigt function using the HighScore Plus software. The individual peaks were fitted then the peak positions were corrected by the offset of the substrate (200) peak from the reference pattern. All peaks could be fit with a single pair of Voigt functions suggesting that they are not comprised of strained layers unlike the films deposited on STO. The fitted peak positions and corresponding d-spacings are given in Table 8.2. Only the position in 2θ corresponding to the Cu $K_{\alpha 2}$ X-rays is given.

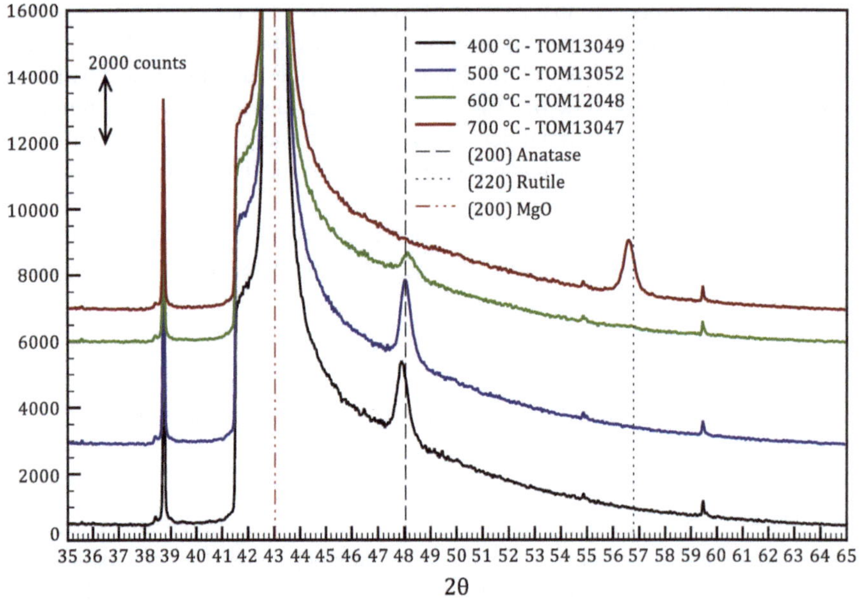

Fig. 8.15 View of anatase (200) XRD peak of undoped TiO_2 films deposited at a range of temperatures on MgO (100) with modified deposition conditions

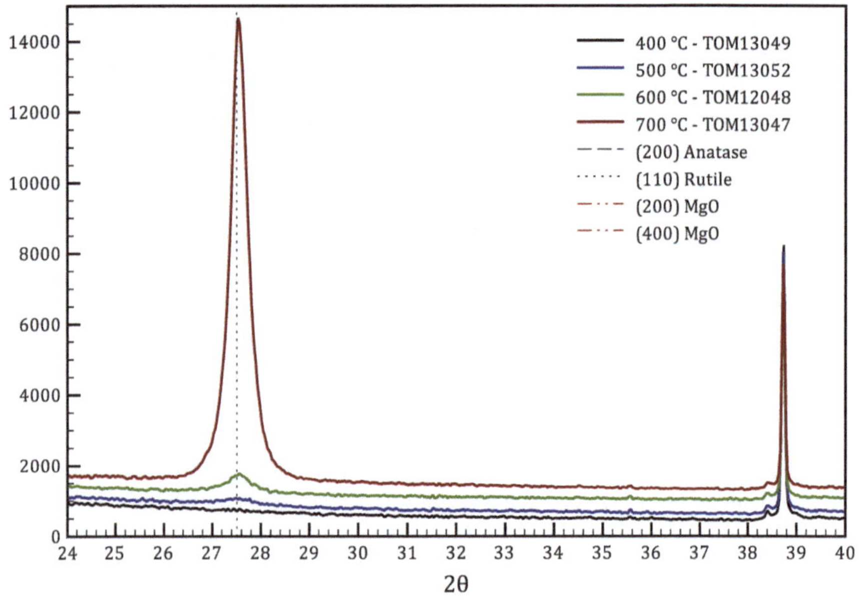

Fig. 8.16 View of rutile (110) XRD peak of undoped TiO2 films deposited at a range of temperatures on MgO (100) with reduced working distance of 50 mm, higher pulse rate of 8 Hz, and 10k pulses

Table 8.2 Peak positions of films deposited on MgO (100)

Temperature (°C)	Anatase (200)		Rutile (110)	
	2θ (°)	d (Å)	2θ (°)	d (Å)
400	47.5630	1.9102	N/A	N/A
500	47.9085	1.8973	N/A	N/A
600	48.0048	1.8937	27.3870	3.2540
700	N/A	N/A	27.4251	3.2495

The reference pattern for anatase gives the (200) reflection at 48.050° suggesting that as the deposition temperature is increased, the film is allowed to relax towards the reference value. Likewise, in the rutile reference pattern the (110) reflection occurs at 27.447°, so the rutile film also relaxes towards the reference with increasing temperature.

The thicknesses of the films were measured by profilometry. The film deposited at 400 °C is measured to be ca. 150 nm, the film deposited at 600 °C is ca. 80 nm thick, and the film deposited at 700 °C is ca. 100 nm thick. An optical measurement gave the thickness of the sample deposited at 700 °C (TOM13047) to be 143 nm. There is a discrepancy between the optical and profilometry measurements, which could be due to the optical measurement being made in the centre of the sample, and the profilometry measurement being made at the edge. Optical transmittance and reflectance measurements of the samples deposited at 400–600 °C show similar spectra that lack interference fringes (Appendix D), suggesting that they are less than 150 nm.

Given that only the temperature was varied, it is surprising that the thickness varies as seen here. Films deposited at a range of temperatures on fused silica (Chap. 7) with similar deposition conditions to those discussed here show thicknesses on the order of 250–300 nm. In fact, the films deposited at higher temperature that form rutile tend to be thinner than the anatase films. They also have fewer interference fringes for a given thickness due to the higher refractive index. That the rutile film deposited on MgO at 700 °C shows interference fringes in the UV-Vis spectra while the anatase films deposited on MgO from 400 to 600 °C do not, indicates that the anatase films are significantly thinner than the rutile ones. This should not be the case, as the same deposition conditions—other than temperature—were used. This suggests that there is some uncontrolled variable in the deposition. The most likely variable is the manual placement of the mirrors that direct the ablating laser beam into the chamber, affecting the spot size and the fluence of the laser pulse.

It could also suggest (but is somewhat unlikely) that the refractive index in the anatase [100] direction is higher than in the rutile [110] direction (and based on the films on fused silica, higher than in the anatase [101] direction).

8.5.3 *Effect of Other Deposition Parameters*

To understand which parameters control whether the film grows as rutile (110) or as anatase (200), the deposition parameters, other than temperature, were varied. This included, pressure, pulse rate, working distance, and the number of pulses. The substrate heating temperature of the substrate during deposition was kept at 400 °C for all samples discussed in this section.

The films discussed are a mixture of undoped and 1 wt% Nb-doped. As shown in Chap. 7, the effect of up to 1 % Nb-doping seemed to have little impact on the growth (at least on fused silica). The effect of varying, distance, pressure, repetition rate and pulse number are shown in Fig. 8.17.

Two films were deposited from the 1 % Nb-doped target with the same deposition conditions to check repeatability. The samples NTOM13255 (black line, Fig. 8.17) and NTOM13271 (blue line, Fig. 8.17) were deposited at 400 °C and 30 mTorr, with 10k pulses at 8 Hz, and a substrate target distance of 50 mm. While the position of the peak maximum is the same (at $2\theta = 47.8°$) it is clear that the intensity of the (200) peak of the first film is significantly higher than that of the second film (about 12000 vs. 6000). This suggests that either the thickness of the

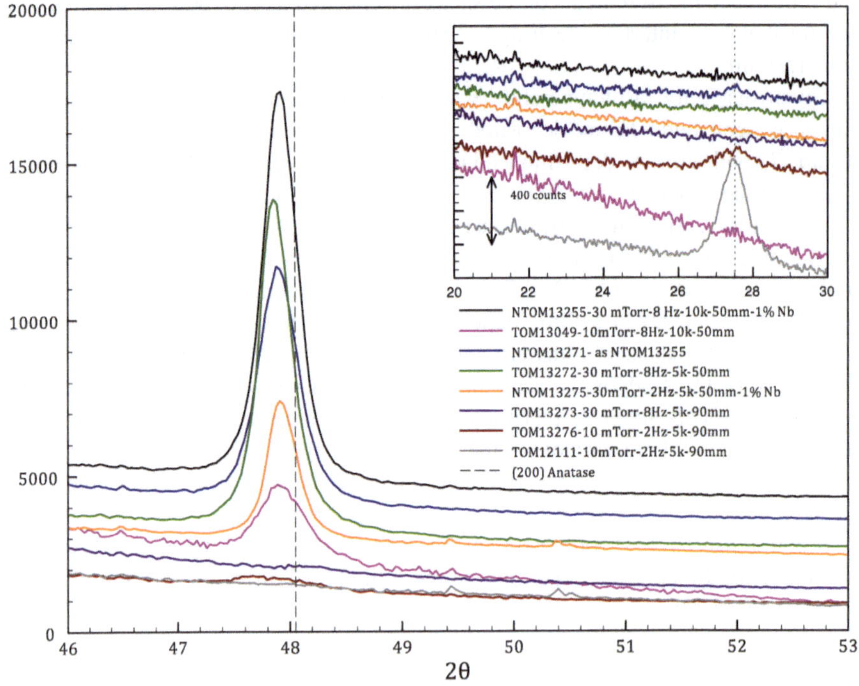

Fig. 8.17 View of anatase (200) XRD peak of undoped TiO_2 films deposited at a range of thicknesses, by altering pulse number, rate, and substrate-target distance

repeated film is less than that of the original, or the crystallinity is less. The thickness of sample NTOM13255 is estimated to be 311 nm (see Appendix D).

This again highlights an issue in the analysis of these films: even with the same deposition conditions, there is still some variability in the growth of the film. It is therefore difficult to make categorical statements about the effect of the deposition parameters when the effects are small. To be certain, it would be necessary to deposit a number of films at each condition and base the comparisons on the statistical properties of the films. It is also interesting to note that the repeated film shows a hint of the rutile (110) peak (blue line, inset of Fig. 8.17).

The (200) XRD peak in Fig. 8.17 for the undoped film deposited with 5k pulses (sample TOM13272, green line) is of similar intensity and width to the lightly doped films deposited with 10k pulses (black and blue lines). The thickness of the undoped film with fewer pulses is estimated to be 164 nm, which is approximately half that of the doped sample. Therefore the thickness is scaling approximately with the number of pulses, and the similar intensity indicates that the undoped film is more crystalline than the doped film. A doped film deposited with only 5k pulses and a lower rate of 2 Hz (sample NTOM13275, yellow line) has a much lower intensity (of about 4000) and width. This is less than half that of the film with 10k pulses (sample NTOM13255) so it seems that some of the reduction in intensity is due to a thinner film (estimated to be 196 nm) due to half as many pulses and some of the reduction is due to a lower pulse rate. It is not totally clear why a reduced pulse rate would result in a thinner film let alone a less crystalline film (which could also account for reduced peak intensity). One possibility it that the lower rate gives the impinging Ti and O ions time to diffuse and results in a less defective lattice, while the higher rate causes the film to grow quickly but with a larger concentration of defects. This could result in a slightly thicker (but less dense) film growing from the same total flux of material. Alternatively, the variation in the flux of material from the target varies significantly from sample to sample. The Nb-doped film deposited with 5k pulses (yellow line) seems to be thinner or less crystalline than the undoped film (green line), and the undoped film is similar in thickness or crystallinity to the doped film with double the number of pulses. Taken together this suggests that the Nb-doping is unfavourable to the growth of an anatase film with the (100) orientation. This could be due to the Nb increasing the temperature at which the film crystallises, for example.

Increasing the substrate-target distance to 90 mm while keeping the other parameters unchanged has a huge impact. The sample deposited with a 50 mm separation (TOM13272, green line) is clearly of the anatase (200) orientation, while that deposited with a 90 mm separation (TOM13273, purple line) shows no crystallinity at all. Both samples were deposited with the same pressure, pulse number and rate (30 mTorr, 5000, 8 Hz). It is possible that the thickness of the samples was reduced with the increased separation, and that the lack of a visible peak is due to fewer planes for diffraction. However, this is the opposite effect of decreasing the substrate-target distance seen for the films grown on STO, where the peak intensity decreased with decreasing distance (albeit counter intuitively). Alternatively, the film may be largely or fully amorphous, indicating that the distance affects the crystallinity significantly.

Reducing the pressure from 30 to 10 mTorr, and the repetition rate from 8 to 2 Hz, the XRD patterns (sample TOM13276, red line) develops a peak at $2\theta = 27.5°$ corresponding to rutile (110). Compared to the original film deposited with these parameters, the intensity of the (110) peak is much lower (red vs. grey line of inset). Comparing sample TOM13049 (pink line) deposited at 10 mTorr to samples NTOM13255/NTOM13271 (black/blue lines) and TOM13272 (green line) deposited at 30 mTorr indicates that the effect of a lower oxygen pressure reduces the intensity of the anatase (200) peak markedly more than decreasing the number of pulses (thickness alone). This suggests that a higher oxygen pressure favours the growth of the (100) surface or orientation.

Without further intermediate samples (and ideally multiple samples at each condition) it is difficult to know precisely which parameter or combination is the key to the growth of rutile rather than anatase. Based on the study to date, it seems that the biggest factor is having a larger substrate-target distance of 90 mm. Although as noted above, this had little effect on the films grown on STO. Reducing the pressure seems to favour a larger rutile (110) peak in conditions that produce rutile (grey and red lines vs. purple) and a smaller anatase (200) peak in conditions that produce anatase (pink line).

On balance, it seems that the combination chamber oxygen pressure and the substrate-target distance are the critical parameters affecting whether anatase or rutile is achieved. Essentially, both these parameters control the interaction of the plume of material ejected from the PLD target, with oxygen gas inside the chamber. This affects the energy of the material in the plume, and the concentration. Higher oxygen pressure should result in greater interaction. This is easy to see during deposition. The plume contracts as oxygen pressure increases. A larger separation of target and substrate should also give greater interaction due to the longer path from target to substrate. It would be reasonable to suppose that increasing pressure would have a similar effect as a larger substrate-target distance. However, the results discussed here seem to show that they have opposite effects. Increasing distance favours rutile, while increasing pressure favours anatase. This is rather counter-intuitive and suggests that something subtler is occurring and warrants further studies. In particular, a study of the effect of varying the substrate-target distance from 50 to 90 mm in 10 mm increments would be the next step to take. It appears that the substrate-target distance is the more sensitive parameter, as increasing the distance from 50 to 90 mm increases the distance by a factor of 1.8 only, while increasing the pressure from 10 to 30 mTorr is a factor of 3 and should otherwise more than compensate unless it is a significantly less important parameter.

Another possibility that could not be confirmed was whether the particular batch of original substrates was slightly (but significantly for film growth) different to the subsequent batches. The original batch was all used, so it was not possible to make a comparison across batches. An obvious difference was that the original substrates were polished on both sides. It is conceivable that quality of polishing is different to that of the one-side polished samples, resulting in an altered surface and growth. That both rutile and anatase films were achieved using substrates from the same batch (as discussed in the preceding paragraphs) through the variation of deposition parameters suggests that batch to batch variability of the substrates is not the principle reason for the growth of rutile rather than anatase films, although it has not been excluded.

It is also possible that instabilities in the laser deposition system may have resulted in misleading conclusions.

In practical terms i.e. the growth of the desired anatase (100) orientation, it is clear that it is achievable using the new configuration of the PLD chamber with the following deposition conditions:

- Substrate-target distance is 50 mm
- At least 5000 pulses
- Pressure is between 10 and 30 mTorr
- Repetition rate between 2 and 8 Hz

8.5.4 *X-ray Texture Analysis—Epitaxy of Films on MgO (100)*

To confirm that the thin films grown on MgO were epitaxial rather than simply preferentially oriented, a number of pole figures were produced for several critical values of 2θ. Pole figures are given for both the rutile (110) and anatase (200) oriented films.

A pole figure with $2\theta = 27.4054°$ is shown in Fig. 8.18 for a rutile film with a known (110) orientation (Sample TOM12114). This pole figure seems of minimal interest other than that it shows a peak centred $\psi = \phi = 0$ that corresponds to the rutile (110) reflection. The original scan (Fig. 8.18a) was done with a resolution of 5° in both ϕ and ψ. A scan with a higher resolution of 2.5° (Fig. 8.18b) shows that the pole is actually comprised of four peaks separated by 90° in ϕ. As these poles occur at $\phi = 45°$ on a (100) oriented cubic substrate, they are aligned with the (111) poles of MgO. This was confirmed by a pole figure with $2\theta = 62.26°$ to identify the (220) poles of MgO. These occurred at $\phi = 0°, 90°, 180°$, and 270° i.e. at 45° to the [100] direction of MgO.

A ψ-scan through the pole (Fig. 8.19) shows that the peak occurs at about $\psi = 3.25°$. Taken as measured it suggests that the (110) grains that make up the film are actually oriented slightly off from the normal to the substrate. It is important to note that, because the film is oriented with the (110) planes parallel to the plane of the substrate, the film has only twofold symmetry. The reference lattice parameters of rutile (ICDD pdf 00-021-1276) are $a = b = 4.5933$ Å and $c = 2.9592$ Å, giving a rectangular lattice in the (110) plane of 6.4959 and 2.9592 Å. As such, it is surprising that there are four off-centre poles for the (110)

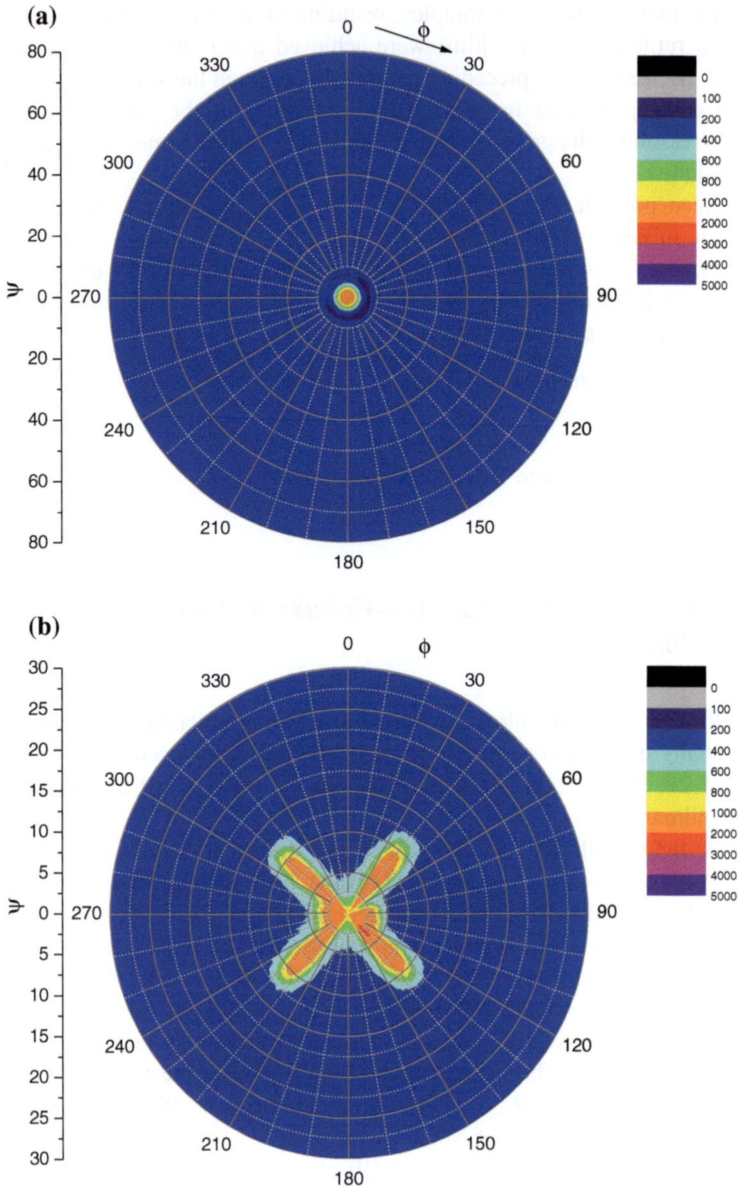

Fig. 8.18 Pole Figure of rutile (110) film on MgO. Scan for peaks at $2\theta = 27.4054°$. 40 kV/40 mA. Resolution of **a** 5° × 5° and **b** 2.5° × 2.5° for view of poles at $\psi = \phi = 0$

planes. The lattice parameter of the reference MgO substrate (ICDD pdf 00-045-0946) is 4.2112 Å. This is not close to either of the lattice parameters of the rutile (110) planes. Compared to the 2.9592 Å of the in-plan*c*-axis of rutile the

Fig. 8.19 ψ-scan of rutile (110) film on MgO. Examination at $2\theta = 27.4054°$, corresponding to the (110) pole of rutile

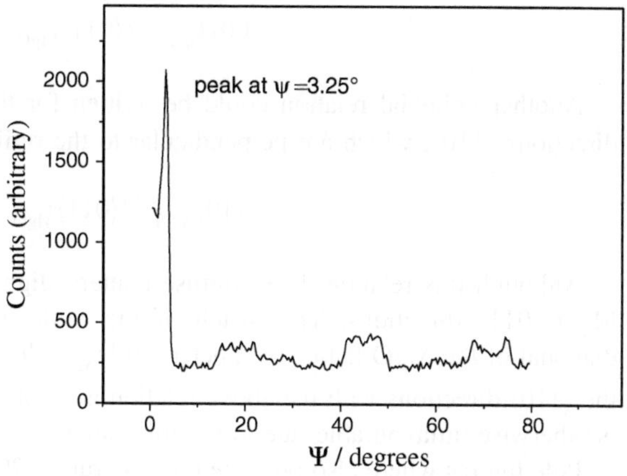

strain would be 42 % (= 4.2112/2.9592 − 1). For half the lattice parameter in the in-plane rutile $\{1\bar{1}0\}$ directions of 3.2480 Å, the strain would be 27 %. The diagonal of the MgO lattice is 5.9556 Å. This approximately twice the *c*-axis lattice parameter of rutile (5.9184 Å) and would give a strain of 1.006. The relation of the rutile 6.4959 Å in plane parameter to the MgO diagonal of 5.9556 Å gives a strain of 91.7 % i.e. the rutile is compressed along the *c*-axis, giving an 8 % mismatch. This seems the most likely fit between the lattices and would correspond to the *c*-axis of rutile being parallel to the $\langle 011 \rangle$ directions of the MgO surface i.e. the rutile *c*-axis is rotated 45° relative the *a* and *b* axes of the substrate. This is also the epitaxy suggested by Silva et al. [1] who note that the rutile (110) plane is rotated 45° with respect to the MgO basal square. Combining the twofold symmetry in the rutile (110) plane with the four directions in the substrate surface—[011], $[0\bar{1}1]$, $[01\bar{1}]$ and $[0\bar{1}\bar{1}]$—it is apparent that there are two possible orientations of the *c*-axis of rutile with respect to the substrate. The grains should grow oriented in plane with two directions rotated by 90° with respect to each other. This helps to explain the appearance of at least two of the four off-centre rutile (110) poles. That the poles are off-centre could be explained if the sample was not mounted completely flat on the goniometer of the diffractometer. This could have given two off-centre poles separated by 180° in ϕ. Taken with the two possible rotational orientations of the grains, this could give another pair of off-centre poles at 90° to the first pair.

As the rutile (110) planes grows on the MgO (100) plane, the first epitaxial relations of the film would be:

$$[110]_{rutile}//[100]_{\mathrm{MgO}}$$

This relation is different to $[110]_{rutile}//[011]_{\mathrm{MgO}}$ as given by Silva et al., which this author believes to be incorrect. The *c*-axis of rutile, the [001] direction grows in plane and is oriented at 45° to both the [010] and [001] directions of the substrate (and all other equivalents). This gives an epitaxial relation of:

$$[001]_{rutile}//\langle 011\rangle_{\mathrm{MgO}}$$

Another epitaxial relation could be written for the other rutile (110) in-plane directions $\langle\bar{1}10\rangle$, which are perpendicular to the rutile [001] direction:

$$\langle\bar{1}10\rangle_{rutile}//\langle 011\rangle_{\mathrm{MgO}}$$

Although this relation does confuse matters slightly as it is also parallel to the MgO $\langle 011\rangle$ directions. The match of the rutile [001] lattice parameter to the diagonal of the MgO lattice parameter $\langle 011\rangle_{\mathrm{MgO}}$ drives the growth. The growth of the $\langle\bar{1}10\rangle$ directions with the above relation is a consequence of the geometry, and is otherwise unfavourable due to the mismatch.

Pole figures where also produced for the rutile (200) and (011) poles, which are shown in Figs. 8.20 and 8.21 respectively. The (110) plane cuts the unit cell across the square *a-b* plane. Therefore, the poles of the equivalent (100) and (010) planes should be at an angle of $\psi = 45°$ to the (110) pole at $\psi = \phi = 0$. The (200) pole figure shows four poles separated by 90° in ϕ at $\psi \sim 55°$. These cannot be the (200) reflections and are attributed to the substrate. It is possible that because the film is thin and strained—even the rutile (110) peak is weak and broad in the 2θ scan—that the (200) reflections are too weak to be detected by the parameters used for the scan.

The pole figure for the rutile (011) plane (Fig. 8.21) shows a number of poles. There are four poles spaced by 90° in ϕ at $\psi = 55°$. There appear to be a further

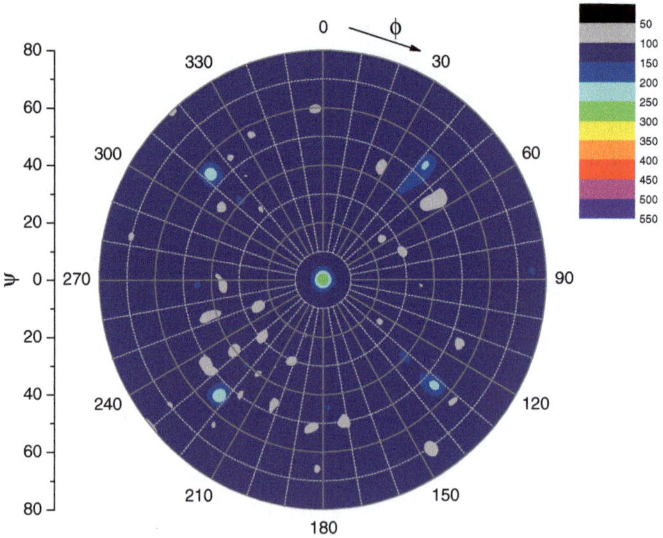

Fig. 8.20 Pole figure of rutile (110) film on MgO(100) at $2\theta = 39.3120°$ for the (200) poles of rutile

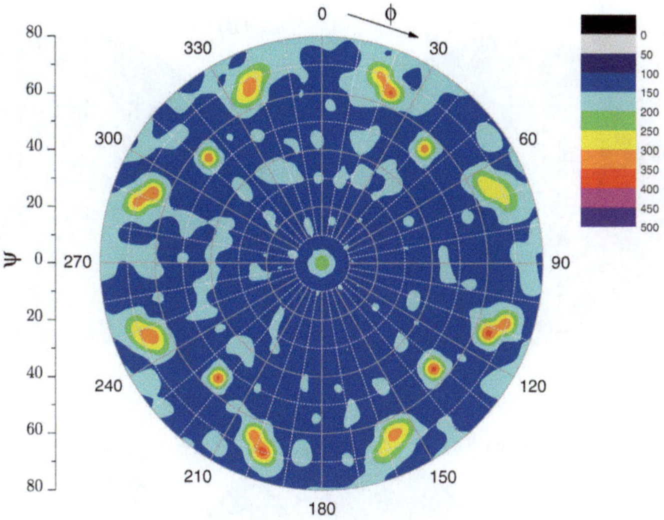

Fig. 8.21 Pole figure of rutile (110) film on MgO(100) for at $2\theta = 36.0410°$

eight poles spaced unevenly in ϕ at $\psi = 65°$ to 70°. As the value of 2θ for the (011) pole figure is similar to that for the (200) pole figure, it is possible that the four poles at $\psi = 55°$ correspond to the same feature of the substrate. This is confirmed by a measurement made on a bare MgO (100) substrate (Fig. 8.22) that shows that the poles at $\psi = 55°$ are indeed from the substrate. In fact, these poles also appear on measurements made on a bare substrate for values of 2θ of 27.4054° and 39.3120° corresponding to the rutile (110) and (200) planes.

It is difficult to use the (200) figure to confirm that the epitaxial relation is indeed the one proposed above, due to the lack of appearance of the simple (200) poles. It might be possible to be more certain if the pole figure was repeated with a thicker or more crystalline film, such as sample TOM13047 deposited at 700 °C. However, the complex (011) pole figure can be analysed by comparison to a pole figure simulated with the software package WinWulff. The simulated pole figure (not shown) indicates that one would expect four poles at about $\psi = 68°$. The poles are the (101), (011), $(10\bar{1})$, and $(01\bar{1})$ planes. They are arranged in ϕ with about 50° between the (101) and (011) poles, which would correspond to the two poles at the top of Fig. 8.21. The $(10\bar{1})$ pole is rotated 180° in ϕ from (101) and $(01\bar{1})$ is rotated 180° relative to the (011) pole. These would correspond to the bottom two poles in Fig. 8.21. This explains four of the eight poles. From the discussion above, the grains should grow with the *c*-axis oriented in two possible perpendicular directions. This gives rise to the second set of $\{011\}$ poles at 90° to the first (left and right pairs in Fig. 8.21) and explains the apparently unequal distribution of the poles in ϕ.

In addition to the two sets of poles, there also appears to be a double spot at each of the $\{011\}$ poles. This could be due to strain or slight variation in the orientation

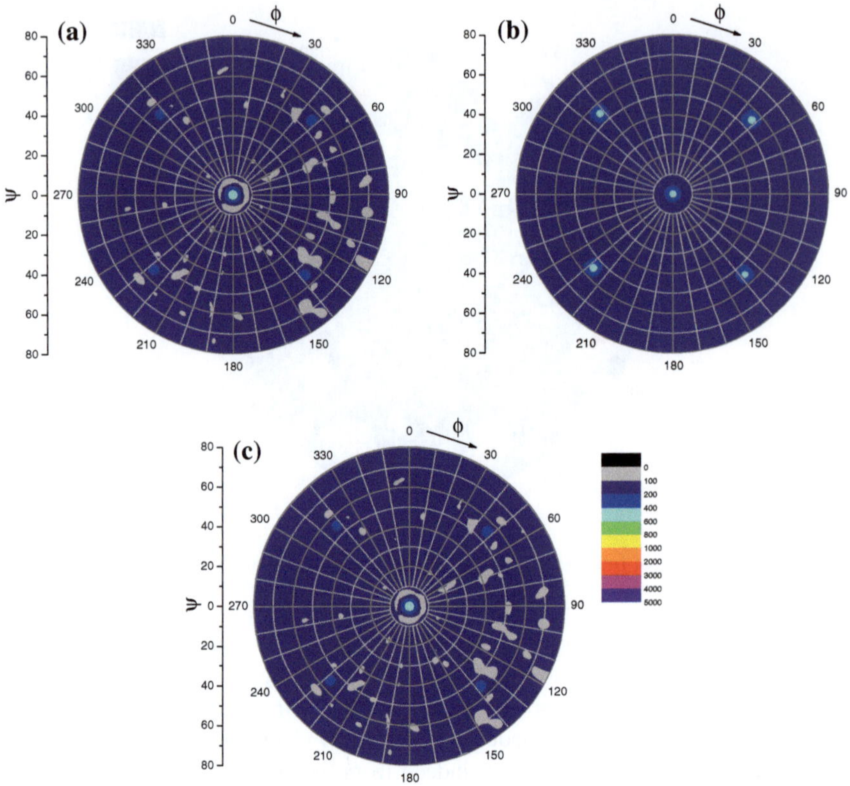

Fig. 8.22 Pole figure of MgO (100) substrate for **a** 2θ = 27.4054°, **b** 2θ = 36.0410°, **c** 2θ = 36.0410°

of each grain, or the hypothesized tilt in the sample mounting that leads to the off-centre poles for the rutile (110) planes. It could also be due to the relatively low resolution (5° × 5°) of the pole figure. There are measurements at ψ = 65° and 70°, which lie either side of the position of the ideal pole (based on the ICDD reference crystal structure). It may be possible to distinguish between these possibilities using a much finer scan, say 1° × 1°.

A pole figure for the anatase (200) sample deposited at 400 °C with a shorter substrate-target distance (Sample TOM13049) is shown in Fig. 8.23. The pole figure was produced with 2θ = 25.2672° to identify the position of the anatase (101) peaks relative to the substrate. The initial 5 × 5 pole figure (Fig. 8.23a) identifies four poles between ψ = 20° and ψ = 25° at a spacing of 90° in ϕ. A higher resolution scan shows they occur at ψ = 22.5°. Based on the reference unit cell parameters of b = 3.7582 Å and c = 9.5139 Å, the (101) planes should be at an angle of $\tan\psi = b/c$, or $\psi = 21.55°$. The ϕ positions of the (101) poles are aligned with the [001] and [010] directions of the substrate (the edges). As the film is oriented with the *a*-axis of the unit cell parallel to the [100] direction of the substrate:

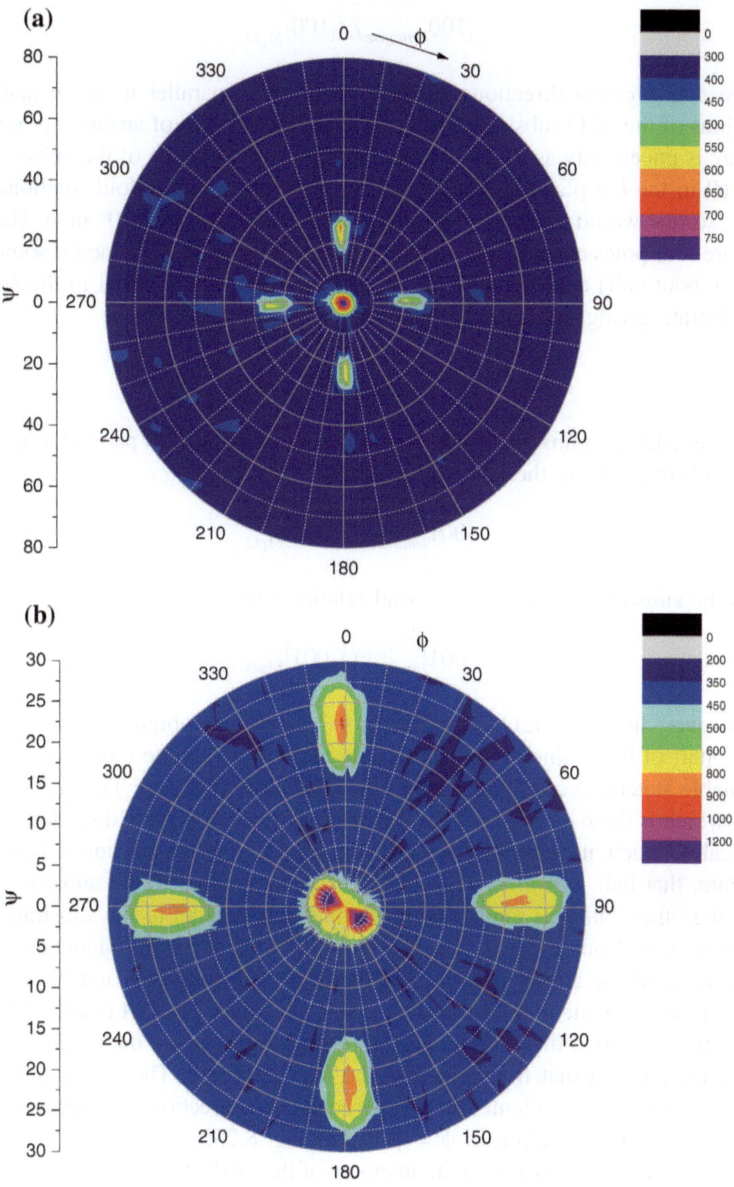

Fig. 8.23 Pole Figure of anatase (200) film on MgO. Scan for peaks at $2\theta = 25.2672°$ corresponding to anatase (101) poles. Resolution of **a** $5° \times 5°$ and **b** $2.5° \times 2.5°$ for view of poles at $\psi = 22.5°$

$$[100]_{anatase}//[100]_{\mathrm{MgO}}$$

The *b* and *c*-axis directions of the film are also parallel to the *b* and *c*-axis directions of the STO substrate. However, the *b* and *c*-axes of anatase are not equal so there is uncertainty as to the epitaxial relations in the plane of the substrate. The unit cell in the *b*-*c* plane is rectangular and thus has only twofold rotational symmetry, so one would expect to see only two poles spaced by 180° in ϕ. However, there are four poles equally spaced by 90° in ϕ. This can be explained if some of the grains (about half) are oriented with the *b*-axis of the film parallel to the *b*-axis of the substrate giving the epitaxial relation:

$$[010]_{anatase}//[010]_{\mathrm{MgO}}$$

The remaining grains are oriented with the *c*-axis of the film parallel to the *b*-axis of the substrate giving the epitaxial relation:

$$[001]_{anatase}//[010]_{\mathrm{MgO}}$$

As the substrate is cubic, the second relation (directly above) is equivalent to

$$[001]_{anatase}//[001]_{\mathrm{MgO}}$$

However, this final relation would not indicate unambiguously that approximately half of the grains of the film are oriented at 90° to the other half. This is a reasonable statement considering that when the film grows, the TiO_2 film does not know which is the *b* direction of the substrate and which is the *c* direction: both are identical. As such, it is obvious that for individual film grains starting to form on the substrate, that half will grow with the film b axis parallel to the substrate b axis, while the other half will grow with the film b axis parallel to the substrate c axis. Any deviation from a 50 to 50 distribution of grains would indicate that the arrangement of the atoms at surface of the MgO substrate does not have fourfold symmetry in the plane of the substrate i.e. it is not cubic. The fact that the intensities of the anatase (101) poles in Fig. 8.23 are equal suggests that there is in fact a 50–50 distribution and that the surface of the MgO is cubic. The statement, that the anatase film grains are oriented in two perpendicular directions, is supported by the presence of two poles close to $\psi = \phi = 0$ in Fig. 8.23b. The intensity of the two central poles is also higher than the intensity of the (101) poles, suggesting that it is either from anatase grains oriented with the (101) planes parallel to the substrate (of which there is no sign in the 2θ scans), or from the substrate. The ICDD reference pattern for MgO (pdf 00-045-0946) does not identify any planes for $2\theta = 25.2672°$. This would correspond to a d-spacing of 3.522 Å. The poles are spaced by 180° at $\phi = 120$ and 300. They do not fall at the mid point between the (101) poles. As there are only two poles, it suggests that they are from the film rather than the substrate (four fold symmetry). They may originate from a slight tilt of 2° to 3° in

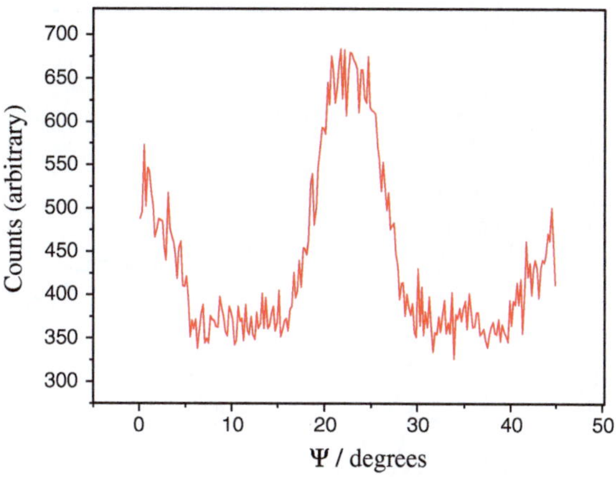

Fig. 8.24 ψ-scan of anatase (200) film on MgO. Examination of the (101) film pole at 2θ = 25.2672° corresponding to anatase (101) poles

the mounted sample, about an axis at an angle of 30° to the sides of the substrate that causes a splitting in the poles.

It is also important to note that because anatase is tetragonal while MgO is cubic, that the 101 directions in anatase are not parallel to the 110 directions of MgO, even thought they are aligned in ϕ (Fig. 8.23).

As in the case of the films on STO, the films on MgO demonstrate significant strain in the ψ-scan through the anatase (101) poles which have a full width half maximum on the order of 10° (Fig. 8.24). This is due to a combination of the lattice mismatch, the two orientations of grains, and the likely tilt of the sample. The in plane lattice parameters of anatase are b = 3.7582 Å and c = 9.5139 Å. The lattice parameter of MgO is 4.2112 Å. This gives a strain of 12 % (=4.2112/3.7582 − 1) for the c axis of the film, and a strain of 13 % for the b axis of the film (=9.5139/2/4.2112 − 1).

8.5.5 Optical Measurements—Films on MgO

The films were characterized using UV-Vis spectroscopy. The transmittance and reflectance data is given in Appendix D. The processed data on the absorption and bandgap is given below. The original depositions that resulted in rutile films are considered first before the subsequent depositions that resulted in anatase and rutile films depending on the deposition temperature.

The absorption coefficient and absorption depth of the rutile films are shown in Fig. 8.25. The absorption coefficients were calculated using the film thickness determined by profilometry. Given that all the films are rutile (110), the variation in the peak absorption is surprising. It suggests that the thickness measurements obtained by profilometry are not accurate. A comparison of the absorption of the films (Fig. 8.26) shows a much smaller variation. The samples deposited at 500–700 °C are remarkably similar, while the sample deposited at 400 °C shows a

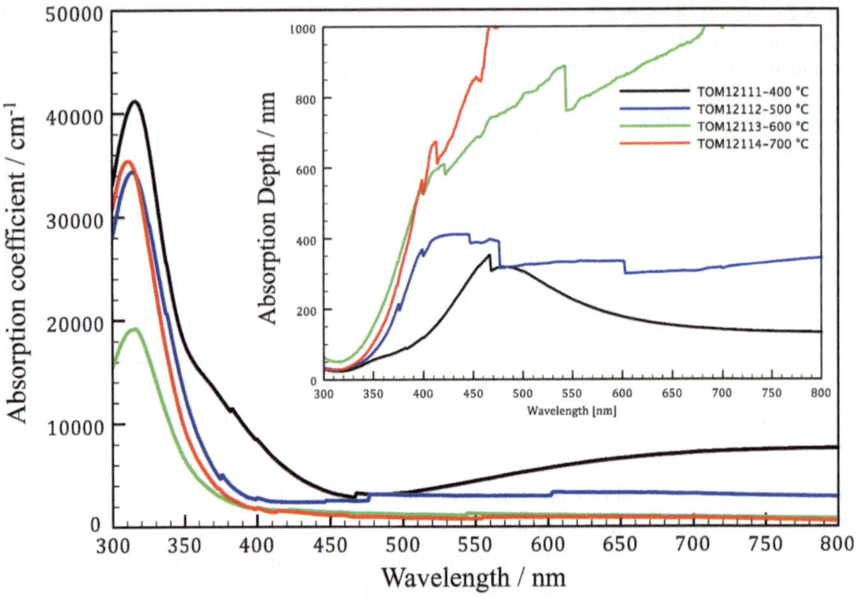

Fig. 8.25 Absorption coefficient of rutile (110) films grown on MgO (100)

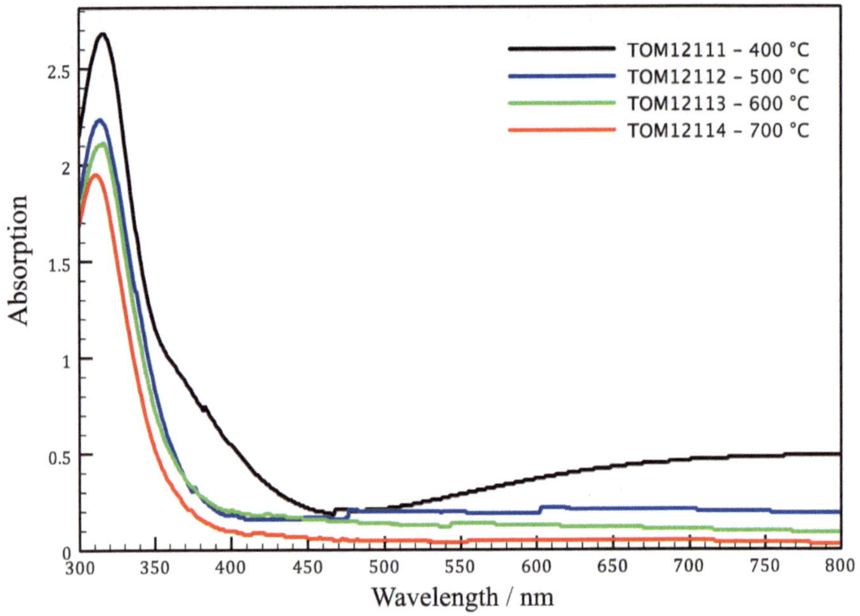

Fig. 8.26 Absorption of rutile (110) films grown on MgO (100)

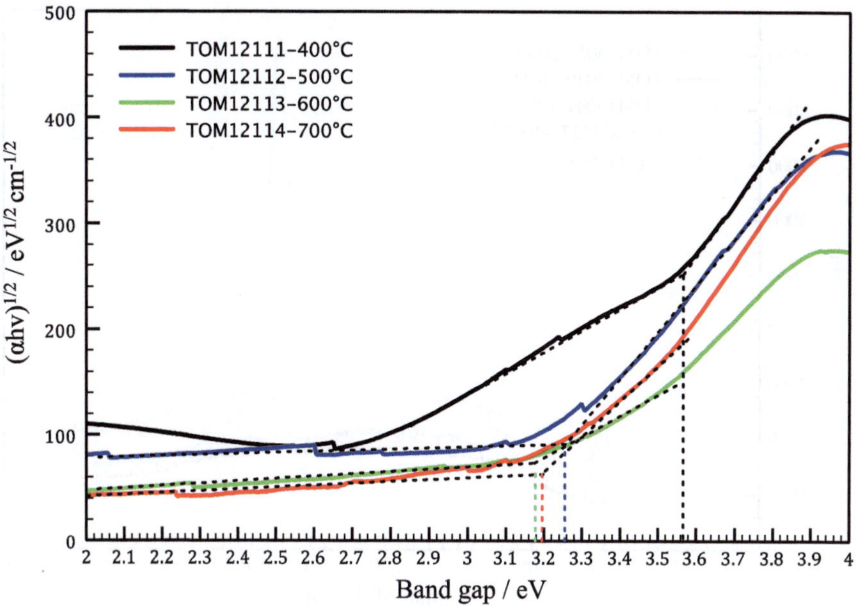

Fig. 8.27 Tauc plot assuming indirect bandgap for rutile (110) films grown on MgO (100)

shoulder in the absorption between 350 and 450 nm. This is probably due to an optical interference fringe close to the band gap. However, that would indicate that the film is much thicker that the 65 nm estimated by profilometry. The shoulder in the absorption data also prevents the bandgap from being estimated properly from a Tauc plot (Fig. 8.27). The bandgap of the samples deposited at 500–700 °C are indirect and in the range 3.15–3.25 eV. This is higher than the typically reported bandgap of rutile [3] at 3.05 eV and may be due to the thinness of the films and the strain due to the substrate.

8.5.6 Raman Spectroscopy—Films on MgO

A small selection of the films deposited on MgO (100) were examined using Raman spectroscopy to confirm the phase of the films, and to try to identify any differences in the Raman spectra that might be due to the orientation of the films, such as activation of different Raman modes.

The Raman spectra of the anatase (200) samples deposited at 400 and 500 °C, and the rutile sample deposited at 700 °C at 10 mTorr oxygen pressure are shown in Fig. 8.28. Interestingly, only the E_g at 445 cm^{-1} and A_{1g} at 610 cm^{-1} vibrations of the rutile sample are visible [4]. There are no visible peaks for the anatase films. For the films deposited on fused silica (Chap. 7), the peaks of the anatase films were

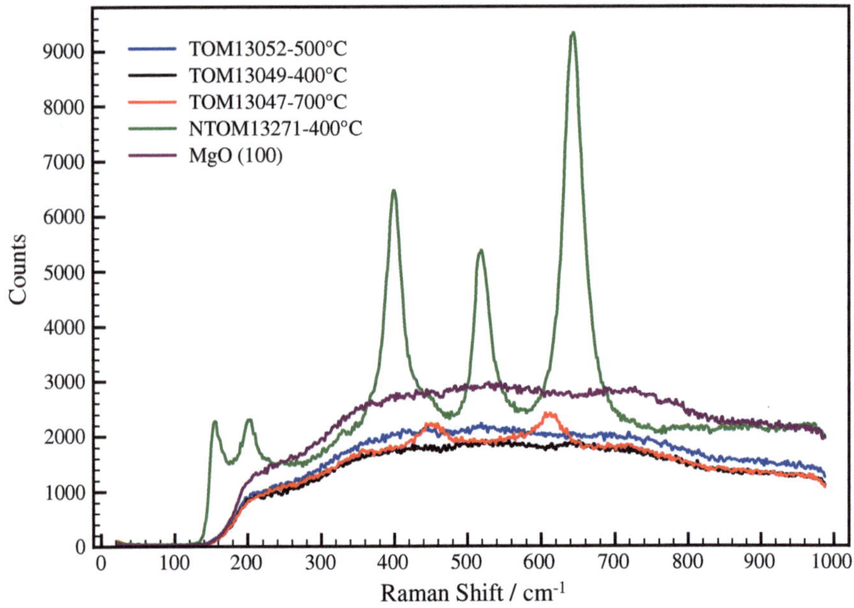

Fig. 8.28 Raman spectra of anatase (200) and rutile (110) films grown on MgO (100). Spectra obtained with 514 nm laser

significantly more intense that those of the rutile for films of similar thickness. This suggests that either the anatase films on MgO are much thinner than the rutile film, less crystalline, or that the (100) orientation of anatase is not Raman active. The Raman spectra of the subsequently deposited film NTOM13271 (with the same conditions except for a higher oxygen pressure of 30 mTorr) shows all the main anatase peaks, B_{1g} at 395 cm^{-1}, A_{1g} at 515 cm^{-1}, and E_g at 635 cm^{-1}. This film also has a more intense anatase (200) peak in the XRD patterns (due to thickness and crystallinity) and so would also be expected to have a more intense Raman spectrum. While a thicker and more crystalline film is expected to give more intense peaks, it is surprising that not even small peaks are visible for the anatase films deposited at 10 mTorr. The presence of peaks may indicate that the Nb doping favours the activation of the Raman modes in the (100) oriented films. Alternatively, the peaks may indicate that the higher oxygen pressure improves the crystallinity of the film.

References

1. Silva, V.F., et al.: Substrate-controlled allotropic phases and growth orientation of TiO_2 epitaxial thin films. J. Appl. Cryst. **43**, 1502–1512 1–11 (2010). doi:10.1107/S0021889810041221

2. Swanepoel, R.: Determination of the thickness and optical-constants of amorphous-silicon. J. Phys. E: Sci. Instrum. **16**, 1214–1222 (1983)
3. Diebold, U.: The surface science of titanium dioxide. Surf. Sci. Rep. **48**, 53–229 (2003)
4. Yan, J., et al.: Understanding the effect of surface/bulk defects on the photocatalytic activity of TiO_2: anatase versus rutile. Phys. Chem. Chem. Phys. **15**, 10978 (2013)

Chapter 9
Conclusions

This thesis describes investigations into the use of titanium dioxide (TiO_2) as a photo-anode for photo-electrochemical water splitting, otherwise termed the production of solar hydrogen.

The thesis reports on the results of four strands of work related to the surface modification and growth of TiO_2 electrodes.

- The surface modification of rutile (110) single crystals by the addition of gold nanoparticles (AuNPs).
- Studies on Nb-doped and reduced rutile (110) single crystals.
- Growth of oriented TiO_2 thin films, for use as electrodes, on amorphous substrates like fused silica, including the effect of Nb doping.
- Growth of oriented TiO_2 thin films on single crystal substrates

The work can be grouped as studies on rutile (110) single crystals, and the growth of TiO_2 thin films by pulsed laser deposition (PLD).

In the investigation of the surface modification of single crystal rutile (110) by AuNPs, several nanoparticle deposition methods were investigated including annealing metal thin films (AMTF), micellar nanolithography (MNL), and nanosphere lithography (NSL). The optical characteristics were determined by UV-Visible spectroscopy (UV-Vis), and the photoelectrochemical performance of Au-TiO_2 photoanodes under UV and visible light was measured using a range of electrochemical techniques.

AuNP arrays, formed by annealing a sputtered gold thin film, demonstrated an optical reflectance spectra characteristic of localized surface plasmon resonance (LSPR) on the AuNPs. Photocurrent measurements demonstrated a small plasmonic photocurrent under illumination by visible light. The incident photon to current efficiency (IPCE) of 0.003 % was vanishingly small but measurable. The variation of the IPCE with wavelength coincided with the reflectance and absorption of the LSPR, with the maximum occurring around 575–600 nm. This photocurrent was not present for the bare rutile (110) electrode. The reasons for the low IPCE were discussed. The proposed mechanism of plasmonic photocurrent generation was

© Springer International Publishing Switzerland 2016

J.C. Alexander, *Surface Modifications and Growth of Titanium Dioxide for Photo-Electrochemical Water Splitting*, Springer Theses,
DOI 10.1007/978-3-319-34229-0_9

discussed critically and in detail in the literature review contained in Chap. 2 of this thesis.

The plasmonic photocurrent was not detected on reduced rutile (110) or a second batch of Nb-doped rutile when either was coated with AuNPs. To investigate the reasons for the lack of plasmonic photocurrent, detailed measurements using electrochemical techniques were made on bare singles crystals from both batches of Nb-doped and reduced rutile (110).

The voltammetric measurements made on the first batch of Nb-doped crystals demonstrated qualitatively different behaviour to the second batch, which had qualitatively similar behaviour to the reduced rutile. The first batch of Nb-doped rutile displays purely capacitive behaviour over a wide potential range, while the second batch and the reduced rutile display current indicating redox processes over the same range. This suggested that the two batches of Nb-doped material had a fundamentally different structure at the surface. Furthermore, this suggested that the Nb-doped rutile that did not display plasmonic photocurrent (when coated in AuNPs) was partly reduced as well as doped. It was suggested that this reduced state either prevented or reduces significantly the mechanism of plasmonic photocurrent.

The parameters describing the space charge region of the rutile electrodes, such as space charge width constant, and charge carrier density, were derived from impedance measurements made on the electrodes. In particular, fitting the capacitance data to the Mott-Schottky equation demonstrated that there is an impurity energy level in the band gap of rutile TiO_2. As the electrode potential was swept positive of the flat band potential, the slope of the plot of C_{SCR}^{-2} against potential changed, indicating an increase in charge carrier concentration. For example, due to the Fermi level passing through a donor energy level. These parameters were used to fit a model to the measured photocurrent data. The model could not be fit in the region close to the flat band potential. Inspection of the voltammetric analysis suggests that, positive of the flat band potential of rutile, there is the expected oxidation of water to oxygen by photo-generated holes, and a simultaneous reduction of a species (presumed to be oxygen) at the surface, which reduces the net anodic current. In a water splitting (or other photo-electrochemical) application, this suggests that the reduction of oxygen could reduce the efficiency of a rutile photoanode.

With the aim of providing a range of oriented surfaces of both anatase and rutile for further investigations, thin films of TiO_2 were grown by pulsed laser deposition (PLD). Films grown on an amorphous fused silica substrate demonstrated sensitivity to the thermal processing. Films deposited at room temperature and subsequently post-annealed in air crystallized with a predominantly (101) orientation. On annealing at successively higher temperatures, the films eventually underwent an anatase to rutile phase transition between 1000 and 1100 °C. With substrate heating during growth, films grew as anatase (101) from 400 to 600 °C, and as rutile with the (110) orientation from 800 °C. Interestingly, films deposited at 700 °C grew as anatase with the (001) orientation dominating. This indicates that it is possible to control the orientation of a growing film by temperature alone, without the need for a single crystal substrate or buffer layer.

Films deposited with 10 wt% Nb-doping demonstrated stability of anatase up to at least 800 °C, also with the (001) orientation i.e. the Nb-doping acted to stabilize anatase with respect to rutile.

Unexpectedly, the Nb-doped thin films did not demonstrate a measurable level of conductivity (as expected from the literature review), even with high doping concentrations. A review of the defect chemistry of TiO_2 and Nb_2O_5 suggested that this was likely due to annealing in at a high oxygen partial pressure. The close to ambient oxygen pressure may have resulted in the formation of ionic, rather than electronic, defects acting to compensate the Nb incorporated into the TiO_2 lattice.

The deposition of TiO_2 thin films on the single crystal substrates STO (100) and MgO (100) was also investigated. Films grown on STO (100) were found to grow as expected with the anatase (001) orientation. Films grown on MgO (100) were found to grow as either anatase (100), the desired orientation, or as rutile (110) at the same temperature. The parameters determining the phase of the film appear to be a combination of the substrate-target working distance and the deposition oxygen pressure. Finally, the epitaxial relations between the TiO_2 films and the substrates were determined by X-ray diffraction texture analysis.

Appendix A
Defect Chemistry of TiO_2

The defect chemistry of TiO_2 and Nb-doped TiO_2 is discussed below. The reasons for including the defect chemistry in such detail was that it was reviewed in detail to try to understand why TiO_2 films doped with Nb failed to demonstrate measurable levels of conductivity. Additionally, it was used to try to understand the differences between Nb-doped and reduced rutile.

The initial theoretical treatment is adapted from 'The Defect Chemistry of Metal Oxides' by Smyth [1], in which TiO_2 doping of Nb_2O_5 is used as an example, and other references [2]. The algorithm used by Smyth was followed to write out the defect chemistry for the case of Nb_2O_5 doping of TiO_2, as done experimentally in this work. The various examples of TiO_2 defects discussed across the text are collected together here for ease of reference. In particular, Smyth (and others) makes no particular distinction between the rutile and anatase phases, as only equilibrium conditions are considered and anatase is only a metastable phase. To distinguish between anatase and rutile defect chemistry requires more sophisticated approaches such as computational chemistry, which calculate the energies of defect formation. It is difficult (if not impossible) to predict which of the defects outlined below will occur without knowledge of the energy of formation of the defect. It is worth considering the defect chemistry of both undoped and doped TiO_2. Ionic (stoichiometric) disorder is considered first, followed by electronic disorder, the limiting case of non-stoichiometry. The defect chemistry is denoted using Kröger-Vink notation as used by Smyth. Other variations of the notation are possible.

The likely defect types are presented for intrinsic and extrinsic defects in both the stoichiometric and non-stoichiometric cases. These were then compared and the defects likely to be present in the reduced and Nb-doped single crystals and thin films described in Chaps. 5–8 were identified.

© Springer International Publishing Switzerland 2016
J.C. Alexander, *Surface Modifications and Growth of Titanium Dioxide for Photo-Electrochemical Water Splitting*, Springer Theses,
DOI 10.1007/978-3-319-34229-0

A.1 Undoped TiO_2—Frenkel Disorder

The first possible defect to be considered is an anion Frenkel defect, where an O^{2-} ion moves from a normally occupied lattice site to a vacant interstitial site according to:

$$O_O + V_I \rightleftharpoons O_I'' + V_O^{\cdot\cdot} \tag{A.1}$$

where O_O denotes an oxygen ion on an oxygen lattice site, V_I a vacant interstitial site, O_I'' an oxygen ion on an interstitial site (with a charge of −2 relative to interstitial lattice), and $V_O^{\cdot\cdot}$ a vacancy on an oxygen lattice site (with a charge of +2 relative to oxygen lattice).

Also possible is a cation Frenkel defect, in which a Ti^{4+} ion moves from a normally occupied lattice site to an interstitial site:

$$Ti_{Ti} + V_I \rightleftharpoons Ti_I^{4\cdot} + V_{Ti}^{4'} \tag{A.2}$$

where Ti_{Ti} denotes Ti^{4+} on a Ti site in the lattice, V_I denotes a vacant interstitial site etc. following the usual conventions.

Smyth points out that the charge of the $Ti_I^{4\cdot}$ defect is higher than that of the O_I'' defect and should be less favourable. However, he also notes that the O lattice sites in rutile, though not anatase, are approximately close-packed, so there is no space for an oxygen interstitial, making the Ti interstitial a more likely defect, despite its higher charge.

The mass-action equation for the cation Frenkel (denoted by subscript 'CF') defect is:

$$\left[V_{Ti}^{4'}\right]\left[Ti_I^{4\cdot}\right] = K_{CF}(T) = K_{CF}^{\circ} e^{\Delta S_{CF}/k} e^{-\Delta H_{CF}/kT} \tag{A.3}$$

where $\left[V_{Ti}{}^{4'}\right]$ indicates the concentration of negatively charged (relative to the lattice) vacancies on Ti lattice sites and $\left[Ti_I^{4\cdot}\right]$ indicates the concentration of positively charged Ti^{4+} ions on interstitial sites. $K_{CF}(T)$ is the equilibrium constant for the formation of the defect at a temperature T, and ΔS_{CF} and ΔH_{CF} are the entropy and enthalpy of formation of the defect.

If it is assumed that the cation Frenkel this is the dominant defect in the material, it is possible to make the approximation:

$$\left[V_{Ti}^{4'}\right] \approx \left[Ti_I^{4\cdot}\right] \tag{A.4}$$

The equation is approximate because other defects such as Schottky disorder and non-stoichiometry may also be present. Substituting for either in the mass-action expression gives:

$$\left[V_{Ti}^{4'}\right] \approx \left[Ti_I^{4\cdot}\right] = K_{CF}^{1/2} \tag{A.5}$$

A.2 Undoped TiO_2—Schottky Disorder

The final intrinsic ionic disorder that may be possible in TiO_2 is the Schottky defect, where an electrically neutral TiO_2 unit is removed from the interior lattice and placed at a surface, leaving one Ti and two O vacancies in the lattice. This is given by:

$$nil \rightleftharpoons V_{Ti}^{4'} + 2V_O^{\cdot\cdot} \tag{A.6}$$

where 'nil' denotes the normal TiO_2 lattice. The mass-action expression for this defect is:

$$\left[V_{Ti}^{4'}\right]\left[V_O^{\cdot\cdot}\right]^2 = K_S(T) = K_S^\circ e^{\Delta S_S/k} e^{-\Delta H_S/kT} \tag{A.7}$$

Assuming that a Schottky Disorder is the dominant defect at some temperature T, then the concentration of Ti vacancies $\left[V_{Ti}^{4'}\right]$ will be approximately half that of oxygen vacancies $\left[V_O^{\cdot\cdot}\right]$:

$$2\left[V_{Ti}^{4'}\right] \approx \left[V_O^{\cdot\cdot}\right] \tag{A.8}$$

Substituting this into the mass-action expression (A.7) gives:

$$\left[V_{Ti}^{4'}\right] \approx \left(\frac{1}{4}\right)^{1/3} K_S^{1/3} \tag{A.9}$$

And

$$\left[V_O^{\cdot\cdot}\right] \approx 2^{1/3} K_S^{1/3} \tag{A.10}$$

A.3 Undoped TiO_2—Non-stoichiometric Disorder

In the region of non-stoichiometry (when it is possible for the oxide to lose oxygen due to low external oxygen partial pressure), electronic disorder due to reduction is possible. The possible reductions are:

$$2O_O \rightarrow O_2 + 2V_O^{\cdot\cdot} + 4e' \tag{A.11}$$

For each molecule of oxygen lost, two oxygen vacancies are created and compensated by four electrons. If a normally occupied Ti_{Ti} and unoccupied interstitial V_I (which have no charge) are added to equation (A.11) substitution for $V_O^{\cdot\cdot}$ can be made to obtain an alternative reaction:

$$Ti_{Ti} + 2O_O + V_{Ti} \rightarrow O_2 + Ti_I^{4\cdot} + 4e' \tag{A.12}$$

This suggests that rather than forming an oxygen vacancy, an interstitial Ti ion may form with the loss of oxygen from the lattice. In both cases, the electrons will either go into the conduction band or reduce a Ti^{4+} ion to a Ti^{3+} ion.

Smyth summarises the evidence for the Ti interstitial being the preferred defect over oxygen vacancies: the $-1/5$ gradient for variation of conductivity with oxygen pressure; blockage of normally vacant octahedral sites in TEM data; and variation of the conductivity minimum with oxygen activity with acceptor doping. While each $Ti_I^{4\cdot}$ is compensated by four electrons, it is not clear what happens to the electrons, which could reduce Ti_{Ti} to Ti_{Ti}' or they might reduce the interstitial Ti and three lattice Ti sites.

The mass-action for the formation of a Ti interstitial compensated by electrons gives:

$$\frac{n^4\left[Ti_I^{4\cdot}\right]P(O_2)}{\left[Ti_{Ti}\right]\left[O_O\right]^2\left[V_{Ti}\right]} \approx n^4\left[Ti_I^{4\cdot}\right]P(O_2) \approx K_{RI} \approx K_{RI}^{\circ}e^{-\Delta H_{RI}/kT} \tag{A.13}$$

where n is the concentration of electrons and $P(O_2)$ is the oxygen activity. The subscript RI is used to indicate an ionic reduction, which here is the formation of a cation interstitial without the formation of a cation vacancy (which otherwise would be a Frenkel disorder). Assuming the non-stoichiometry is insignificant means the terms in the denominator are unity.

A.4 Nb-Doped TiO_2—Extrinsic Ionic Disorder

For extrinsic ionic disorder introduced by doping with Nb, there are two main possibilities for the compensating ionic defect. As was done experimentally (as described in Chaps. 7 and 8), both defects are based on doping TiO_2 with Nb_2O_5:

$$Nb_2O_5 \xrightarrow{(2TiO_2)} 2Nb^{\cdot}_{Ti} + 4O_O + O''_I \tag{A.14}$$

And

$$2Nb_2O_5 \xrightarrow{(5TiO_2)} 4Nb^{\cdot}_{Ti} + 10O_O + V^{4'}_{Ti} \tag{A.15}$$

Either an interstitial oxygen O''_I or a Ti vacancy $V^{4'}_{Ti}$ are possible as the compensating defect of the $Nb^{\cdot}_{Ti}$. Smyth suggests that $V^{4'}_{Ti}$ is the most likely defect for rutile, due to the close-packed anion sub-lattice. The anatase phase does not have a close-packed anion sub-lattice, so interstitial oxygen may be more likely than in rutile, and could even be the dominant defect. This is not known at present.

The extrinsic disorder is now considered together with the intrinsic, as it can explain the behaviour of defects at high oxygen pressure $P(O_2)$, which is the most relevant to the thin films described in this thesis.

A.5 Nb-Doped TiO_2—Schottky Disorder as Ionic Defect

The following example is presented for Nb doping of TiO_2 and a Schottky disorder. Other examples with Frenkel defects are also possible.

If Schottky disorder were assumed to be the dominant intrinsic ionic defect, then combined with the equation $2\left[V^{4'}_{Ti}\right] \approx \left[V^{\cdot\cdot}_O\right]$ and assuming that $V^{4'}_{Ti}$ is the dominant compensating mechanism of the extrinsic defect, the charge neutrality condition would be:

$$4\left[V^{4'}_{Ti}\right] \approx \left[Nb^{\cdot}_{Ti}\right] + 2\left[V^{\cdot\cdot}_O\right] \tag{A.16}$$

At low Nb-doping levels, $4\left[V^{4'}_{Ti}\right] \approx 2\left[V^{\cdot\cdot}_O\right]$ and at high doping, $\left[V^{\cdot\cdot}_O\right]$ can be neglected compared to $\left[Nb^{\cdot}_{Ti}\right]$ and the condition becomes $4\left[V^{4'}_{Ti}\right] \approx \left[Nb^{\cdot}_{Ti}\right]$, i.e. there will be four interstitial Ti for each Nb incorporated into the lattice. Substituting for $\left[V^{4'}_{Ti}\right]$ in $\left[V^{4'}_{Ti}\right]\left[V^{\cdot\cdot}_O\right]^2 = K_S$ (A.7) gives:

$$\left(\frac{1}{4}\right)\left[Nb^{\cdot}_{Ti}\right]\left[V^{\cdot\cdot}_O\right]^2 = K_S \tag{A.17}$$

And

$$\left[V_{\mathrm{O}}^{\cdot\cdot}\right] = \left(\frac{4}{\left[\mathrm{Nb}_{\mathrm{Ti}}^{\cdot}\right]}\right)^{1/2} K_S^{1/2} \tag{A.18}$$

So the concentration of oxygen vacancies $\left[V_{\mathrm{O}}^{\cdot\cdot}\right]$ would decrease with a gradient of $-1/2$ with increasing $\left[\mathrm{Nb}_{\mathrm{Ti}}^{\cdot}\right]$ on a log-log concentration plot. Therefore, the extrinsic defect concentration at which the Nb-doping would become dominant is dependent on the temperature, embedded in the temperature dependence of the equilibrium coefficient K_S. At higher temperature, the concentration of intrinsic defects is higher, so a higher Nb-content would be required for extrinsic defects to dominate.

A.6 Non-stoichiometry

In conditions typical of vacuum deposition, such as during pulsed laser deposition (as used to grow films for this thesis), it is expected that oxide materials are able to lose oxygen and be reduced. This can act as a compensation mechanism for the incorporation of $\mathrm{Nb}_{\mathrm{Ti}}^{\cdot}$ defects and should result in conductivity. As the material is then free to lose non-metal like oxygen rather than have to incorporate it interstitially, O_I'' can be replaced in the equilibrium condition (Eq. A.14) by the reaction:

$$\mathrm{O}_I'' \rightarrow \frac{1}{2}\mathrm{O}_2 + V_I + 2e' \tag{A.19}$$

This results in the non-stoichiometric defect compensation mechanism:

$$\mathrm{Nb_2O_5} \xrightarrow{(2\mathrm{TiO_2})} 2\mathrm{Nb}_{\mathrm{Ti}}^{\cdot} + 4\mathrm{O_O} + \frac{1}{2}\mathrm{O}_2 + 2e' \tag{A.20}$$

For every two units of Nb_2O_5, eight oxygen atoms are incorporated, and one O_2 molecule is emitted, leaving behind four electrons in the lattice that provide conductivity.

The preceding is also true if the $V_{\mathrm{Ti}}^{4'}$ term is substituted in (A.15) by:

$$V_{\mathrm{Ti}}^{4'} + 2\mathrm{O_O} \rightarrow \mathrm{O}_2 + 4e' \tag{A.21}$$

This suggests that in vacuum conditions, Nb-doped films can result in the loss of oxygen from the lattice, resulting in the addition of four electrons for every molecule of oxygen lost. These additional charge carriers can either be injected into the conduction band or reduce reducible species or lattice sites in the material.

Whether oxygen is lost and the material becomes electronically conductive, or whether the Nb is compensated by ionic disorder will depend on the relative enthalpy changes, and to a lesser extent entropy changes, of each process. However, experimental data suggests that electronic defects are indeed the favoured mechanism, due to the detected conductivity. It is noted that the films produced in this project and discussed in Chaps. 7 and 8 were not detectably conductive, and as such seem to have been ionically rather than electronically compensated. This is likely to have been due to the oxygen partial pressure used during annealing just below atmospheric conditions.

The next question is whether the electrons are free to move in the conduction band or are localized on a charge lattice site, and so do not result in conductivity. Smyth [1] and Arbiol et al. [3] note that Ti^{4+} is a reducible ion, with both Ti^{3+} and Ti^{2+} being achievable oxidation states, found as Ti_2O_3 and TiO. As such, the following modification to the compensation mechanism can be written to denote the reduction of a Ti^{4+} ion to Ti^{3+}:

$$2\mathrm{Ti}_{\mathrm{Ti}} + 2e' \rightarrow 2\mathrm{Ti}'_{\mathrm{Ti}} \tag{A.22}$$

The modified full expression is then:

$$\mathrm{Nb_2O_5} \xrightarrow{(2\mathrm{TiO_2})} 2\mathrm{Nb}^{\cdot}_{\mathrm{Ti}} + 4\mathrm{O_O} + \frac{1}{2}\mathrm{O_2} + 2\mathrm{Ti}'_{\mathrm{Ti}} \tag{A.23}$$

Furthermore, Nb^{5+} is also reducible to Nb^{4+} as evidenced by the existence of NbO_2 that has the rutile structure, and to Nb^{2+} for NbO [4, 5]. The reduction of Nb^{5+} defects could be written as:

$$2\mathrm{Nb}^{\cdot}_{\mathrm{Ti}} + 2e' \rightarrow 2\mathrm{Nb}_{\mathrm{Ti}} \tag{A.24}$$

And the defect compensation mechanism of (A.23) can be re-written as:

$$\mathrm{Nb_2O_5} \xrightarrow{(2\mathrm{TiO_2})} 2\mathrm{Nb}_{\mathrm{Ti}} + 4\mathrm{O_O} + \frac{1}{2}\mathrm{O_2} \tag{A.25}$$

These expressions show that the conductivity of the material, even allowing for non-stoichiometry, should depend on whether it is energetically favourable for the electrons to be in the conduction band or reduce either the Ti and Nb ions on the Ti sub-lattice. As discussed in Chap. 5, it seems that an electron should occupy the 4s band of Ti preferentially over the 5s band of Nb.

There have been a number of publications that use the methods of computational chemistry, such as density functional theory, to try to calculate the energy of these defects relative to that of the conduction band [6–8]. There have also been many experimental papers that give an indication of those energies, though these papers have focussed on rutile in the main [9]. Studies are usually done on rutile, as it is the stable phase and can be heated to high temperatures for conductivity measurements without risk of a phase change, unlike anatase.

Conductivity in rutile is generally reported to be low, even with high concentrations of doping. The conductivity has also been seen temperatures well above room temperature. This suggests that the electronic disorder introduced by say Nb-doping in non-stoichiometric conditions, results in the reduction of either the Ti^{4+} ion or the donor ion, and that the donor energy level sits several tenths of an electronvolt below the conduction band (based on a value greater than k_BT of ca. 0.025 mV at room temperature). This is also evidenced by reduced TiO_2 single crystals being blue in colour [10]. Fully oxidized pure rutile is fully transparent to visible light due to the ca. 3 eV bandgap, so the blue colouring indicates the presence of states in the bandgap, or colour-centres [11]. Nb-doped rutile is also blue in colour (see Results Chaps. 5 and 6), which also suggests that the electrons that should compensate the Nb are localized to a state in the bandgap.

Even though the blue colour is formed by reducing TiO_2, and assumed to induce electronic conductivity, the equilibrium defect at ambient $P(O_2)$ should in fact be ionic disorder, such as Ti vacancies or oxygen interstitials. If this were true in practice, it would suggest that these defects also result in states within the bandgap. However, it seems more likely that the reduced samples are not truly in equilibrium under ambient conditions, i.e. the defects have been fixed by cooling.

Smyth points out that point defects in high dielectric materials such as TiO_2 have a tendency to coalesce and form crystallographic shear planes. These are the Magnéli phases of Ti_nO_{2n-1}. A study by Blanchin et al. [12] referenced by Smyth apparently showed that these shear planes were present mainly for samples annealed either at low temperature or cooled slowly. Samples quenched from high temperature exhibited only point defects. This suggests that the point defects are the stable distribution of defects at high temperature, and are frozen in when the material is quenched. If the material is cooled slowly or annealed, the defects coalesce into a lower energy configuration.

To this author's knowledge, the studies on the Magnéli phases and sub-stoichiometric TiO_2 in general, have all been done on rutile. Whether the same phenomenon occurs with anatase is unknown, but it has not been reported, suggesting that is does not occur. Although this may be due to the difficulty (and expense) of obtaining single crystals of anatase and the stability of the phase when heated, i.e. it is difficult to study the conductivity of anatase at elevated temperature due to its propensity to transform to the more stable rutile phase. Hence, it would be difficult to measure the equilibrium conductivity of anatase as a function of oxygen partial pressure (the standard technique for investigating the defect mechanism), as the low temperatures required would oppose diffusion of defects.

An interesting point to note about the reduction of TiO_2 is that the compound Ti_2O_3 is a metallic conductor. Reports by Harrison et al. [13, 14] show that the Magnéli phases can be viewed as planes of Ti_2O_3 cutting a matrix of TiO_2, i.e. from the perspective of a volume spanning approximately one unit cell either side of the shear plane, the material seems to have the Ti_2O_3 structure, so this shear plane should be conductive. It is possible that some of the measured conductivity of reduced or doped TiO_2, especially at room temperature, is due to charge transport along these shear planes. For example, this effect is exploited in the application of TiO_2 to memristor devices, in which the resistivity is tuned by the reduction or oxidation of the TiO_2 by the application of a potential [15].

Appendix B
Phase Diagrams of TiO_2 and H_2O

The potential-pH (Pourbaix) phase diagrams are introduced below and used subsequently to try to determine the identity of oxidized and reduced species in the voltammetry. Phase diagrams are presented for:

- O–H species expected in the electrolyte
- Ti–H_2O species at the interface of the electrode and electrolyte

The Ti–H_2O diagram is required to understand the stability of the TiO_2 electrode in aqueous solutions, while the O–H diagram is useful in understanding the possible reactants and product that may be formed in addition to the oxidation of water.

The phase diagrams are constructed from tabulated data of the standard Gibbs free energy of formation. The reference book used was "Standard Potentials in Aqueous Solutions" [16]. In some cases, the source of the data for the tabulated species it is not always clear. For example, both TiO_2 and TiO_2 (hydrated) appear in the data with values of the Gibbs free energy that differ enough to significantly alter the phase diagram. This will be discussed further below.

In addition to the tabulated data, an assumption on the activity coefficient of soluble species is required (for example Ti^{3+}). The phase diagrams presented in this section are based on the activity of soluble species estimated at an operational pH of 13.6 (for 1 M NaOH). The partial pressure of oxygen and hydrogen are also required. These were assumed to be unity in constructing the diagrams, but as discussed in Sect. B.2, it is likely that the partial pressure of hydrogen during measurements was much lower with the associated implications for hydrogen production.

B.1 Ti–H_2O Phase Diagram

In the case of the Ti–H_2O diagram at pH 13.6, the only soluble species expected is $HTiO_3^-$. This is equivalent to TiO_2 gaining a hydroxide ion (or loss of a proton from H_2O–TiO_2.) The reaction may be written as:

$$TiO_2(\text{hydrous}) + H_2O \rightarrow HTiO_3^- + H^+ \tag{B.1}$$

© Springer International Publishing Switzerland 2016

J.C. Alexander, *Surface Modifications and Growth of Titanium Dioxide for Photo-Electrochemical Water Splitting*, Springer Theses,
DOI 10.1007/978-3-319-34229-0

where ΔG^0 for the reaction is 102,578 J mol^{-1}. The equilibrium constant, log K = 17.97 can be used to derive an equation that describes the pH dependence of the soluble species:

$$\log\left(HTiO_3^-\right) = pH - 17.97 \quad \text{(B.2)}$$

For pH 13.6, this predicts the activity to be 4×10^{-5}. The pH dependence of the activity of dissolved titanium (III) and titanium (IV) soluble species is shown in Fig. B.1. If the activity of is calculated based on anhydrous TiO_2 then the activity of $HTiO_3^-$ is predicted to be 4×10^{-17} (and is below 10^{-10} over the full range of pH).

Tabulated data is given for hydrous material in addition to anhydrous, and results in a diagram with smaller regions of stability. The phase diagrams for both are given. It is not entirely clear which diagram is most appropriate, as it could be argued that the surface is always hydrated, while the bulk of a single crystal is not.

There is some uncertainty as to whether or not TiO_2 (hydrous) should be written H_2O–TiO_2 i.e. already containing the water molecule. The Gibbs energy of formation of TiO_2 is −888,400 J mol^{-1}, while for TiO_2 (hydrous) it is −821,300 J mol^{-1}. Interestingly, Bard et al. [16] also tabulates data for $TiO_2.H_2O$ at −1,058,500 J mol^{-1}. It is not clear what the difference is between $TiO_2.H_2O$ and TiO_2 (hydrous). Repeating the calculation $TiO_2.H_2O \rightarrow HTiO_3^- + H^+$, results in a pH dependence $\log\left(HTiO_3^-\right) = pH - 17.98$, which is almost identical to the original calculation, suggesting that $TiO_2.H_2O$ and TiO_2(hydrous) are merely different ways of writing the same phase.

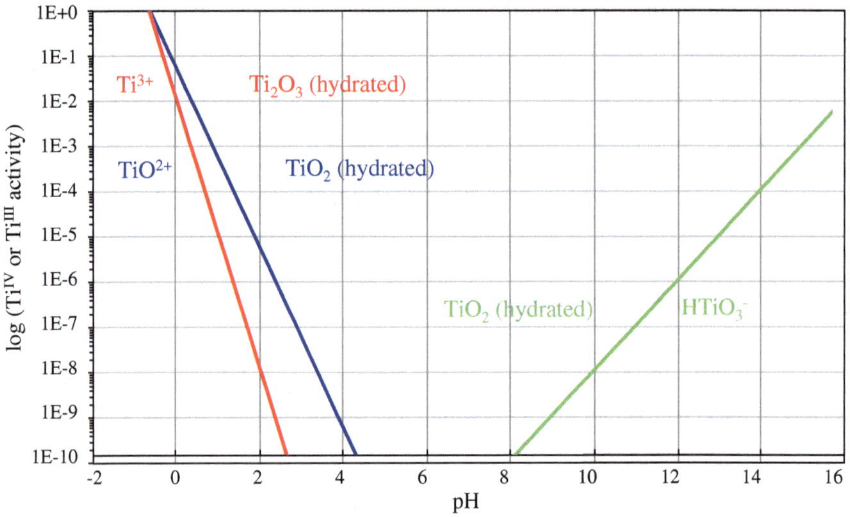

Fig. B.1 Predicted pH dependence of activities of dissolved Ti^{III} and Ti^{IV} species in equilibrium with their respective hydrated oxides

If the calculation is done for $TiO_2(c) + H_2O \rightarrow HTiO_3^- + H^+$, then the pH dependence is $\log(HTiO_3^-) = pH - 29.73$

The potential-pH diagrams are shown below in Figs. B.2, B.3, B.4 and B.5. The first diagram, in Fig. B.2, shows the stability of anhydrous TiO_2 in water. At the activity assumed for dissolved Ti^{IV} (4×10^{-5}), TiO_2 is stable over a wide range of potential and pH. In particular, at the operational pH of 13.6, there should be no solubility of Ti^{IV} in the electrolyte. Oxidation of TiO_2 to $TiO_3.2H_2O$ may be expected to occur at about +1.4 V (SHE). However, this is about 1 V positive of the water oxidation potential, and so unlikely to occur. Reduction of TiO_2 to Ti_4O_7, Ti_3O_5, Ti_2O_3, and eventually TiO is not predicted to occur between about −1.3 V and −1.6 V (SHE). Again, this is about 0.5 V negative of the water reduction potential.

Given this diagram, redox processes are not expected between the water reduction and oxidation potentials. In the case of the HgO|Hg electrode, this corresponds to −0.9 to +0.33 V. However, the voltammetry on the bare TiO_2 single crystals demonstrates several features in this range. As the TiO_2 electrode is immersed in H_2O, it might be reasonable to suppose that the TiO_2 at the interface with the electrolyte is hydrated. The phase diagram for hydrous TiO_2 is shown in Fig. B.3 with a detailed view of the diagram from pH 10 to 16 shown in Fig. B.4.

The phase diagram for TiO_2 (hydrous) (Fig. B.3) predicts that at the operational pH (13.6), $HTiO_3^-$ may be present at dissolved activities up to 4×10^{-5}. However, there was no indication of either pH changes in the electrolyte or optical absorption by dissolved species following experiments. This was determined by UV-Vis (results not shown).

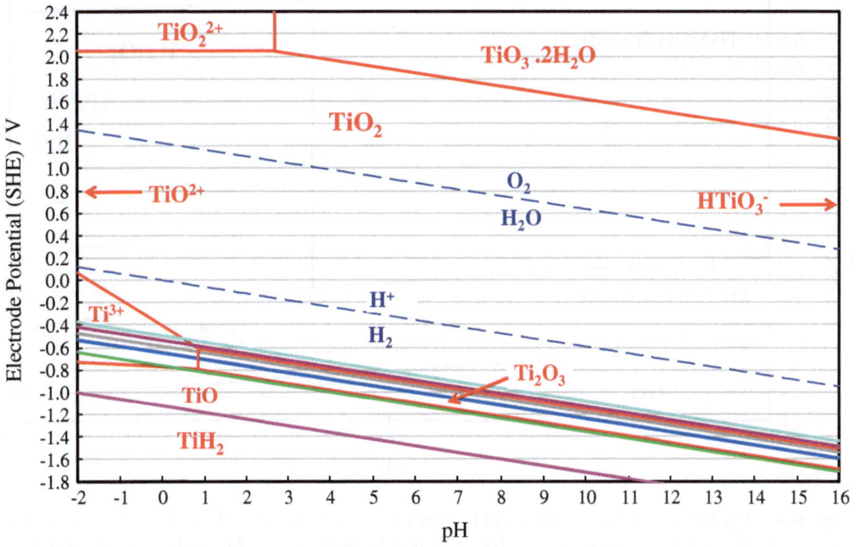

Fig. B.2 TiO_2–H_2O potential-pH diagram at 298 K, 0.1 MPa and for a dissolved titanium activity of 4×10^{-5}. Hydrous TiO_2 was not considered

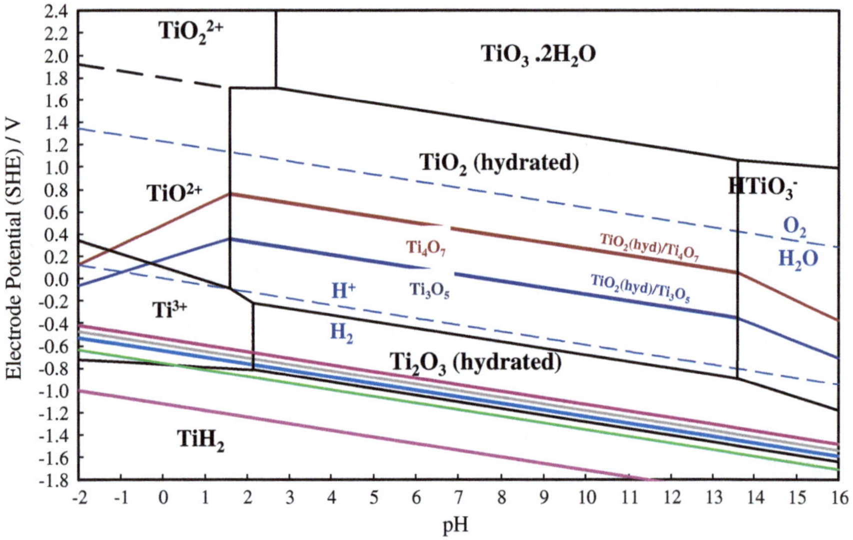

Fig. B.3 TiO_2–H_2O potential-pH phase diagram considering hydrated TiO_2 (rather than the anhydrous phase) and hydrated Ti_2O_3, for 298 K, 0.1 MPa and a dissolved titanium activity of 4×10^{-5}

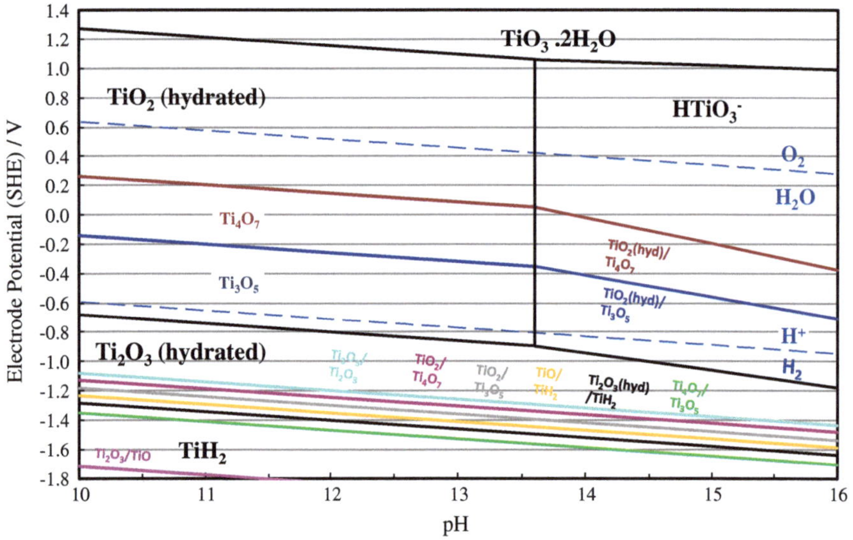

Fig. B.4 (Partial) Electrode potential-pH diagram for 298 K, 0.1 MPa, dissolved activity of dissolved Ti^{III} and Ti^{IV} species of 4×10^{-5} for the pH range 10–16. The stabilities of the reduced phases are colour co-ordinated with the text

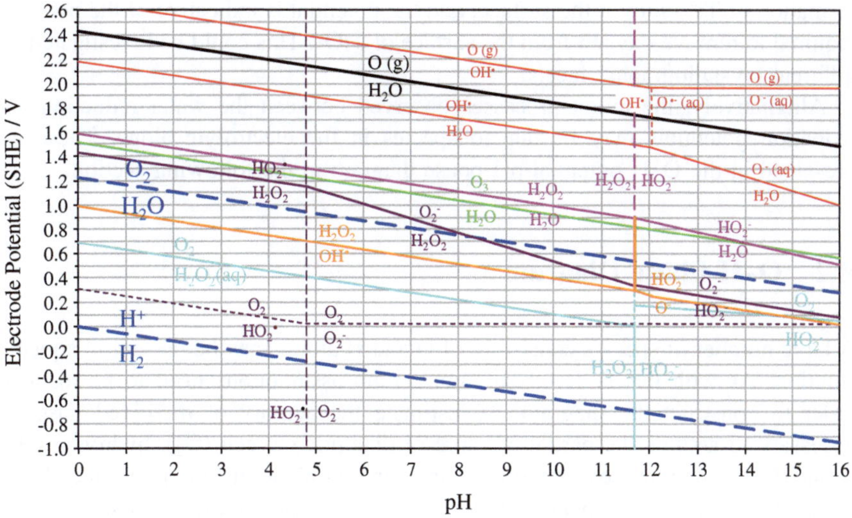

Fig. B.5 O–H potential-pH phase diagram. The partial pressure of oxygen and hydrogen gas is assumed to be 1 atm. The dissolved activities of other aqueous species are assumed to be 10^{-6}

Figure B.3 predicts that it may be possible to reduce TiO_2 to Ti_4O_7 at about +0.05 V (SHE) or to Ti_3O_5 at about −0.35 V (SHE), corresponding to −0.05 V (HgO|Hg) and +0.25 V (HgO|Hg), respectively. This suggests that any redox processes seen between the water reduction and oxidation potentials may be due to these phase changes.

However, the lack of any identifiably reduced phases in the TiO_2 electrodes in X-ray diffraction (XRD) measurements suggests that any reduction was minimal and may have been confined to a small layer at the surface. This would be supported by the fact that the peaks seen in the voltammetry plateau after a few hundred mV. However, if the reduced phase accounts for anything less than a few percent of the sample volume, XRD is unlikely to detect a change in the surface of a rutile single crystal due to the large signal from the underlying rutile.

Ti_4O_7 would be expected to form first. The potential for the reduction of Ti_4O_7 to Ti_3O_5 is at about −1.55 V (SHE) or −1.65 V (HgO|Hg). Interestingly, the potential to further reduce Ti_3O_5 to Ti_2O_3 is less negative, at −1.3 V (SHE). If the potential were negative enough to reduce Ti_4O_7 to Ti_3O_5, then it might be assumed that this would be further reduced immediately to Ti_2O_3, implying that Ti_3O_5 is at best only meta-stable.

It is not necessarily the case that TiO_2 can be reduced directly to Ti_4O_7 by electrochemical means. In fact, there are known to be numerous Magnéli phases of the form Ti_nO_{2n-1} expected between TiO_2 and Ti_4O_7. Hence, it is not clear how much help the potential-pH diagram is in truly understanding the electrochemical stability of TiO_2. This can be illustrated by asking the question: at what potential does TiO_2 become $Ti_{50}O_{99}$ or similar? Perhaps the redox processes are the formation of Ti^{3+} sites at the surface, and the potential is indicative of the potential required to remove oxygen from the surface of TiO_2. This could be investigated for different crystallographic facets.

Alternatively, the redox processes observed between the water reduction and oxidation potentials may be due to species in the electrolyte alone. These are presented in the O–H potential-pH phase diagram in the following section.

B.2 O–H Phase Diagram

A large number of species consisting of O and H atoms may be present in aqueous solution. As with the Ti–H_2O diagram, certain assumptions need to be made in order to construct the diagram. To estimate the activity of dissolved gases, it necessary to assume the partial pressure of diatomic oxygen (O_2), hydrogen (H_2), ozone (O_3), and atomic oxygen (O). In an operating photo-electrochemical cell with the anode and cathode separated by a membrane, the partial pressures should be in the region of 1 atm for both O_2 and H_2, assuming that the cell is working at atmospheric pressure.

In the case of a voltammetric experiment in a single small volume cell (without a membrane), this may be more complicated. If the evolution of O_2 and H_2 is negligible (which it was), then the partial pressures should be on the order of $P(O_2) = 0.2$ atm. and $P(H_2) = 10^{-6}$ atm. This affects the potential of water reduction and oxidation as shown in Eqs. (B.3) and (B.4).

$$U(H^+/H_2) = U^0 - \ln 10.\frac{RT}{F}pH - \ln 10.\frac{RT}{2F}\log P(H_2) \tag{B.3}$$

$$U(H_2O/O_2(g)) = U^0 - \ln 10.\frac{RT}{F}pH - \ln 10.\frac{RT}{4F}\log P(O_2(g)) \tag{B.4}$$

In the case of water reduction at pH 13.6, the effect of $P(H_2)$ shifts the potential from −0.8046 V (SHE) to −0.6271 V(SHE), a shift of 0.1775 V. In the case of water oxidation, at pH 13.6, the potential shifts from +0.4245 V (SHE) to +0.4142 V (SHE), which is a shift of −0.0103 V, and is essentially negligible.

The diagram presented below is for standard conditions with the partial pressure assumed to be 1 atm for all gases. This gives a dissolved activity of oxygen of 1.4×10^{-3} (i.e. saturated solution). The dissolved activity of other species was assumed to be negligible 10^{-6}.

It is clear that the phase diagram for O–H species in the electrolyte is extremely complex. In particular, at pH 13.6 there are a number of competing processes between the water reduction and oxidation potentials. The concentration of most of these species is expected to be very low compared to the concentration of Ti species at the surface. The kinetics of each of these reactions on TiO_2 is not explicitly known (to the knowledge of the author) and it is hard to be certain. It is noted that some of the lines for reactions as calculated (and assuming activities of dissolved species) do not meet correctly in the diagram, specifically O_2/HO_2^- with O_2/H_2O_2.

The possible presence of species might be predicted by assuming that an anodic scan is carried out starting just positive of the H^+/H_2 potential, and assuming that the dominant species in solution is H_2O or OH^-. The resulting oxidation current density An anodic scan of potential would be expected to first pass through the O_2^-/O_2 potential at 0.02 V (SHE), oxidizing any superoxide ions (O_2^-) present at the interface. At a slightly more positive potential, HO_2^- would be oxidized to hydroperoxyl ($HO_2\cdot$) or superoxide (O_2^-). The O_2^- would be expected to be unstable and be further oxidized to O_2, as the formation potential is positive of its oxidation potential. $HO_2\cdot$ would be oxidized to O_2 at a similar potential. Beyond the water oxidation potential H_2O/O_2, there is also the possibility of the oxidation producing HO_2^- (which would be further oxidized), ozone (O_3), or O^-, which should also oxidize to HO_2^-. It is clear that the only long-lived species will be H_2O and O_2. Hydrogen peroxide (H_2O_2) is not expected to be present above pH 11.7.

On the subsequent negative-going potential scan, the species expected would be H_2O and O_2. It is possible that both $HO_2\cdot$ and O_2^- could form due to the reduction of O_2. However, at the potentials at which they are formed, they would be unstable and be further reduced to H_2O.

B.3 Species Expected in Voltammetry from Phase Diagrams

By reviewing the phase diagrams for Ti–H_2O and O–H, the possible species in solution can be estimated. Based on the O–H diagram it is unlikely that anything other than H_2O, O_2, and H_2 would be present and result in repeatable processes in the voltammetry. If the processes were detected, the current would be small and expected to be mass transport limited due to the low concentration of reactant. From the Ti–H_2O potential-pH diagram, it seems that there are a number of possible reduction and oxidation processes that may change the surface of the electrode. It is possible that hydrated TiO_2 may even be unstable between the water reduction and oxidation potentials. At potentials negative of the water reduction potential, there appears to be a complex phase diagram demonstrating reduction from TiO_2 to TiO between −1.2 and −1.6 V (SHE).

Adsorption of contaminants, dissolved Ti (III) species, and a range of O–H species could also provide measurable currents in the voltammetry.

Given the complex structure of both the Ti–H_2O and O–H phase diagrams, it is extremely difficult to identify with certainty what reaction is responsible for the measured current at any given potential. This is complicated further by the space charge region, which provides a current when charging and discharging. The presence of the space charge region can also result in a shift in the onset of oxidation currents: the width of the space charge region prevents electrons tunnelling from solution species into the conduction band. This is demonstrated in Fig. 6.2 (dark blue line) where even at +2.0 V (HgO|Hg) no water oxidation occurs. These shifts in onset of processes mean that a process may become apparent only at potentials significantly positive or negative or the reversible potentials shown in Sect. 6.5.

Appendix C
Optical Measurements—Films on Fused Silica

The optical transmittance and reflectance data of TiO_2 thin films deposition on fused silica and MgO are presented in the following sections.

The film transmittance is shown in blue, and the reflectance in red. Solid and dashed black lines show the substrate transmittance and reflectance respectively. The green circles show the estimated refractive index, obtained using the Swanepoel method. The estimate of the film thickness is also given.

C.1 TiO_2—Effect of Deposition Pressure

See Figs. C.1, C.2, C.3 and C.4.

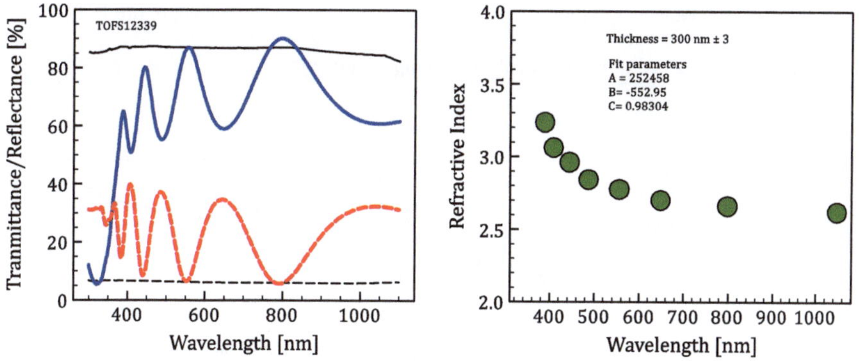

Fig. C.1 Film deposited at 10 mTorr at room temperature

© Springer International Publishing Switzerland 2016

J.C. Alexander, *Surface Modifications and Growth of Titanium Dioxide for Photo-Electrochemical Water Splitting*, Springer Theses,
DOI 10.1007/978-3-319-34229-0

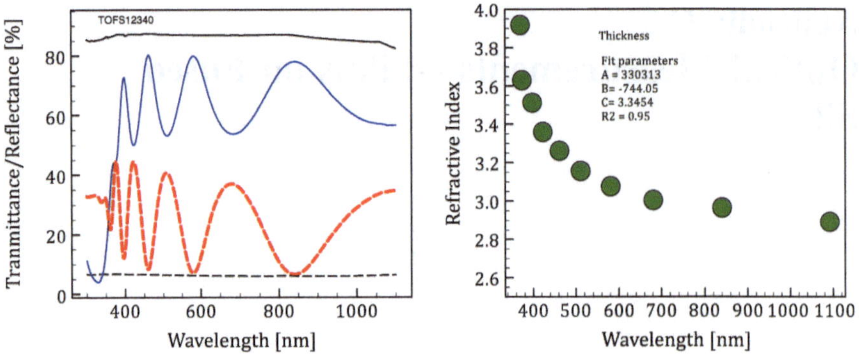

Fig. C.2 Film deposited at 30 mTorr at room temperature

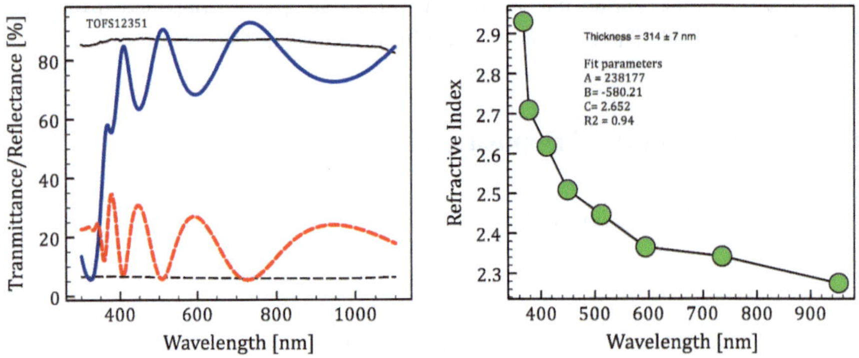

Fig. C.3 Film deposited at 40 mTorr at room temperature

Fig. C.4 Film deposited at 50 mTorr at room temperature. Refractive index could not be modelled

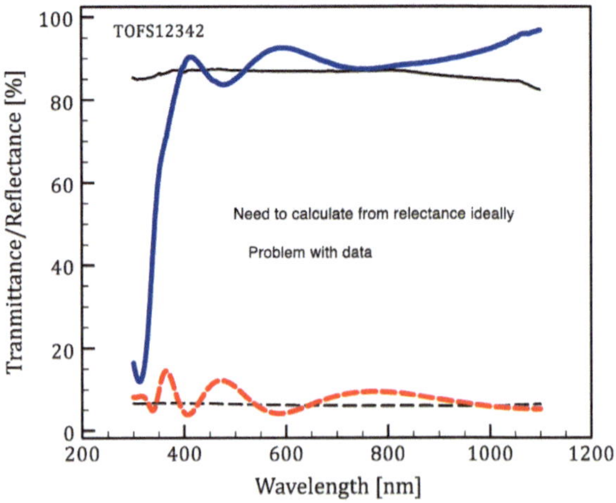

C.2 TiO_2—Effect of Deposition Temperature

See Figs. C.5, C.6, C.7, C.8 and C.9.

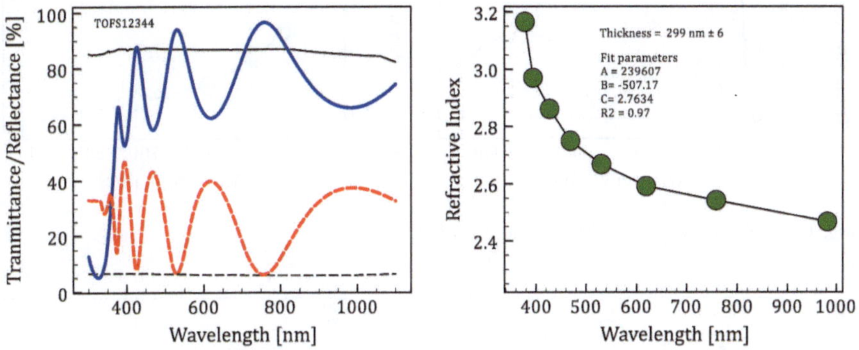

Fig. C.5 Film deposited at 400 °C and 30 mTorr

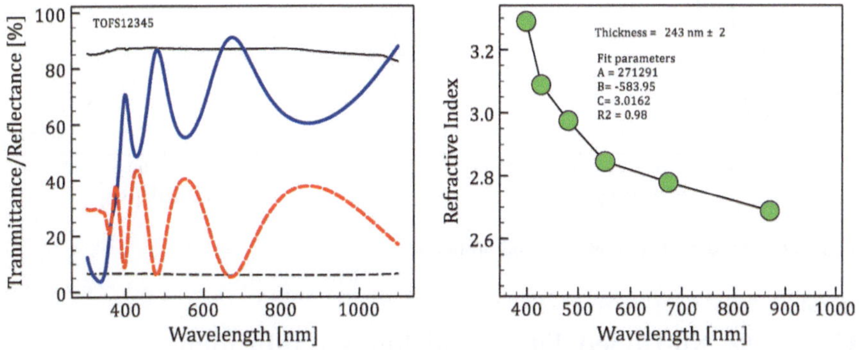

Fig. C.6 Film deposited at 500 °C and 30 mTorr

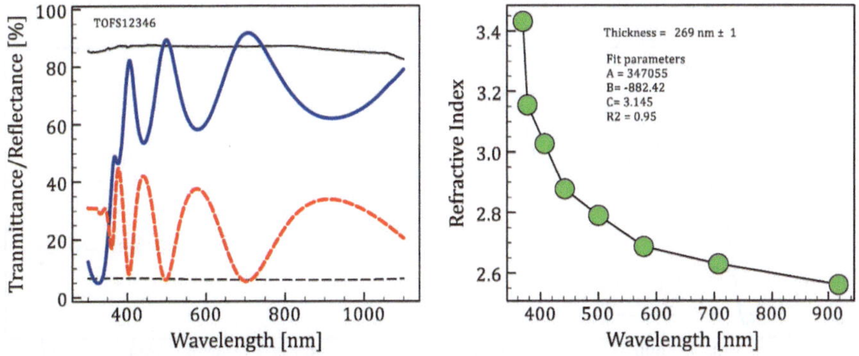

Fig. C.7 Film deposited at 600 °C and 30 mTorr

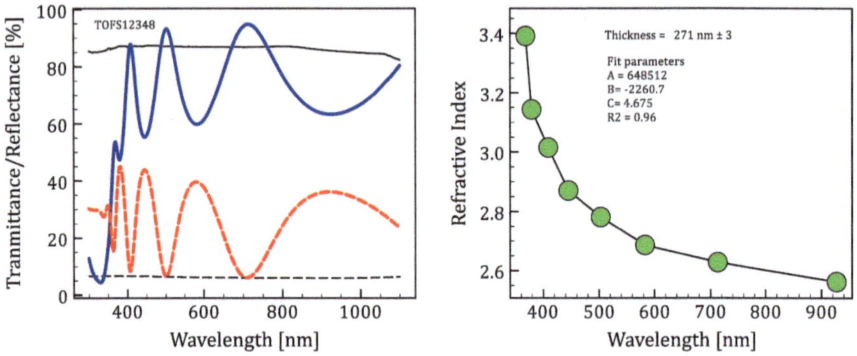

Fig. C.8 Film deposited at 700 °C and 30 mTorr

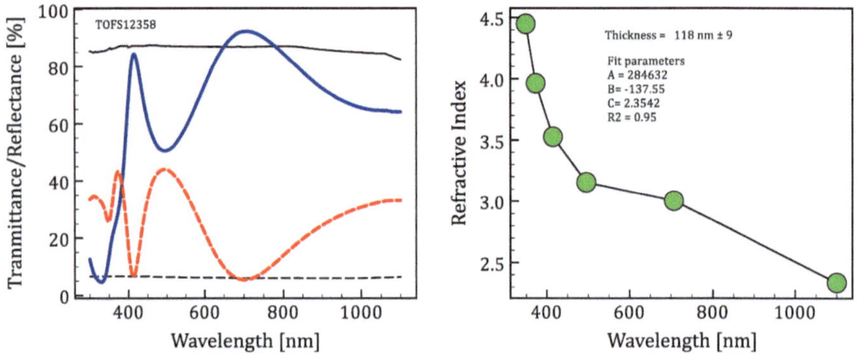

Fig. C.9 Film deposited at 800 °C and 30 mTorr

C.3 1 % Nb-Doped TiO_2—Thickness Calibration

See Figs. C.10, C.11, C.12 and C.13.

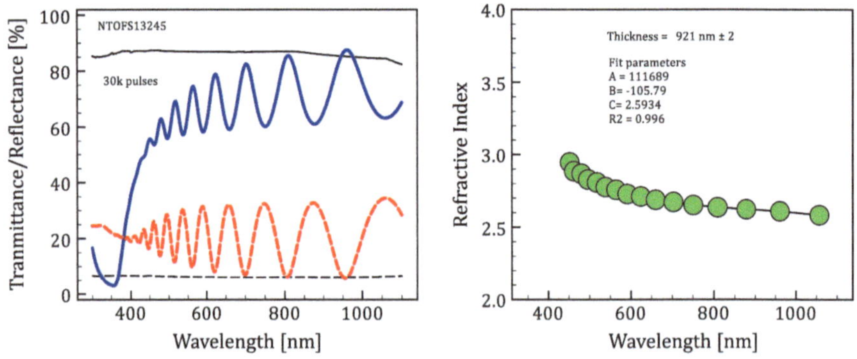

Fig. C.10 Film deposited with 30k laser pulses at 400 °C and 30 mTorr

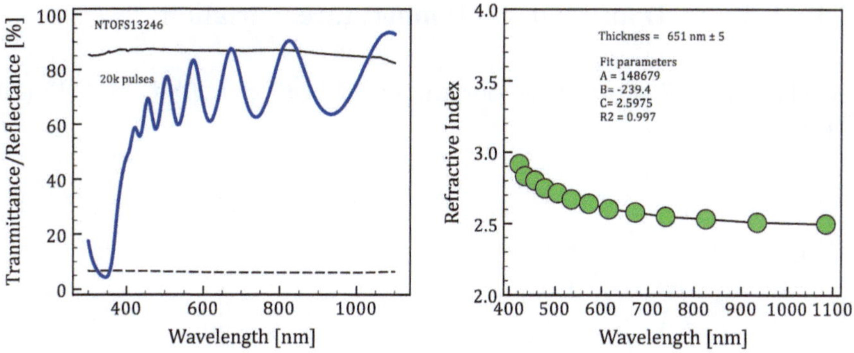

Fig. C.11 Film deposited with 20k laser pulses at 400 °C and 30 mTorr

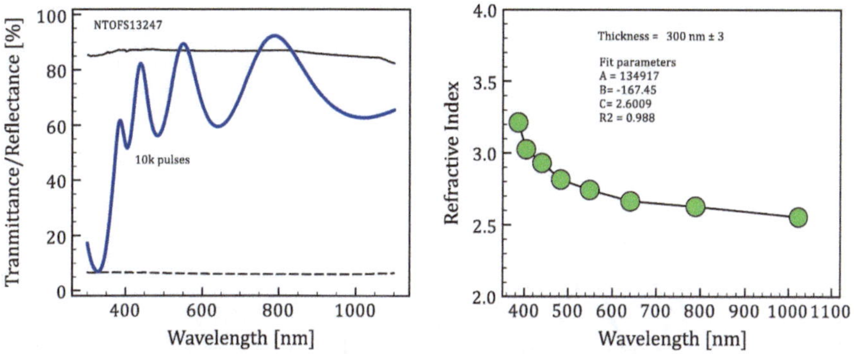

Fig. C.12 Film deposited with 10k laser pulses at 400 °C and 30 mTorr

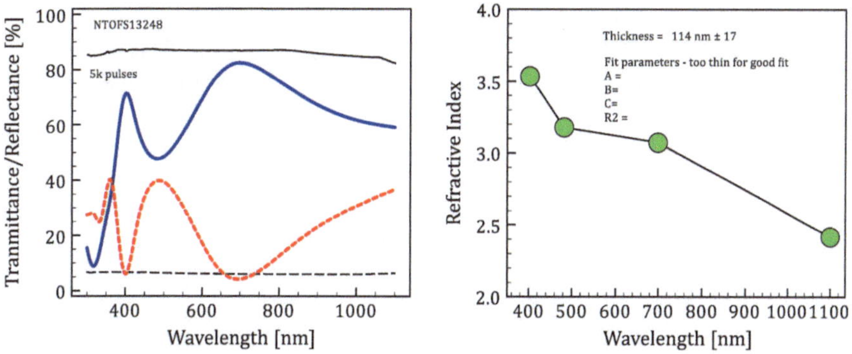

Fig. C.13 Film deposited with 5k laser pulses at 400 °C and 30 mTorr

C.4 1 % Nb-Doped TiO_2—Temperature Variation

The film deposited at 400 °C is shown in Fig. C.12 (Figs. C.14, C.15, C.16 and C.17).

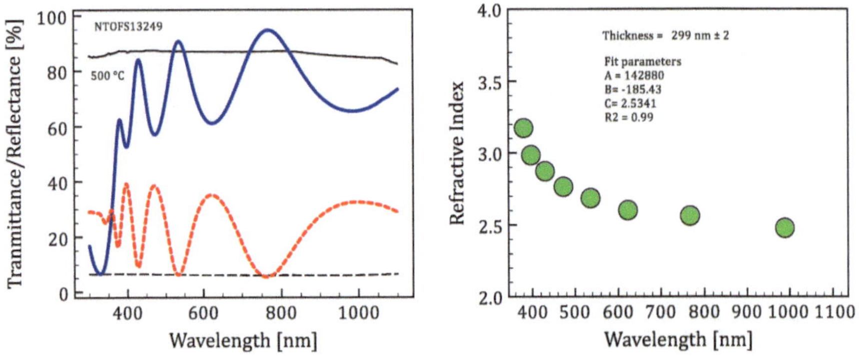

Fig. C.14 Film deposited at 500 °C and 30 mTorr

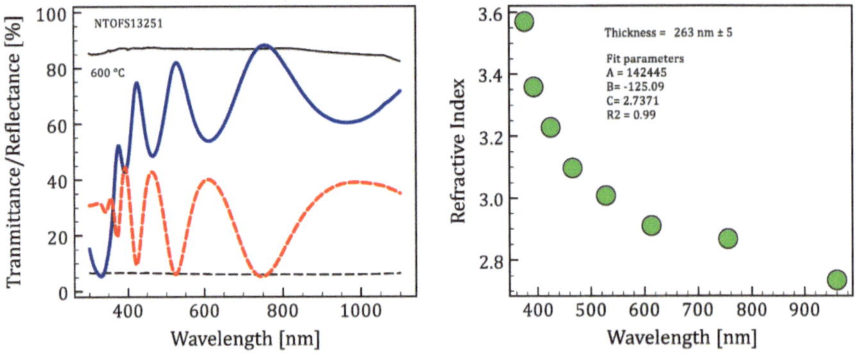

Fig. C.15 Film deposited at 600 °C and 30 mTorr

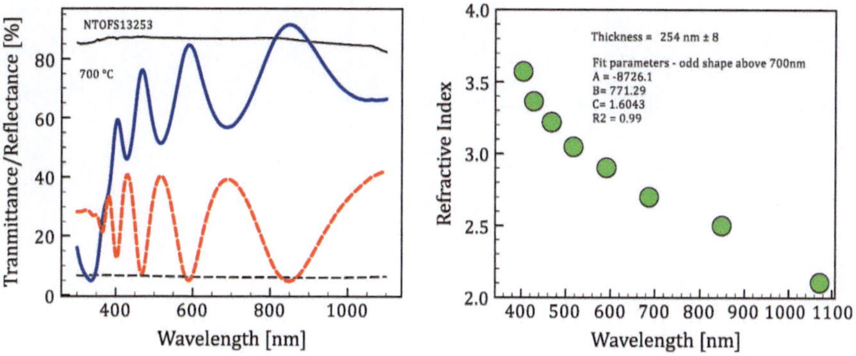

Fig. C.16 Film deposited at 700 °C and 30 mTorr

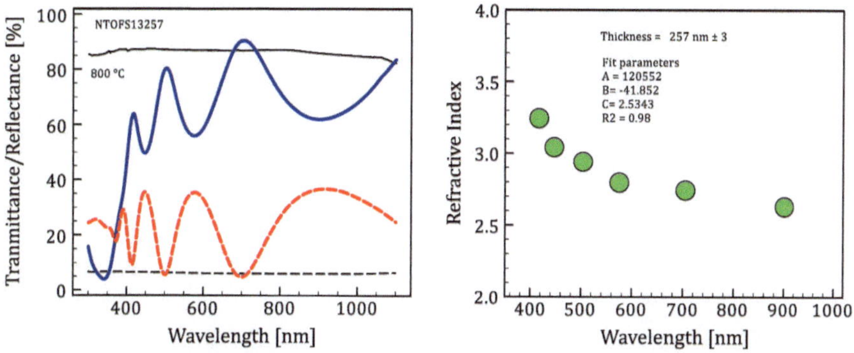

Fig. C.17 Film deposited at 800 °C and 30 mTorr

Appendix D
Optical Measurements—Films on MgO (100)

D.1 Effect of Deposition Temperature

The figure below shows the transmittance and reflectance of samples TOM13047, TOM13048, TOM13049, and TOM13052. Apart from sample TOM13047 the films do not exhibit interference fringes that can be used to estimate the film thickness and refractive index (Fig. D.1).

The thickness of sample TOM13047 is estimated to be 134 ± 7 nm (Fig. D.2).

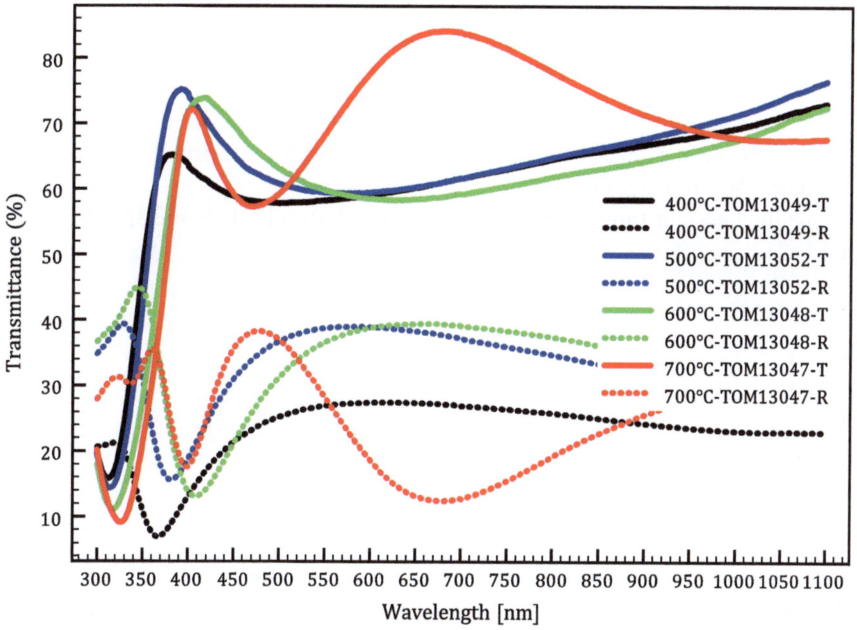

Fig. D.1 Transmittance and reflectance measurements for films deposited at substrate heating temperatures of 400–800 °C and 30 mTorr

© Springer International Publishing Switzerland 2016
J.C. Alexander, *Surface Modifications and Growth of Titanium Dioxide for Photo-Electrochemical Water Splitting*, Springer Theses,
DOI 10.1007/978-3-319-34229-0

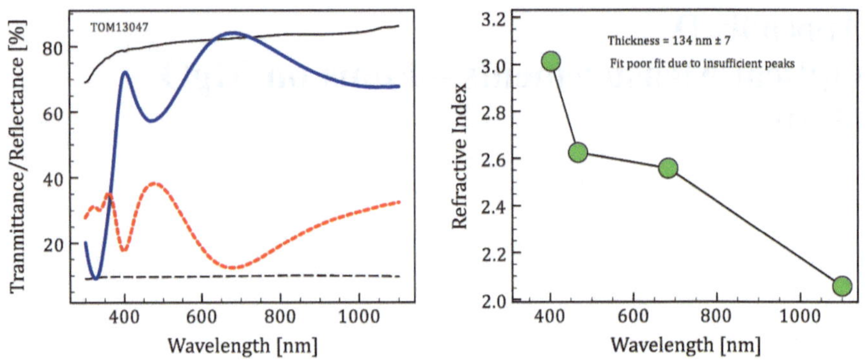

Fig. D.2 Film deposited at 800 °C and 30 mTorr

D.2 Effect of Deposition Parameters Other than Temperature

The following samples were deposited on one-side polished substrates. As a result, it was not possible to make a transmittance measurement. The reflectance data is presented here.

The thickness of sample TOM13255 is estimated to be 311 ± 13 nm (Fig. D.3).
The thickness of sample TOM13272 is estimated to be 164 ± 12 nm (Fig. D.4).

The thicknesses of samples TOM13273 and TOM13276 could not be estimated, as the films are not thick enough to give optical interference fringes. It should be less than 150 nm (Fig. D.5).

The thickness of sample TOM13275 is estimated to be 196 ± 2 nm (Figs. D.6 and D.7).

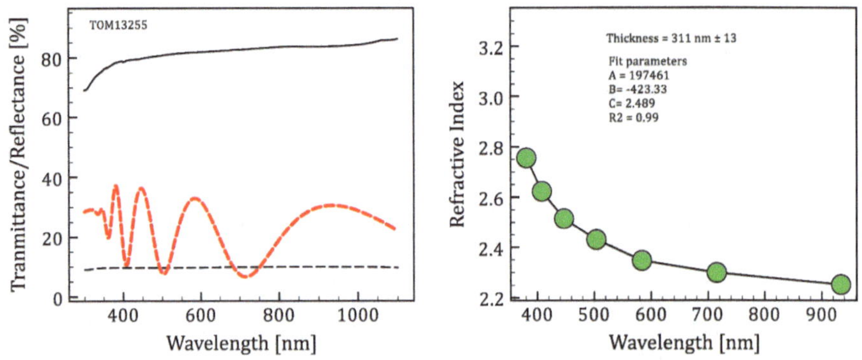

Fig. D.3 Film deposited at 400 °C and 30 mTorr with 5k laser pulses

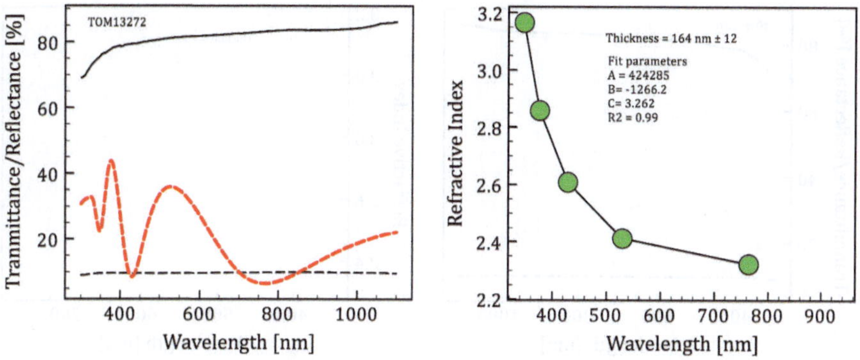

Fig. D.4 Film deposited at 400 °C and 30 mTorr with 5k laser pulses

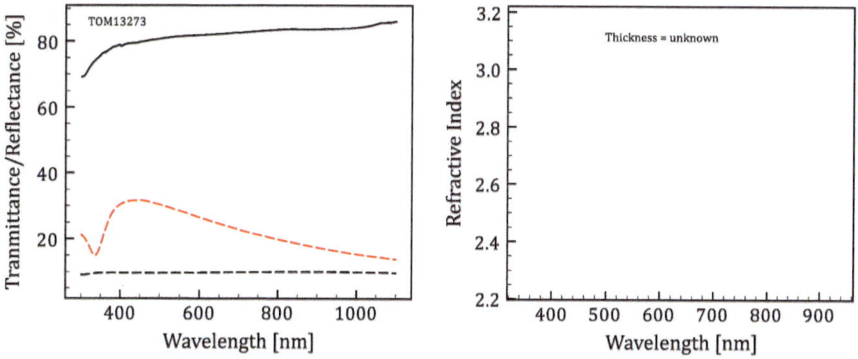

Fig. D.5 Film deposited at 400 °C and 30 mTorr with 5k laser pulses

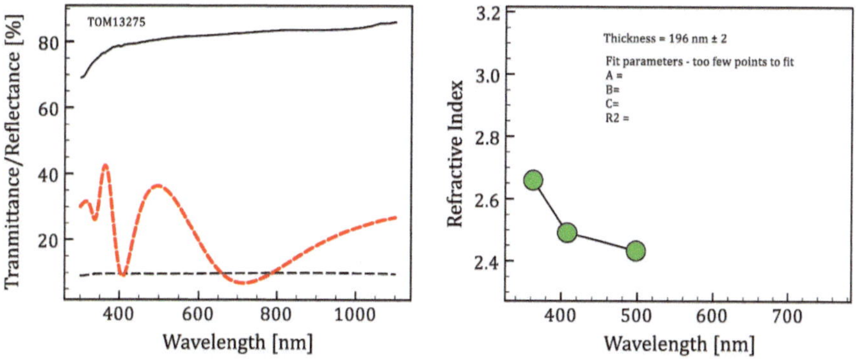

Fig. D.6 Film deposited at 400 °C and 30 mTorr with 5k laser pulses

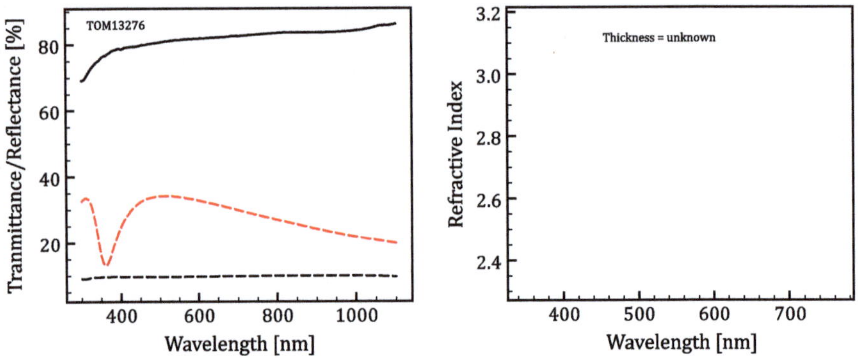

Fig. D.7 Film deposited at 400 °C and 10 mTorr with 5k laser pulses

Appendix E
Additional XRD Measurements

The results of X-ray Diffraction (XRD) analysis of a range of TiO_2 samples are presented in this appendix.

Figure E.1 shows XRD patterns of the PLD targets fabricated with undoped TiO_2 (red line) and 1 wt% Nb doped TiO_2 (blue line). The 1 % Nb-doped target was formed by mixing pure rutile TiO_2 powder with pure Nb_2O_5 powder before sintering to produce a ceramic target. As shown in the figure, all peaks occurring in the patterns for both samples coincide. While only a portion of the data is shown, the peaks are completely aligned i.e. there is no shift in the peaks evident due to the

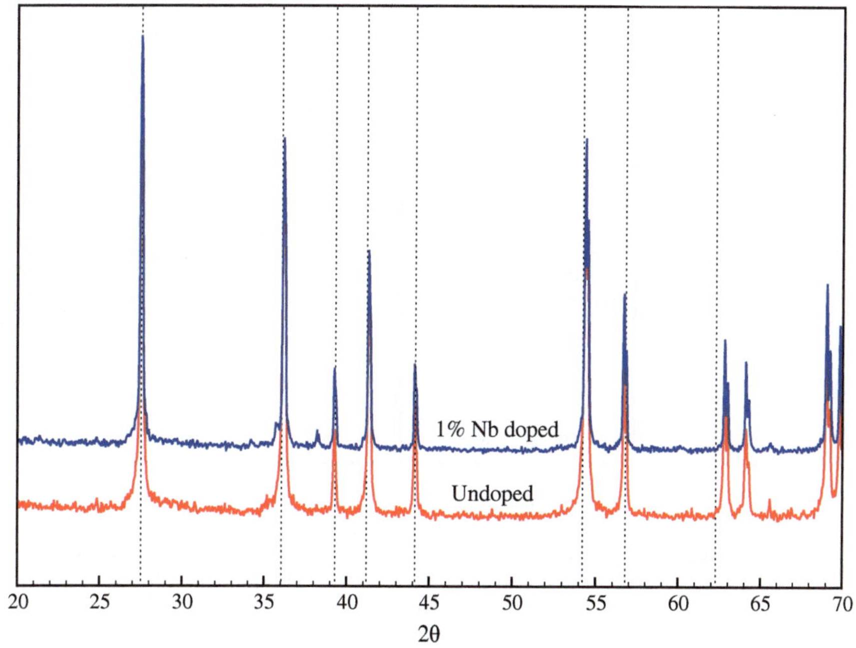

Fig. E.1 XRD patterns of undoped and 1 % Nb doped TiO_2 pulsed laser deposition targets. Standard rutile reference pattern shown by dotted lines

© Springer International Publishing Switzerland 2016

J.C. Alexander, *Surface Modifications and Growth of Titanium Dioxide for Photo-Electrochemical Water Splitting*, Springer Theses,
DOI 10.1007/978-3-319-34229-0

presence of Nb altering the lattice parameter. This suggests the addition of only 1 % Nb has not had a measurable impact on the lattice. This might be expected when substituting Nb, with and ionic radius of 0.64 Å, for Ti with an ionic radius of 0.605 Å. There are, however, two additional peaks in the pattern of the 1 % Nb doped sample, at ca. 35.5° and 37°. By comparison to standard reference patterns, neither peak corresponds to any rutile, anatase, or Nb_2O_5 peaks. It is possible the peaks correspond to a different oxidation state of TiO_2 or a different Ti-Nb-O compound.

Additional pellets were also produced with Nb-doping of 5, 10, and 30 % to begin to investigate the solubility limit of Nb in TiO_2. XRD patterns are shown in Fig. E.2. At Nb-doping up to 10 % only the rutile TiO_2 phase is present. For the sample produced with 30 % Nb, additional phases are present in addition to rutile TiO_2. The peaks occur at broadly similar positions as both monoclinic and tetragonal Nb_2O_5, but are not aligned with any one reference pattern examined. This suggests that the pellet is comprised of several T–Nb–O phases.

XRD patterns of the undoped and 0.05 wt% Nb-doped rutile (110) single crystals were also measured and are presented in Fig. E.3. There is no shift in the relative peak positions between samples. As such, neither the degree of doping nor

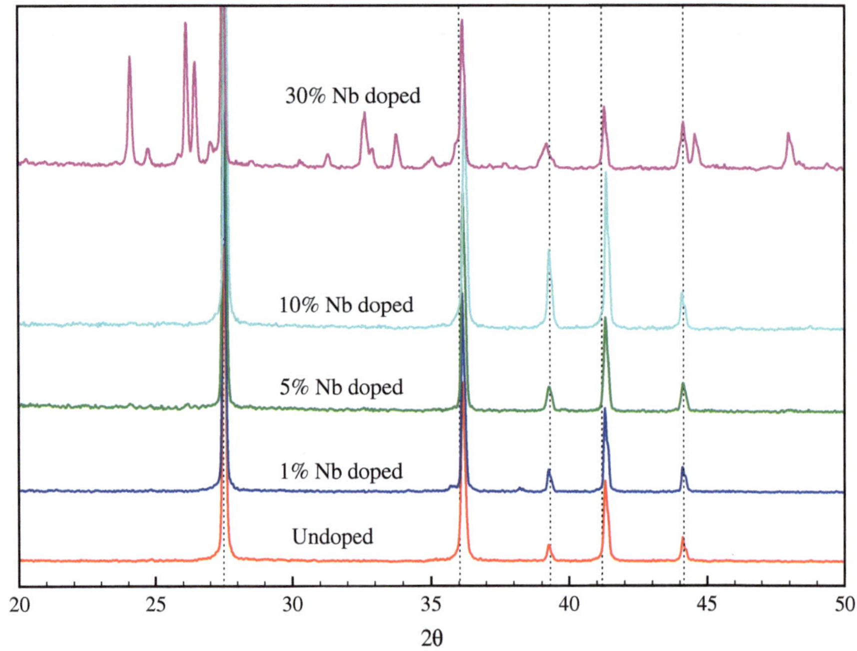

Fig. E.2 XRD patterns of Nb-doped TiO_2 pellets for Nb-doping of 0, 1, 5, 10, and 30 %

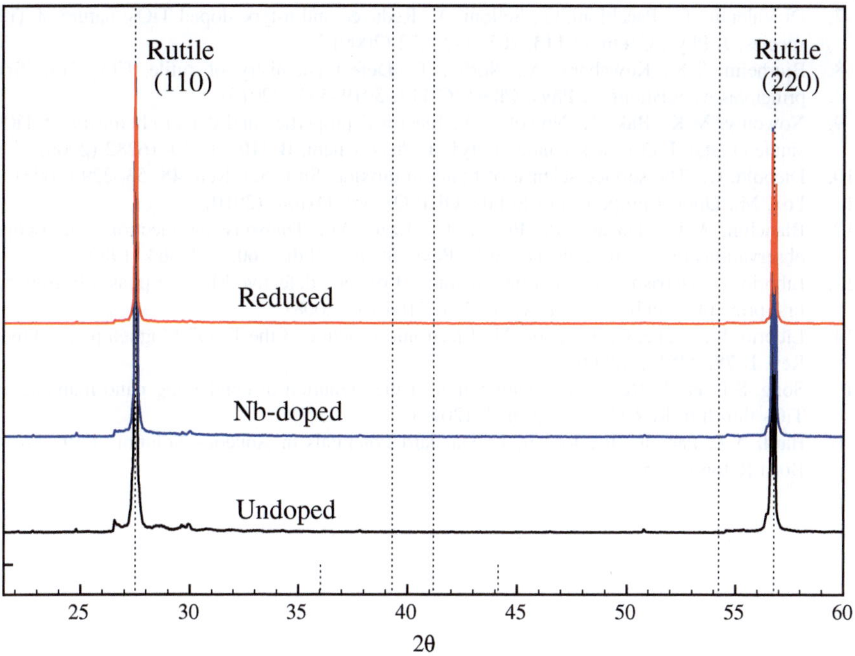

Fig. E.3 XRD patterns of undoped, Nb-doped, and reduced rutile (110) single crystals

reduction could be ascertained from the XRD patterns. This may be due to the extremely low level of doping and the close match between the ionic radius of Ti and Nb. In the case of the reduced sample, it is suggested that the degree of reduction is minimal and is insufficient for a change in the bulk lattice parameter i.e. due to reduction close to the surface only.

References

1. Smyth, D.M.: The Defect Chemistry of Metal Oxides. Oxford University Press, Oxford (2000)
2. Sheppard, L.R.: Defect Chemistry and Charge Transport in Niobium-Doped Titanium Dioxide, pp. 1–139. University of New South Wales, Sydney (2007)
3. Arbiol, J., et al.: Effects of Nb doping on the TiO_2 anatase-to-rutile phase transition. J. Appl. Phys. **92**, 853 (2002)
4. Weibin, Z., et al.: The investigation of NbO_2 and Nb_2O_5 electronic structure by XPS, UPS and first principles methods. Surf. Interface Anal. **45**, 1206–1210 (2013)
5. Morris, D., et al.: Photoemission and STM study of the electronic structure of Nb-doped TiO_2. Phys. Rev. B **61**, 13445 (2000)
6. Di Valentin, C., et al.: N-doped TiO_2: theory and experiment. Chem. Phys. **339**, 44–56 (2007)

7. Di Valentin, C., Pacchioni, G., Selloni, A.: Reduced and n-type doped TiO_2: nature of Ti^{3+} species. J. Phys. Chem. C **113**, 20543–20552 (2009)
8. Bjørheim, T.S., Kuwabara, A., Norby, T.: Defect chemistry of rutile TiO_2 from first principles calculations. J. Phys. Chem. C **117**, 5919–5930 (2013)
9. Nowotny, M.K., Bak, T., Nowotny, J.: Electrical properties and defect chemistry of TiO_2 single crystal. I. electrical conductivity†. J. Phys. Chem. B **110**, 16270–16282 (2006)
10. Diebold, U.: The surface science of titanium dioxide. Surf. Sci. Rep. **48**, 53–229 (2003)
11. Fox, M.: Optical Properties of Solids. OUP Oxford, Oxford (2010)
12. Blanchin, M.G., Faisant, P., Picard, C., Ezzo, M.: Transmission electron microscope observations of slightly reduced rutile. Phys. Status Solidi *A* **60**, 357–364 (1980)
13. Liborio, L., Harrison, N.: Thermodynamics of oxygen defective Magnéli phases in rutile: a first-principles study. Phys. Rev. B **77**, 104104 EP (2008)
14. Liborio, L., Mallia, G., Harrison, N.: Electronic structure of the Ti4O7 Magnéli phase. Phys. Rev. B **79**, 245133 (2009)
15. Song, S.J., et al.: Real-time identification of the evolution of conducting nano-filaments in TiO_2 thin film ReRAM. Sci. Rep. **3**, (2013)
16. Bard, A.J., Parsons, R., Jordan, J.: Standard Potentials in Aqueous Solution. CRC Press, Boca Raton (1985)

MIX
Papier aus verantwortungsvollen Quellen
Paper from responsible sources
FSC® C105338

If you have any concerns about our products,
you can contact us on
ProductSafety@springernature.com

In case Publisher is established outside the EU,
the EU authorized representative is:
Springer Nature Customer Service Center GmbH
Europaplatz 3, 69115 Heidelberg, Germany

Printed by Libri Plureos GmbH
in Hamburg, Germany